METHODS IN MOLECULAR BIOLOGY™

Series Editor
John M. Walker
School of Life Sciences
University of Hertfordshire
Hatfield, Hertfordshire, AL10 9AB, UK

For other titles published in this series, go to
www.springer.com/series/7651

Therapeutic Oligonucleotides

Methods and Protocols

Edited by

John Goodchild

Department of Chemistry, Worcester State University, Worcester, MA, USA

Humana Press

Editor
John Goodchild
Department of Chemistry
Worcester State University
Worcester, MA 01602, USA
jgoodchild1@worcester.edu

ISSN 1064-3745 e-ISSN 1940-6029
ISBN 978-1-4939-5822-1 ISBN 978-1-61779-188-8 (eBook)
DOI 10.1007/978-1-61779-188-8
Springer New York Dordrecht Heidelberg London

Softcover re-print of the Hardcover 1st edition 2011

Printed on acid-free paper

Humana Press is part of Springer Science+Business Media (www.springer.com)

Preface

It seems fitting that this book on therapeutic oligonucleotides should begin by mentioning the name of Paul Zamecnik. A great gentleman of science, Paul died in 2009, thirty-one years after reporting that an oligonucleotide complementary to the end of Rous sarcoma virus could interfere with its replication (1, 2). Despite his suggestion that oligonucleotides might be used to treat various diseases, the publication aroused little interest at the time. Even for those who shared his vision, it was prohibitively difficult or expensive for most workers to acquire sufficient amounts of oligonucleotides in the 1970s. In any case, few RNA sequences were known that could be used as drug targets.

During the 1980s and 1990s, improvements in synthesis and sequencing of nucleic acids led to an increased use of "antisense" oligonucleotides complementary to "sense" mRNAs to prevent their translation into proteins. This helped to assign functions to the growing number of identified genes. There were also a handful of people who, like Paul Zamecnik, believed that antisense oligonucleotides offered a general approach for treating disease by modulating the levels of relevant proteins.

These early aspirations did not anticipate the problems that emerged during attempted implementation. Antisense inhibition of translation is not as straightforward as was hoped and has still not produced any important drugs. Yet, 30 years on, there is still intense interest in therapeutic oligonucleotides that justifies the present volume. Why is that? I would like to propose two reasons.

The first is that the nucleic acids and RNA, in particular, continue to surprise us. New discoveries have refreshed enthusiasm for oligonucleotide therapeutics by suggesting new applications. Most recently, small RNAs such as siRNA and miRNA have become the focus of interest.

The second reason is the unique predictability of base pairing that offers an alluring shortcut to drug design. There is ample proof of principle for the therapeutic potential of oligonucleotides but this has not yet been reduced to practice in a generally applicable manner. However, it is still reasonable to suggest that one success could lead to many others using the same approach.

Why has this not happened already? Early hopes were naïve due to ignorance of the chemical and biological complexity of nucleic acids. Only the base pairing properties of oligonucleotides were considered and not the fact that they are complex molecules in their own right. The icon of the double helix illustrates base pairing beautifully but it ignores the effects of sequence on structure. It was tempting to assume that all antisense oligonucleotides would behave similarly and would differ only in hybridizing to different sequences of RNA. However, a single-stranded nucleic acid is densely packed with functional groups that can participate in secondary and tertiary structure formation. For example, the nucleoside guanosine alone contains 15 lone pairs of electrons and six hydrogen atoms able to form hydrogen bonds. Its heterocyclic base can stack with other bases and is also capable of tautomerization. In addition, each nucleotide has an ionized phosphate group that can recruit metal ions of varying charge. Even small oligonucleotides have abundant capacity to form higher order structures.

This book illustrates many different ways that oligonucleotides might be applied as therapeutics. It contains a selection of established and emerging methods that have been chosen for their potential to change the field. Often, the same problems are encountered in different approaches; cellular uptake is probably the single most important problem that must be addressed in all of them. So, for example, a method for getting siRNA into cells of one type might be used with ribozymes or aptamers in other types of cells. All types of oligonucleotide must survive the same onslaught of nucleases long enough to be effective; modifications that stabilize antisense oligonucleotides might be used with aptamers or any of the other classes of therapeutics.

Similarly, several protocols describe methods for optimizing or improving cell uptake that might be tried with other classes of oligonucleotide. They include photochemical internalization, modified cell penetrating peptides, antibody conjugates, and nanoparticles. Other contributions address quantitation of RNA therapeutics in cells, assaying gene knockdown, selecting the best target site, and synthesis of various modified oligonucleotides. The chapter on agRNA contains much useful information on control experiments and off-target effects, selection of RNA targets, validation of experiments, preparation of cells for successful transfection, and the effects of stress due to transfection. This is valuable reading for anybody using oligonucleotides.

The hope remains that approaches for uptake and delivery of one oligonucleotide might work with others, that developing one successful drug might lead to others. It is possible that we already know all the methods needed to solve the problem of oligonucleotide delivery. It could be that we just need to find the right combination of some of the methods in the literature – such as those reported here. Maybe a 3′-*S*-phosphorothioate oligonucleotide combined with an antibody or cell penetrating polypeptide or light activated uptake would be ideal. Nobody has yet tried such combinations. Perhaps drawing together some of the most promising new approaches in a book such as the one you are reading will help someone to find a winning combination.

John Goodchild

References

1. Zamecnik, P. C., and Stephenson, M. L. (1978) Inhibition of Rous sarcoma virus replication and cell transformation by a specific oligodeoxynucleotide. *Proceedings of the National Academy of Sciences USA* 75, 280–284.
2. Stephenson, M. L., and Zamecnik, P. C. (1978) Inhibition of Rous sarcoma viral RNA translation by a specific oligodeoxynucleotide. *Proceedings of the National Academy of Sciences USA* 75, 285–288.

Contents

Contributors

LUIS MARAT ALVAREZ-SALAS • *Laboratorio de Terapia Génica, Departamento de Genética y Biología Molecular, CINVESTAV, México D.F., México*
SAMIR EL ANDALOUSSI • *Department of Neurochemistry, Stockholm University, Stockholm, Sweden*
MARÍA LUISA BENÍTEZ-HESS • *Laboratorio de Terapia Génica, Departamento de Genética y Biología Molecular, CINVESTAV, México D.F., México*
RITA BHARDWAJ • *Division of Infectious Diseases, Department of Biomedical Sciences, Cummings School of Veterinary Medicine, Tufts University, North Grafton, MA, USA*
SIGURD BØE • *Department of Tumor Biology, Institute for Cancer Research, The Norwegian Radium Hospital, Montebello, Oslo, Norway*
PRISCA BOISGUERIN • *UMR 5235 CNRS, Université Montpellier 2, Montpellier, France; Institute of Medical Immunology, Charité-Universitätsmedizin Berlin, Berlin, Germany*
BRAD BRYAN • *Department of Biology, Ghosh Science and Technology Center, Worcester State University, Worcester, MA, USA*
TAO CHEN • *Departments of Chemistry, Department of Physiology and Functional Genomics, Shands Cancer Center, Moffitt Cancer Center, Center for Research at Bio/Nano Interface, UF Genetics Institute and McKnight Brain Institute, University of Florida, Gainesville, FL, USA*
ANGIE CHENG • *Molecular and Cell Biology Division, Life Technologies, Austin, TX, USA*
DAVID R. COREY • *Departments of Pharmacology and Biochemistry, University of Texas Southwestern Medical Center at Dallas, Dallas, TX, USA*
RICHARD COSSTICK • *Department of Chemistry, University of Liverpool, Liverpool, UK*
LAURENCE CROMBEZ • *Department of Molecular Biophysics and Therapeutics, Centre de Recherches de Biochimie Macromoléculaire, Montpellier, France*
ERIC DAUSSE • *University of Bordeaux, ARNA laboratory, Bordeaux, France; INSERM, U869, IECB ARNA laboratory, Pessac, France*
EMILY DENNSTEDT • *Department of Biology, Ghosh Science and Technology Center, Worcester State University, Worcester, MA, USA*
CARMELO DI PRIMO • *University of Bordeaux, ARNA laboratory, Bordeaux, France; INSERM, U869, IECB ARNA laboratory, Pessac, France*
GILLES DIVITA • *Department of Molecular Biophysics and Therapeutics, Centre de Recherches de Biochimie Macromoléculaire, Montpellier, France*
MICHAEL J. DONOVAN • *Departments of Chemistry, Department of Physiology and Functional Genomics, Shands Cancer Center, Moffitt Cancer Center, Center for Research at Bio/Nano Interface, UF Genetics Institute and McKnight Brain Institute, University of Florida, Gainesville, FL, USA*
BIRGIT ENGESÆTER • *Department of Tumor Biology, Institute for Cancer Research, The Norwegian Radium Hospital, Montebello, Oslo, Norway*
LANCE P. FORD • *Bioo Scientific Corporation, Austin, TX, USA*
JAMES W. GAYNOR • *Department of Chemistry, University of Liverpool, Liverpool, UK*

JOHN GOODCHILD • *Department of Chemistry, Worcester State University, Worcester, MA, USA*
YOSHIYUKI HARI • *Graduate School of Pharmaceutical Sciences, Osaka University, Suita, Osaka, Japan*
FATOUMA SAID HASSANE • *UMR 5235 CNRS, Université Montpellier 2, Montpellier, France*
LAURA L. HAUCK • *AVI Biopharma, Inc., Corvallis, OR, USA*
FRÉDERIC HEITZ • *Department of Molecular Biophysics and Therapeutics, Centre de Recherches de Biochimie Macromoléculaire, Montpellier, France*
DONALD J. HNATOWICH • *Division of Nuclear Medicine, Department of Radiology, University of Massachusetts Medical School, Worcester, MA, USA*
EIVIND HOVIG • *Department of Tumor Biology, Institute for Cancer Research, The Norwegian Radium Hospital, Montebello, Oslo, Norway*
REN-JUN HSU • *Institute of Molecular and Cellular Biology, National Taiwan University, Taipei, Taiwan*
TAKESHI IMANISHI • *Graduate School of Pharmaceutical Sciences, Osaka University, Suita, Osaka, Japan*
EKAMBAR R. KANDIMALLA • *Idera Pharmaceuticals, Inc., Cambridge, MA, USA*
TETSUYA KODAMA • *Graduate School of Pharmaceutical Sciences, Osaka University, Suita, Osaka, Japan*
GREICE KRAUTZ-PETERSON • *Division of Infectious Diseases, Department of Biomedical Sciences, Cummings School of Veterinary Medicine, Tufts University, North Grafton, MA, USA*
TAO LAN • *Idera Pharmaceuticals, Inc., Cambridge, MA, USA*
ÜLO LANGEL • *Department of Neurochemistry, Stockholm University, Stockholm, Sweden*
BERNARD LEBLEU • *UMR 5235 CNRS, CC107, Université Montpellier 2, Montpellier Cedex 5, France*
LIANDE LI • *Department of Physiology, The University of Texas Southwestern Medical Center, Dallas, TX, USA*
XINRONG LIU • *Division of Nuclear Medicine, Department of Radiology, University of Massachusetts Medical School, Worcester, MA, USA*
YI LIU • *Department of Physiology, The University of Texas Southwestern Medical Center, Dallas, TX, USA*
SUSAN MAGDALENO • *Molecular and Cell Biology Division, Life Technologies, Austin, TX, USA*
NIKKI B. MARSHALL • *Department of Microbiology, Oregon State University, Corvallis, OR, USA*
LING MENG • *Departments of Chemistry, Department of Physiology and Functional Genomics, Shands Cancer Center, Moffitt Cancer Center, Center for Research at Bio/Nano Interface, UF Genetics Institute and McKnight Brain Institute, University of Florida, Gainesville, FL, USA*
MAY C. MORRIS • *Centre de Recherches de Biochimie Macromoléculaire, Interactions and Molecular Mechanisms Regulating the Cell Cycle, Montpellier, France*
DAN V. MOURICH • *Department of Microbiology, Oregon State University, Corvallis, OR, USA; AVI Biopharma, Inc., Corvallis, OR, USA*
SATOSHI OBIKA • *Graduate School of Pharmaceutical Sciences, Osaka University, Suita, Osaka, Japan*

LINA PRASMICKAITE • *Department of Tumor Biology, Institute for Cancer Research, The Norwegian Radium Hospital, Montebello, Oslo, Norway*
MALLIKARJUNA REDDY PUTTA • *Idera Pharmaceuticals, Inc., Cambridge, MA, USA*
PABLO REYES-GUTIÉRREZ • *Laboratorio de Terapia Génica, Departamento de Genética y Biología Molecular, CINVESTAV, México D.F., México*
JACOB C. SCHWARTZ • *Departments of Pharmacology and Biochemistry, University of Texas Southwestern Medical Center at Dallas, Dallas, TX, USA*
KWAME SEFAH • *Departments of Chemistry, Department of Physiology and Functional Genomics, Shands Cancer Center, Moffitt Cancer Center, Center for Research at Bio/Nano Interface, UF Genetics Institute and McKnight Brain Institute, University of Florida, Gainesville, FL, USA*
RANNAR SILLARD • *Department of Neurochemistry, Stockholm University, Stockholm, Sweden*
PATRICK J. SKELLY • *Division of Infectious Diseases, Department of Biomedical Sciences, Cummings School of Veterinary Medicine, Tufts University, North Grafton, MA, USA*
WEIHONG TAN • *Departments of Chemistry, Department of Physiology and Functional Genomics, Shands Cancer Center, Moffitt Cancer Center, Center for Research at Bio/Nano Interface, UF Genetics Institute and McKnight Brain Institute, University of Florida, Gainesville, FL, USA*
MASOUD M. TOLOUE • *Bioo Scientific Corporation, Austin, TX, USA*
JEAN-JACQUES TOULMÉ • *University of Bordeaux, ARNA laboratory, Bordeaux, France; INSERM, U869, IECB ARNA laboratory, Pessac, France*
HUAI-JEN TSAI • *Institute of Molecular and Cellular Biology, National Taiwan University, Taipei, Taiwan*
ALEXANDER V. VLASSOV • *Molecular and Cell Biology Division, Life Technologies, Austin, TX, USA*
YI WANG • *Department of Environmental Science and Engineering, Tsinghua University, Beijing, China*
DONG YU • *Idera Pharmaceuticals, Inc., Cambridge, MA, USA*
YUNFEI ZHANG • *Departments of Chemistry, Department of Physiology and Functional Genomics, Shands Cancer Center, Moffitt Cancer Center, Center for Research at Bio/Nano Interface, UF Genetics Institute and McKnight Brain Institute, University of Florida, Gainesville, FL, USA*

Chapter 1

Therapeutic Oligonucleotides

John Goodchild

Abstract

A brief historical introduction describes early attempts to silence specific genes using the antisense oligonucleotides that flourished in the 1980s. Early aspirations for therapeutic applications were almost extinguished by the unexpected complexity of oligonucleotide pharmacology. Once the biochemistry and molecular biology behind some of the pharmacology was worked out, new approaches became apparent for using oligonucleotides to treat disease. The biochemistry of small nucleic acids is outlined in **Section 2**. Various approaches employing oligonucleotides to control cellular functions are reviewed in **Section 3**. These include antisense oligonucleotides and siRNA that bind to RNA, antigene oligonucleotides that bind to DNA, and aptamers, decoys, and CpG oligonucleotides that bind to proteins.

Key words: Antisense oligonucleotides, siRNA, miRNA, antigene, aptamer, decoy, CpG, ribozyme, DNAzyme.

1. Introduction

The first suggestion that oligonucleotides might be used therapeutically was by Zamecnik and Stephenson in 1978 (1, 2). They reported that oligonucleotides complementary to terminal repeat sequences of the Rous sarcoma virus could inhibit its replication. This finding generated little interest at the time. It was not until the introduction of automated DNA synthesizers that Zamecnik extended his studies to include HIV (as it became called later) in 1986 (3). This paper fell on more receptive ears, in part because of the urgency of the emerging AIDS problem. Soon, a rash of companies appeared to develop modified oligonucleotides (phosphorothioates and methylphosphonates) as drugs. These "antisense" oligonucleotides were complementary to sequences in either viral or messenger RNAs. Unlike

J. Goodchild (ed.), *Therapeutic Oligonucleotides*, Methods in Molecular Biology 764,
DOI 10.1007/978-1-61779-188-8_1, © Springer Science+Business Media, LLC 2011

conventional drugs, which bind directly to proteins, antisense oligonucleotides block the action of proteins by suppressing translation. However, few of the early companies survive in their original form; the task of silencing genes in vivo was too challenging for the technology of the time and has still not been solved satisfactorily.

As individual antisense oligonucleotides were studied, it became apparent that they do not all behave similarly. Like proteins, single-stranded nucleic acids can adopt complex secondary and tertiary structures. These can bind to other molecules and are even able to mimic the functions of antibodies and enzymes. The variability of oligonucleotide secondary structures can complicate the pharmacology of these compounds, leading to unexpected biological consequences. Also, it transpired that nucleic acids, especially small RNA molecules, have important natural functions in cells that were unknown in the 1980s and 1990s. These functions are described in **Section 2** of this chapter and in **Chapter 11** by Li and Liu (this volume). They increase the number of pharmacological effects that might be exhibited by exogenous oligonucleotides, making development of drugs more difficult. They also suggested new ways for using oligonucleotides that gave rise to a new generation of compounds and companies. Translation of RNA may now be suppressed by siRNAs, ribozymes, and DNAzymes in addition to the original antisense oligonucleotides. Antigene oligonucleotides bind directly to DNA while aptamers, decoys, and CpG oligonucleotides bind to proteins or other types of molecules. All these are discussed in **Section 3**.

2. The Role of RNA in Cells

Recent discoveries have led to a reevaluation of the role of RNA in cells. To the classic functions of messenger, transfer, and ribosomal RNAs have been added control of gene expression and other activities previously assumed to be carried out by proteins.

Over 20 years ago it was found that some bacteria and other cells use endogenous antisense RNA to suppress translation of mRNA. A few years later, the catalytic activity of ribozymes was recognized in ribonuclease P and various self-cleaving RNAs and later in rRNA (4). More recently, control elements called riboswitches were discovered in mRNA. By binding to the appropriate metabolite, riboswitches sense whether or not translation should proceed (5).

Through the discovery of RNA interference (RNAi), the importance of previously undetected RNA molecules with only 18–30 nucleotides became apparent (6). They may directly regulate over 30% of genes in cells (7). Found in both the cytoplasm and the nucleus, small RNAs can suppress translation of mRNAs involved in cell differentiation, growth and proliferation, migration, apoptosis, metabolism, and defense. Often, these activities involve an RNA-induced silencing complex (RISC) in the cytoplasm (8, 9). Under the right circumstances, small RNAs may also activate translation (10, 11).

Currently, three major classes of small RNAs are recognized (12, 13). These are microRNAs (miRNA), small interfering RNAs (siRNA), and PIWI interacting RNAs (piRNA). Other categories have been proposed (14) and others may remain to be discovered.

Over 850 miRNAs have been found in human cells. They comprise 21 or 22 nucleotides excised from larger RNA precursors. In the cytoplasm, RISC anneals miRNA to partly or completely complementary sequences in an mRNA target. This blocks translation of the mRNA in one of two ways. If the sequences are completely complementary, the mRNA is destroyed by nucleases. If the sequences contain mismatches, translation is blocked but the mRNA is not degraded (15).

miRNAs are involved in the pathogenesis of diverse diseases including cancer, stroke, diabetes, diseases of the liver, kidney, and cardiovascular system as well as neurodegenerative and infectious diseases. This suggests new roles for therapeutic oligonucleotides that might mimic or else block miRNAs (16). Two recent examples will be given. In mice, the growth of new blood vessels is controlled by an miRNA (17). A corresponding miRNA in humans might be a target for oligonucleotide antagonists in ischemic disease. In mice, an miRNA in skeletal muscle promotes regeneration of neuromuscular synapses after injury (18). Oligonucleotide agonists of this RNA might be of interest in neurodegenerative disease.

siRNAs, essential players in RNA interference, are formed from precursor double-stranded RNAs. These arise during viral replication or when endogenous antisense RNA is transcribed. Large RNA duplexes are trimmed down to 21–23 base pairs with overhangs of two nucleotides at the 3′-ends. In the cytoplasm, RISC directs the antisense strand of the siRNA duplex to hybridize with the target mRNA to prevent translation or induce degradation depending on the degree of complementarity.

Of more restricted occurrence are piRNAs that were discovered more recently. Containing 25–30 nucleotides, they are methylated on 2′-hydroxyl groups at their 3′-ends. piRNAs are required for development of germ cells in animals where they bind to the PIWI clade of Argonaute proteins.

Not all the new discoveries about RNA involve small molecules. Much of the human genome is transcribed into noncoding RNA (19). This takes place on both strands of DNA in both directions and gives both sense and antisense RNA (13). The function of all this RNA is still not well understood. It includes RNA precursors of siRNA and miRNA that affect translation. It also controls transcription in various ways. Many of the long, non-coding RNAs are conserved between species. A new category has been suggested for such RNA that may transport repressors to promoters to regulate cell cycling, immunity, and stem cell differentiation. Called large intervening non-coding RNA (lincRNA), they have 2300–17,200 nucleotides (20). Over 4000 may exist in human cells although not everybody agrees that they represent a true subclass of RNA.

Perhaps even more surprising is that small RNAs such as siRNA and miRNA and long non-coding antisense RNAs are responsible for epigenetic changes within cells. This topic has been reviewed recently (21). Various mechanisms are involved, often involving methylation of histones or DNA at promoters or other sites targeted by the RNA (22). Resulting long-term changes in gene expression are passed on to daughter cells. In other cases, it seems that blocking of transcription may occur without methylation.

These mechanisms are being unraveled and involve proteins from the same Argonaute family found in RISC. The endogenous epigenetic regulator in human cells is long, non-coding antisense RNA (asRNA) that directs a complex of proteins, including a DNA methylase, to the complementary site on DNA. siRNA or miRNA may substitute for endogenous asRNA in this mechanism. Much remains to be learned about these activities of ribooligonucleotides. For example, different results can be obtained by targeting DNA sites upstream of the RNA polII promoter, at the RNA polII binding site, or even at a splice junction. There also appear to be examples of small RNAs blocking the large asRNA, resulting in derepression of transcription. Non-translated antisense RNA is the target for antigeneRNA (agRNA) that can increase or decrease transcription (23).

In animals, many early antisense deoxyoligonucleotides caused unwanted biological effects due to the presence of CG motifs. These often trigger an innate immune response intended to combat bacterial infections. Prokaryotic DNA is richer in CG elements while the corresponding sequences in eukaryotes are generally methylated. This was the first of several mechanisms involving members of the toll-like receptor (TLR) family of proteins that respond to pathogenic nucleic acids (24).

Each site of action of nucleic acids in cells presents an opportunity for therapeutic intervention. In the last few years it has become apparent that these are more numerous than had been realized and the various options are discussed in **Section 3**. Conversely, each site of action increases the chance of unwanted side effects by oligonucleotide drugs. We already know how to avoid some of these. Safer molecules will be designed with better understanding of their natural functions in cells.

3. Therapeutic Oligonucleotides

This section presents the various approaches for using oligonucleotides in therapy. As each of these could easily merit a chapter of its own, citations are restricted mainly to existing reviews or recent reports of significant new departures.

Oligonucleotides are suited particularly as antagonists of cellular RNA molecules. Given the sequence of the target, complementary oligonucleotides can be selected immediately and winnowed using various guidelines based on past experience. The most common RNA targets for this approach are mRNA or pre-mRNA, viral RNA, miRNA, or the telomerase-associated RNA in cancer cells. DNA is a more challenging target due to the presence of the complementary strand. Proteins and other molecules can also serve as targets for oligonucleotides but not, of course, through Watson–Crick base pairing.

Within nuclease-rich cells or serum, unprotected oligonucleotides survive only briefly and so must be modified chemically or sequestered within a protective particle. Alternatively, they may be continuously generated from a vector following transfection. The two approaches present different sets of problems and the present review is concerned primarily with the use of modified or packaged oligonucleotides.

Because their physicochemical properties are so different from conventional drugs, oligonucleotides presented a new challenge for drug development. Cost and scale of synthesis, toxicity, rapid degradation, delivery, and pharmacokinetics of large polyanions were all problematic. However, numerous clinical trials have been conducted and two oligonucleotides have received FDA approval. The consensus among many workers is that the major obstacle to more widespread success is cellular uptake. Transfection agents are commonly used and are the subject of intense research (25, 26). Surprisingly, it seems possible that uptake may be more efficient with cells in vivo than in vitro. A recent breakthrough reports improved uptake of at least some types of antisense oligonucleotide without transfection agents (27).

3.1. Oligonucleotides That Hybridize to RNA

Oligonucleotide antagonists of RNA are commonly classified according to their structure and mechanism of action.

Antisense oligonucleotides are primarily from the DNA family and the hybrid they form with RNA is a substrate for cellular ribonuclease H that destroys the RNA component. However, some antisense oligonucleotides bind so strongly to RNA that they can prevent translation or splicing without the need for nucleases. These oligonucleotides are said to act by the steric block mechanism.

siRNA oligonucleotides hybridize to RNA under the direction of the cellular complex, RISC, which was mentioned earlier. RISC has an associated nuclease activity that also destroys the target RNA.

The optimum target sequence within an RNA must be determined experimentally or with the help of computational methods that attempt to predict the structure of the target RNA or the thermodynamics of binding of any particular oligonucleotide (for some of the many published approaches, see (28–32)). Site selection is a perennial problem for all methods using oligonucleotides that bind to RNA.

3.1.1. Antisense Oligonucleotides

Although Zamecnik originally proposed the term "hybridon," it was "antisense" that became the accepted designation for oligonucleotides complementary to coding RNA. Of late, antisense oligonucleotides have been overshadowed by the more recently discovered siRNAs. The two approaches are opposite sides of the same coin; they differ primarily in which nucleases they recruit to destroy the target RNA. Most direct comparisons in the literature find siRNA to be more potent. In a recent example (28), cleavage of RNA in Drosophila embryo lysates was three times faster with antisense oligonucleotide as with siRNA directed to same sequence. In vivo, however, siRNA was twice as efficient. The authors suggest that binding of antisense oligonucleotides may be blocked by proteins bound to RNA. Certainly, it is probably in the favor of siRNAs that they utilize cells' endogenous sequence-based pathway for culling RNA.

However, antisense oligonucleotides are still a viable option. Many are reported to be in development with expectations for a new drug application from Isis during 2010 (33). Recently, a modified antisense oligonucleotide was shown to produce long-term suppression of hepatitis C virus in chronically infected chimpanzees (34). Also, unlike siRNA, antisense oligonucleotides can be directed at non-coding RNAs such as that in telomerase. This is considered a target in cancer cells (35). Even miRNA has emerged as a new target for antisense oligonucleotides. This could lead to modulation of multiple genes in a particular pathway rather than ablation of a single gene product when targeting an mRNA (33, 36–38). Various forms of antisense oligonucleotides complementary to miRNA are called antagomirs, antimirs, miRNA decoys, or miRNA sponges by different workers and are being tested in the clinic.

miRNAs are Nature's own antisense oligonucleotides used to suppress translation. In some situations, it would be desirable to reinforce their activity. For example, a newly discovered role of miRNA is that of tumor suppressor and exogenously added miRNA can inhibit tumors (39). Therapeutic applications for antisense oligonucleotides, then, should include that of miRNA agonist as well as antagonist.

Despite problems of nuclease degradation, cell uptake, cytotoxicity, and other off-target effects, reports abound of antisense activity in cells and cell extracts. These use a variety of chemical modifications and uptake strategies that have been widely studied along with delivery, pharmacokinetics, pharmacology, and toxicity of oligonucleotides. A useful source of reviews on these topics is the journal *Current Opinion in Molecular Therapeutics* that produces an annual issue on oligonucleotide therapeutics. As drug candidates, oligonucleotides have the advantage that base pairing is remarkably tolerant of chemical changes in the sugar–phosphate backbone and even in the bases themselves. Most modifications increase resistance to nucleases and some stabilize the duplex with RNA.

Replacing phosphate groups with thiophosphate (phosphorothioate oligonucleotides) not only slows down nuclease degradation but also destabilizes the duplex with RNA. Despite various side effects, this modification is useful as it is one of the few tolerated by ribonuclease H.

Modifications promoting a sugar conformation optimum for duplex formation increase duplex stability. This can be achieved by fusing a second ring onto the deoxyribose to restrict its conformation as in locked or bridged nucleic acids (40–42). Alternatively, ribonucleotides have long been alkylated at the 2′-hydroxyl group for nuclease resistance (43). Uncharged oligonucleotide analogues that are not subject to charge–charge repulsion and hybridize particularly strongly with RNA are phosphorodiamidate morpholino oligomers (44) and peptide nucleic acids (45).

These are some of the most frequently used modifications but there are many others and finding new ones is a favorite task of nucleoside chemists. A common approach is to combine different modifications to achieve a superior balance of pharmacokinetic, pharmacodynamic, and toxicological properties. To activate ribonuclease H, phosphorothioates are often included in the center of the oligonucleotide. These may be flanked by sequences with another modification that confers nuclease resistance and tighter binding to the RNA target (27). Alternatively, the most strongly hybridizing oligonucleotides may be used to prevent translation or splicing by steric blocking. This obviates the need for ribonuclease H and phosphorothioates.

3.1.2. siRNA

Cells make their own antisense ribooligonucleotides. They occur in siRNA or miRNA duplexes. In the RISC particle, the antisense strand (or guide strand) hybridizes to the target RNA and directs nuclease cleavage. Synthetic siRNA introduced into cells performs the same function. The synthetic RNA can be introduced in three forms.

1. A duplex of around 22 base pairs with 3′-end overhangs of two bases. This is the form of natural siRNA.
2. A short RNA hairpin (shRNA) that mimics the natural precursor of siRNA. Once in cells, this is processed to siRNA of the same form as just described.
3. A single antisense strand.

Ribooligonucleotides are broken down rapidly by ribonucleases. Opportunities for protection of siRNA by chemical modification are greatly restricted by the requirements of recognition by RISC. Guidelines have been developed for design of synthetic siRNA incorporating 2′-*O*-methyl, phosphorothioate or 2-thiouracil modifications at certain selected positions (46–48). Unmodified siRNA might be used directly for therapy (49) but it is generally protected from nucleases by encapsulation in a liposome or some other delivery agent. siRNA can also be transcribed from vectors in cells.

As with antisense oligonucleotides, delivery of siRNA is the biggest obstacle to overcome. It involves survival in the circulation, cell entry, and escape from any packaging (such as liposomes) or from endosomes where uptake often leads. In vivo, modified siRNA may enter at least some cells directly (50). A variety of packaging technologies are in various stages of development all the way up to clinical trials. These involve lipids, fusion proteins, and other macromolecules (33).

The problem of delivery is mitigated if the drug can be delivered locally. In a corporate statement, Alnylam claimed the first antiviral proof of principle in humans by intranasal delivery of naked, modified siRNA against respiratory syncytial virus. TransDerm has reported activity with one patient in a trial of naked, unmodified siRNA injected to treat a skin disease of the foot (pachyonychia congenita) (33).

Similar to earlier experiences with antisense oligonucleotides, unexpected biological properties have emerged during the development of siRNA therapeutics. These include effects on the immune system and on endogenous RNA processing (33, 51–54). For example, angiogenesis in the eye was ascribed to innate immune response to double-stranded RNA (55). Immunotoxic effects have been avoided without loss of desired activity by chemical modification (52). Further, transfection of siRNA upregulates genes under the control of miRNA. Possibly exogenous RNA swamps intracellular RNA-processing pathways (56). In mice, oversaturation of miRNA pathways with shRNA is fatal (57).

3.1.3. Ribozymes and DNAzymes

It is fairly straightforward to design a ribozyme to cleave a sequence found in a target RNA (58). The ribozyme must contain antisense sequences that will bind to the target. It must also contain a sequence that will fold into a structure with ribonuclease activity. Such sequences are found in natural hammerhead or hairpin ribozymes. Unlike siRNA or many antisense oligonucleotides, ribozymes need not depend on cellular nucleases for activity.

Composite synthetic ribozymes have been studied extensively in vitro. The antisense regions may be readily modified for protection against nucleases, but nucleosides required for catalytic activity cannot be changed without loss of activity. Although cleavage of target can be demonstrated in cell-free systems, the performance of ribozymes in vivo has been less encouraging and interest in them as therapeutics has waned of late. However, they may still be contenders for expression from vectors in transformed cells. A recent study compared siRNA, the current yardstick for oligonucleotides, with a ribozyme directed at the same site on HIV RNA (59). The ribozyme suppressed the virus throughout the 56-day observation period while the siRNA was inactive.

A surprising development in the study of ribozymes was the discovery of analogous deoxyoligonucleotides with ribonuclease activity (60). These DNAzymes have catalytic domains unrelated to any yet found in nature and may be remarkably simple (61). Also, they may be more amenable to modification than ribozymes. A modified LNA DNAzyme was recently shown to block human papilloma virus replication in cells and might be a candidate for use against cervical cancer (62).

3.2. Oligonucleotides That Hybridize to DNA

At least two applications can be envisaged for antigene oligonucleotides that act on DNA directly rather than RNA. One is to modulate gene expression at its earliest stage. The other is to correct harmful sequences arising from mutations. The problem is in finding oligonucleotides that can bind to DNA that is already double stranded.

One solution utilizes oligonucleotides that give particularly stable duplexes and displace one of the strands in DNA forming a “D loop.” Examples are locked nucleic acids (LNA), described earlier, and peptide nucleic acids (PNA). The latter are not true nucleic acids as the sugar–phosphate backbone is replaced by an uncharged polypeptide. LNAs or PNAs complementary to a sequence in a promoter can inhibit gene expression in human cells (63–65). A peptide was conjugated to PNA in these studies so as to assist uptake into cells.

Rather than trying to separate the two DNA strands, earlier work used oligonucleotides that can form triple-stranded structures (triplex-forming oligonucleotides or TFOs) that were reviewed recently (66). TFOs take up residence in the major groove of the double helix and form Hoogsteen or reverse Hoogsteen base pairs with one of the strands. To form a helix under physiological conditions, this strand must be very A/G rich, which still restricts any therapeutic potential despite extensive research using modified oligonucleotides. TFOs have proved useful as research tools and there are now many reports of their activity in cells. They can inhibit elongation of the growing RNA or, surprisingly, may stimulate transcription if they bind upstream of a promoter. When conjugated with psoralen, a substance that cross-links with DNA on irradiation, TFOs can induce site-specific recombination and gene correction in mammalian cells (67).

3.3. Oligonucleotides That Bind Targets Other Than Nucleic Acids

Oligonucleotides may bind to other types of molecules such as proteins or small molecules. This does not involve Watson–Crick base pairing but may sometimes be sufficiently selective and robust for therapeutic applications.

3.3.1. Aptamers

Single-stranded oligonucleotides may fold into complex secondary and tertiary structures. This is illustrated by the familiar structures adopted by tRNA molecules. A peptide exposed to a library of oligonucleotides with many different sequences is likely to find a few that have architectures to which it can bind. If these oligonucleotides can be separated and amplified using PCR, the library can be enriched in sequences that bind to the peptide. By reiterating the process many times with increasing stringency, oligonucleotides that bind selectively and strongly can be identified, sequenced, and synthesized. This process is known by the acronym SELEX (Systematic Evolution of Ligands by EXponential enrichment) and the oligonucleotide products are called aptamers. Just as ribozymes and DNAzymes are nucleic acids analogous to enzymes, so aptamers are nucleic acids analogous to antibodies. Aptamers can be raised against proteins or small molecules and manufactured without the use of live animals. An example is Macugen® (pegaptanib) for treatment of wet macular degeneration (68). This is only the second oligonucleotide to receive FDA approval.

Aptamers were reviewed recently (47). Like antibodies, they might be used to inhibit the action of harmful proteins such as those overexpressed on the surface of cancer cells (69). By binding to surface proteins, aptamers might also be used to target cells for drug delivery. For example, aptamers have been conjugated to drugs (70), photosensitizers (71), liposomes (72), and nanoparticles (73). They have even been used to target siRNAs to HIV-infected cells (74). Through choice of the right target, an aptamer has been used to induce gene expression (75).

Like antibodies, the action of aptamers is a function of their tertiary structure. It would not be surprising if changes to the sequence or chemical nature of an aptamer abrogated its activity so any modifications are best introduced during the SELEX procedure. This restricts chemical modifications of the starting nucleotides to those accepted by the polymerase used in the PCR step (76).

3.3.2. Decoys

Proteins such as transcription factors recognize certain sequences in DNA. They can also bind to the same sequence in a double-stranded decoy oligonucleotide. Decoy oligonucleotides are used to lure proteins away from their binding sites in DNA and a variety of therapeutic applications have been suggested (77). Any modifications introduced to improve nuclease resistance must not prevent recognition by the protein. Partial protection against exonucleases may be achieved by restricting modifications to the ends of the decoy or by using circular oligonucleotides with self-complementary sequences giving a dumb-bell configuration (78). To protect against endonucleases as well, locked nucleic acids have been used as decoys (79).

Currently, decoy oligonucleotides are not pursued as aggressively as other forms of therapeutic oligonucleotides. However, in a recent development, a decoy oligonucleotide was used as a suicide inhibitor of DNA methyltransferase (80). The oligonucleotide contained three hemimethylated CpG motifs that are recognized by the enzyme. Within this binding site were incorporated 5-fluoro-2′-deoxycytidine residues that link covalently to the enzyme when it tries to methylate them. The oligonucleotide was further partially modified with phosphorothioates.

3.3.3. CpG Oligonucleotides

The most familiar defense of the body against pathogens involves highly specific antibodies produced by the adaptive immune system. Less well known is the innate immune system that responds to bacteria or viruses in general. A family of proteins, the toll-like receptors (TLR), searches for molecules originating in pathogens. This early-warning system stimulates the innate immune system and other responses. Triggering or blocking the innate immune system is a possible strategy for treatment of infections, cancer, allergies, and autoimmune diseases.

Several TLRs (TLR3, 7, 8, and 9) recognize foreign nucleic acids (24). For example, a marker that distinguishes mammalian from bacterial DNA is the dimer CG. This occurs more frequently in bacterial DNA and, furthermore, in mammalian DNA, the cytidine is usually methylated. The function of TLR 9 is to detect the presence of bacterial DNA. By binding to unmethylated CG in certain DNA sequence contexts, it initiates an antibacterial response.

Synthetic oligonucleotides that mimic bacterial DNA bind to TLR9 leading to various immunological responses. This was originally discovered through investigating unexpected side effects of antisense oligonucleotides. Once the cause of these effects was understood, they could be avoided or induced as desired through manipulation of sequences. As has often happened with oligonucleotides, unwanted effects suggested new therapeutic opportunities. Oligonucleotides that stimulate TLR9 have been used for immunotherapy all the way to clinical trials (81), and modified oligonucleotides have recently been discovered that can antagonize the action of TLR9 (81).

4. Conclusion

A few years ago, a leading US politician tied himself in knots talking about the "things we know we don't know and the things we don't know we don't know." The wording caused some amusement at the time. But the story of therapeutic oligonucleotides contains one incidence after another of "things we didn't know we didn't know." That has helped make the development of oligonucleotides more difficult. In addition to the problems we knew about, there were others we did not know about. We did not know, for example, that the innate immune system responds to exogenous oligonucleotides and they would have to be designed to avoid this effect. However, the discovery of unsuspected phenomena has proven to be a wellspring of inspiration for those working in oligonucleotides. Ribozymes and RNA interference are two examples. Such discoveries have been responsible for increasing the number of therapeutic approaches from one (Zamecnik's antisense oligonucleotides) to all those described above. There may be more to come but we don't know what we don't know!

The enduring appeal of oligonucleotides is that the same approach may be used to design drugs against many diseases. All oligonucleotide must successfully overcome the same obstacles if they are to reach an intracellular target. These include crossing cell membranes and facing an onslaught of nucleases. Both of these contribute to the biggest problem: achieving a high-enough concentration of drug in the vicinity of the target. Although this review has often stressed differences between individual oligonucleotides, they are all still from the same family of polyanions. When it comes to getting them across cell membranes and resisting nucleases, it may be that what works for one oligonucleotide would work for many. The rewards of success in one case may yet be enormous.

References

1. Zamecnik, P. C., & Stephenson, M. L. (1978) Inhibition of Rous sarcoma virus replication and cell transformation by a specific oligodeoxynucleotide. *Proceedings of the National Academy of Sciences USA* **75**, 280–284.
2. Stephenson, M. L., & Zamecnik, P. C. (1978) Inhibition of Rous sarcoma viral RNA translation by a specific oligodeoxynucleotide. *Proceedings of the National Academy of Sciences USA* **75**, 285–288.
3. Zamecnik, P., Goodchild, J., Taguchi, Y., & Sarin, P. S. (1986) Inhibition of replication and expression of human T-cell lymphotropic virus type III in cultured cells by exogenous synthetic oligonucleotides complementary to viral RNA. *Proceedings of the National Academy of Sciences USA* **83**, 4143–4146.
4. Wilson, T. J., & Lilley, D. M. (2009) The evolution of ribozyme chemistry. *Science* **232**, 1436–1438.
5. Breaker, R. R. (2008) Complex riboswitches. *Science* **319**, 1795–1797.
6. Ghildiyal, M., & Zamore, P. D. (2009) Small silencing RNAs: An expanding universe. *Nature Reviews Genetics* **10**, 94–108.
7. Lewis, B. P., Burge, C. B., & Bartel, D. P. (2005) Conserved seed pairing, often flanked by adenosines, indicates that thousands of human genes are microRNA targets. *Cell* **120**, 15–20.
8. Moazed, D. (2009) Small RNAs in transcriptional gene silencing and genome defence. *Nature* **457**, 413–420.
9. Carthew, R. W., & Sontheimer, E. J. (2009) Origins and mechanisms of miRNAs and siRNAs. *Cell* **136**, 642–655.
10. Vasudevan, S., Tong, Y., & Steitz, J. A. (2007) Switching from repression to activation: MicroRNAs can up-regulate translation. *Science* **318**, 1931–1934.
11. Dolgin, E. (2009) Now showing: RNA activation. *The Scientist* **23**, 34–39.
12. Farazi, T. A., Juranek, S. A., & Tuschl, T. (2008) The growing catalog of small RNAs and their association with distinct Argonaute/Piwi family members. *Development* **135**, 1201–1214.
13. Olejniczak, M., Galka, P., & Krzyzosiak, W. J. (2010) Sequence-non-specific effects of RNA interference triggers and microRNA regulators. *Nucleic Acids Research* **38**, 1–16.
14. Lee, H.-C., et al. (2009) qiRNA is a new type of small interfering RNA induced by DNA damage. *Nature* **459**, 274–277.
15. Ku, G., & McManus, M. T. (2008) Behind the scenes of a small RNA gene-silencing pathway. *Human Gene Therapy* **19**, 17–26.
16. Marquez, R. T., & McCaffrey, A. P. (2008) Advances in MicroRNAs: Implications for gene therapists. *Human Gene Therapy* **19**, 27–38.
17. Bonauer, A., et al. (2009) MicroRNA-92a controls angiogenesis and functional recovery of ischemic tissues in mice. *Science* **324**, 1710–1713.
18. Williams, A. H., et al. (2009) MicroRNA-206 delays ALS progression and promotes regeneration of neuromuscular synapses in mice. *Science* **326**, 1549–1554.
19. Amaral, P. P., Dinger, M. E., Mercer, T. R., & Mattick, J. S. (2008) The eukaryotic genome as an RNA machine. *Science* **319**, 1787–1789.
20. Pennisi, E. (2009) Some RNA may play key role in repressing genes, slowing cancer. *Science* **324**, 1252–1253.
21. Morris, K. V. (2009) RNA-directed transcriptional gene silencing and activation in human cells. *Oligonucleotides* **19**, 299–305.
22. Hawkins, P. G., Santoso, S., Adams, C., Anest, V., & Morris, K. V. (2009) Promoter targeted small RNAs induce long-term transcriptional gene silencing in human cells. *Nucleic Acids Research* **37**, 2984–2995.
23. Schwartz, J., et al. (2008) Antisense transcripts are targets for activating small RNAs. *Nature Structural & Molecular Biology* **15**, 842–848.
24. Panter, G., Kužnik, A., & Jerala, R. (2009) Therapeutic applications of nucleic acids as ligands for toll-like receptors. *Current Opinion Molecular Therapeutics* **11**, 133–145.
25. Juliano, R., Alam, M. R., Dixit, V., & Kang, H. (2008) Mechanisms and strategies for effective delivery of antisense and siRNA oligonucleotides. *Nucleic Acids Research* **36**, 4158–4171.
26. Juliano, R., Bauman, J., Kang, H., & Ming, X. (2009) Biological barriers to therapy with antisense and siRNA oligonucleotides. *Molecular Pharmaceutics* **6**, 686–695.
27. Stein, C. A., et al. (2010) Efficient gene silencing by delivery of locked nucleic acid antisense oligonucleotides, unassisted by transfection reagents. *Nucleic Acids Research* **38**, e3.
28. Rudnick, S. I., Swaminathan, J., Sumaroka, M., Liebhaber, S., & Gewirtz, A. M. (2008) Effects of local mRNA structure on posttranscriptional gene silencing. *Proceedings of the National Academy of Sciences* **105**, 13787–13792.

29. Ichihara, M., et al. (2007) Thermodynamic instability of siRNA duplex is a prerequisite for dependable prediction of siRNA activities. *Nucleic Acids Research* **35**, e123.
30. Kauffmann, A. D., Campagna, R. J., Bartels, C. B., & Childs-Disney, J. L. (2009) Improvement of RNA secondary structure prediction using RNase H cleavage and randomized oligonucleotides. *Nucleic Acids Research* **37**, e121.
31. Li, Z., Fortin, Y., & Shen, S.-H. (2009) Forward and robust selection of the most potent and noncellular toxic siRNAs from RNAi libraries. *Nucleic Acids Research* **37**, e8.
32. Gifford, L. K., et al. (2005) Identification of antisense nucleic acid hybridization sites in mRNA molecules with self-quenching fluorescent reporter molecules. *Nucleic Acids Research* **33**, e28.
33. Perkel, J. M. (2009) RNAi therapeutics: A two-year update. *Science* **326**, 454–456.
34. Lanford, R. E., et al. (2010) Therapeutic silencing of MicroRNA-122 in primates with chronic hepatitis C virus infection. *Science* **327**, 198–201.
35. Hashizume, R., et al. (2008) New therapeutic approach for brain tumors: Intranasal delivery of telomerase inhibitor GRN163. *Neuro-oncol* **10**, 112–120.
36. Lu, Y., et al. (2009) A single anti-microRNA antisense oligodeoxyribonucleotide (AMO) targeting multiple microRNAs offers an improved approach for microRNA interference. *Nucleic Acids Research* **37**, e24.
37. Medina, P. P., & Slack, F. J. (2009) Inhibiting microRNA function in vivo. *Nature Methods* **6**, 37–38.
38. Horwich, M. D., & Zamore, P. D. (2008) Design and delivery of antisense oligonucleotides to block microRNA function in cultured drosophila and human cells. *Nature Protocols* **3**, 1537–1549.
39. Trang, P., et al. (2009) Regression of murine lung tumors by the let-7 microRNA. *Oncogene* **29**, 1580–1587.
40. Koizumi, M. (2006) ENA® oligonucleotides as therapeutics. Current Opinion Molecular Therapeutics **8**, 144–149.
41. Mitsuoka, Y., et al. (2009) A bridged nucleic acid, 2′,4′-BNACOC: Synthesis of fully modified oligonucleotides bearing thymine, 5-methylcytosine, adenine and guanine 2′,4′-BNACOC monomers and RNA-selective nucleic-acid recognition. *Nucleic Acids Research* **37**, 1225–1238.
42. Ørum, H., & Wengel, J. (2001) Locked nucleic acids: A promising molecular family for gene-function analysis and antisense drug development. *Current Opinion Molecular Therapeutics* **3**, 239–243.
43. Rottman, F., Dunlap, B. E., & Friderici, K. H. (1971) 2′-O-methyl polynucleotides as templates for cell-free amino acid incorporation. *Biochemistry* **10**, 2581–2587.
44. Iversen, P. L., & Newbry, S. (2005) Manipulation of zebrafish embryogenesis by phosphorodiamidate morpholino oligomers indicates minimal non-specific teratogenesis. *Current Opinion Molecular Therapeutics* **7**, 104–108.
45. Boutimah-Hamoudi, F., et al. (2007) Cellular antisense activity of peptide nucleic acid (PNAs) targeted to HIV-1 polypurine tract (PPT) containing RNA. *Nucleic Acids Research* **35**, 3907–3917.
46. Bramsen, J. B., et al. (2009) A large-scale chemical modification screen identifies design rules to generate siRNAs with high activity, high stability and low toxicity. *Nucleic Acids Research* **37**, 2867–2881.
47. Thiel, K. W., & Giangrande, P. H. (2009) Therapeutic applications of DNA and RNA aptamers. *Oligonucleotides* **19**, 209–222.
48. Collingwood, M. A., et al. (2008) Chemical modification patterns compatible with high potency dicer-substrate small interfering RNAs. *Oligonucleotides* **18**, 187–200.
49. Hickerson, R. P., et al. (2008) Stability study of unmodified siRNA and relevance to clinical use. *Oligonucleotides* **18**, 345–354.
50. Querbes, W., et al. (2009) Direct CNS delivery of siRNA mediates robust silencing in oligodendrocytes. *Oligonucleotides* **19**, 23–29.
51. Kalluri, R., & Kanasaki, K. (2008) Generic block on angiogenesis. *Nature* **452**, 543–544.
52. Robbins, M., Judge, A., & MacLachlan, I. (2009) siRNA and innate immunity. *Oligonucleotides* **19**, 89–101.
53. Kenworthy, R., et al. (2009) Short-hairpin RNAs delivered by lentiviral vector transduction trigger RIG-I-mediated IFN activation. *Nucleic Acids Research* **37**, 6587–6599.
54. Olejniczak, M., Galka, P., & Krzyzosiak, W. J. (2009) Sequence-non-specific effects of RNA interference triggers and microRNA regulators. *Nucleic Acids Research* **38**, 1–16.
55. Kleinman, M. E., et al. (2008) Sequence- and target-independent angiogenesis suppression by siRNA via TLR3. *Nature* **452**, 591–597.
56. Khan, A. A., et al. (2009) Transfection of small RNAs globally perturbs gene regulation by endogenous microRNAs. *Nature Biotechnology* **27**, 549–555.
57. Grimm, D., et al. (2006) Fatality in mice due to oversaturation of cellular

microRNA/short hairpin RNA pathways. *Nature* **441**, 537–541.
58. Goodchild, J. (2000) Hammerhead ribozymes: Biochemical and chemical considerations. *Current Opinion Molecular Therapeutics* **2**, 272–281.
59. Müller-Kuller, T., Capalbo, G., Klebba, C., Engels, J. W., & Klein, S. A. (2009) Identification and characterization of a highly efficient anti-HIV pol hammerhead ribozyme. *Oligonucleotides* **19**, 265–271.
60. Schlosser, K., & Li, Y. (2009) Biologically inspired synthetic enzymes made from DNA. *Chemistry & Biology* **16**, 311–322.
61. Schlosser, K., & Li, Y. (2009) DNAzyme-mediated catalysis with only guanosine and cytidine nucleotides. *Nucleic Acids Research* **37**, 413–420.
62. Reyes-Gutiérrez, P., & Alvarez-Salas, L. M. (2009) Cleavage of HPV-16 E6/E7 mRNA mediated by modified 10-23 deoxyribozymes. *Oligonucleotides* **19**, 233–242.
63. Beane, R., Gabillet, S., Montaillier, C., Arar, K., & Corey, D. (2008) Recognition of chromosomal DNA inside cells by locked nucleic acids. *Biochemistry* **57**, 13147–13149.
64. Beane, R., et al. (2007) Inhibiting gene expression with locked nucleic acids (LNAs) that target chromosomal DNA. *Biochemistry* **46**, 7572–7580.
65. Hu, J., & Corey, D. R. (2007) Inhibiting gene expression with peptide nucleic acid (PNA)-peptide conjugates that target chromosomal DNA. *Biochemistry* **46**, 7581–7589.
66. Duca, M., Vekhoff, P., Oussedik, K., Halby, L., & Arimondo, P. B. (2008) The triple helix: 50 years later, the outcome. *Nucleic Acids Research* **36**, 5123–5138.
67. Liu, Y., Nairn, R. S., & Vasquez, K. M. (2009) Targeted gene conversion induced by triplex-directed psoralen interstrand crosslinks in mammalian cells. *Nucleic Acids Research* **37**, 6378–6388.
68. Fraunfelder, F. W. (2005) Pegaptanib for wet macular degeneration. *Drugs of Today* **41**, 703–709.
69. Blake, C. M., Sullenger, B. A., Lawrence, D. A., & Fortenberry, Y. M. (2009) Antimetastatic potential of PAI-1–Specific RNA aptamers. *Oligonucleotides* **19**, 117–128.
70. Huang, Y.-F., et al. (2009) Molecular assembly of an aptamer-drug conjugate for targeted drug delivery to tumor cells. *ChemBioChem* **10**, 862–868.
71. Mallikaratchy, P., Tang, Z. W., & Tan, W. H. (2008) Cell specific aptamer-photosensitizer conjugates as a molecular tool in photodynamic therapy. *ChemMedChem* **3**, 425–428.
72. Cao, Z., et al. (2009) Reversible cell-specific drug delivery with aptamer-functionalized liposomes. *Angewandte Chemie International Edition English* **48**, 6494–6498.
73. Levy-Nissenbaum, E., Radovic-Moreno, A. F., Wang, A. Z., Langer, R., & Farokhzad, O. C. (2008) Nanotechnology and aptamers: Applications in drug delivery. *Trends in Biotechnology* **26**, 442–449.
74. Zhou, J., et al. (2009) Selection, characterization and application of new RNA HIV gp 120 aptamers for facile delivery of dicer substrate siRNAs into HIV infected cells. *Nucleic Acids Research* **37**, 3094–3109.
75. Hunsicker, A., et al. (2009) An RNA aptamer that induces transcription. *Chemistry & Biology* **16**, 173–180.
76. Kang, J., Lee, M. S., Copland, J. A., Luxon, B. A., & Gorenstein, D. G. (2008) Combinatorial selection of a single standed DNA thioaptamer from whole cell selection. *Chemistry – A European Journal* **14**, 1769–1775.
77. Tomita, N., Morishita, R., Tomita, T., & Ogihara, T. (2002) Potential therapeutic applications of decoy oligonucleotides. *Current Opinion Molecular Therapeutics* **4**, 166–170.
78. Tomita, N., et al. (2003) Development of novel decoy oligonucleotides: Advantages of circular dumb-bell decoy. *Current Opinion Molecular Therapeutics* **5**, 107–112.
79. Crinelli, R., Bianchi, M., Gentilini, L., Palma, L., & Magnani, M. (2004) Locked Nucleic Acids (LNA): Versatile tools for designing oligonucleotide decoys with high stability and affinity. *Current Drug Targets* **5**, 745–52.
80. Warncke, S., Gégout, A., & Carell, T. (2009) Phosphorothioation of oligonucleotides strongly influences the inhibition of bacterial (M.HhaI) and human (Dnmt1) DNA methyltransferases. *ChemBioChem* **10**, 728–734.
81. Krieg, A. M. (2006) Therapeutic potential of toll-like receptor 9 activation. *Nature Reviews Drug Discovery* **5**, 471–484.

Chapter 2

Dinucleotides Containing 3′-*S*-Phosphorothiolate Linkages

James W. Gaynor and Richard Cosstick

Abstract

The 3′-*S*-phosphorothiolate (3′-SP) linkage has proven to be a very useful analogue of the phosphodiester group in nucleic acid derivatives; it is achiral and also shows good resistance to nucleases. Whilst oligonucleotides containing a 3′-SP linkage are best prepared using phosphoramidite chemistry, the corresponding dinucleotides are most efficiently synthesised using a Michaelis–Arbuzov reaction between a nucleoside 5′-phosphite and a nucleoside 3′-*S*-disulphide. The method described here is for a thymidine dinucleotide and is based on the use of a silyl phosphite, which is more reactive than simple alkyl phosphites and also simplifies the deprotection strategy. Full experimental details and spectroscopic data for the synthetic intermediates and the target dinucleotide are provided.

Key words: Phosphorothiolate, dinucleotide, oligonucleotide, Michaelis–Arbuzov reaction, silyl phosphate, chemical synthesis.

1. Introduction

Historically, dinucleotides (or more correctly dinucleoside phosphates) have been prepared as model compounds to evaluate the potential of different chemical approaches to the synthesis of oligonucleotides (1) and have not therefore attracted much attention in their own right. However, in the last decade dinucleotides have become recognised as a distinct group of anti-viral agents that can inhibit the activity of HIV integrase (2, 3) and show activity against hepatitis B (4, 5).

As a potential means to enhance biological activity of nucleotides, the phosphodiester bond is usually modified in order to increase resistance to nucleases and phosphodiesterases. The phosphorothioate modification, in which sulphur replaces

J. Goodchild (ed.), *Therapeutic Oligonucleotides*, Methods in Molecular Biology 764,
DOI 10.1007/978-1-61779-188-8_2, © Springer Science+Business Media, LLC 2011

a non-bridging oxygen atom, is an effective analogue for this purpose (4) and is convenient to prepare by chemical procedures (6, 7). This modification results in an asymmetric phosphorus atom and therefore produces mixtures of diastereoisomers, unless the stereochemistry is controlled (8). Replacement of one of the bridging oxygen atoms leads to an achiral phosphorothiolate linkage that can be prepared with either the 3′-oxygen (9) or the 5′-oxygen (9, 10) substituted by sulphur.

The 3′-*S*-phosphorothiolate (3′-SP) linkage (**Fig. 2.1**) initially attracted interest as a probe to obtain detailed mechanistic information about enzymes that process nucleic acids (9). However, subsequently it has been shown through NMR studies on both dinucleotides (11) and oligonucleotide duplexes (12) that a single 3′-*S*-phosphorothiolate modification shifts the conformation of its attached 2′-deoxyribose sugar almost completely to what is described as the north or (C-3′ *endo*) conformation, which is the sugar conformation found in RNA. Thus, oligodeoxynucleotides containing a phosphorothiolate linkage are conformationally good mimics of RNA and as a result they form thermodynamically more stable duplexes with complementary RNA than the corresponding unmodified oligodeoxynucleotides (13). This property suggests that oligonucleotides containing 3′-SP linkages may be of use as both antisense agents and in RNA interference, where enhanced binding to an RNA target is required. It is also interesting to note that dinucleotides containing sugars that are expected to exist predominantly in the north conformation also show good activity against hepatitis B (4).

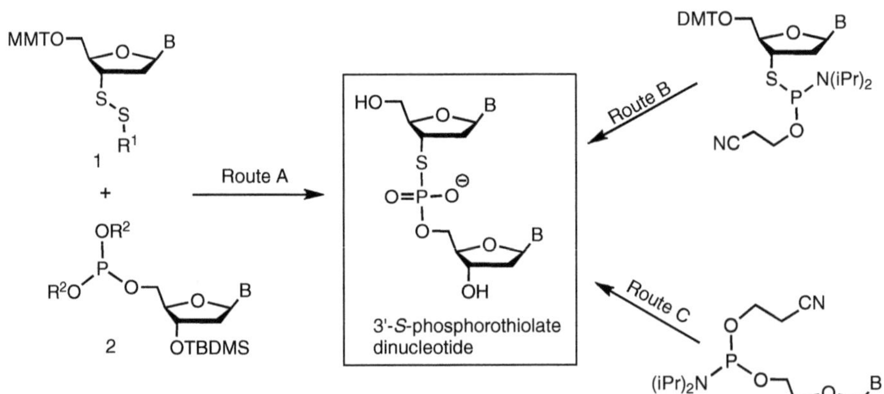

Fig. 2.1. Synthetic routes for the preparation of phosphorothiolate dinucleotides and oligonucleotides. Route A shows the Michaelis–Arbuzov reaction, which is the preferred method for the preparation of dinucleotides containing a phosphorothiolate linkage. Routes B and C show methods based on phosphoramidite chemistry, which have mainly been used for the automated synthesis of oligonucleotides with a 3′-SP modification. B = nucleobase, R^1 = aryl, R^2 = alkyl or trimethyl silyl, MMT = monomethoxytrityl, DMT = dimethoxytrityl, TBDMS = *tertiary*-butyldimethylsilyl.

The chemical synthesis of oligonucleotides containing 3′-SP linkages requires manipulation of the deoxyribose sugar and is therefore more challenging than the synthesis of the phosphorothioate analogues. A number of different chemical approaches have been used to prepare dinucleotides and oligonucleotides containing 3′-SP linkages. Our own research has shown that the most efficient route to prepare phosphorothiolate dinucleotides is through a Michaelis–Arbuzov reaction between a nucleoside 3′-*S*-disulphide (1, **Fig. 2.1**) and a nucleoside 5′-*O*-phosphite (2, **Fig. 2.1**) (Route A, **Fig. 2.1**) (14, 15). An attractive feature of this reaction is that a number of different alkyl or silyl substituents can be used in the phosphite component and this provides a means to control reactivity and vary the protecting group that remains on the phosphorothiolate linkage at the completion of the reaction. In this respect the silyl phosphites are particularly valuable in that they are more reactive than simple alkyl phosphites and are easily prepared from the corresponding *H*-phosphonates (14, 15). In addition, the silyl group on the phosphorothiolate linkage is removed when the reaction is quenched by the addition of water (Step 8 of **Section 3.3**). For these reasons the Michaelis–Arbuzov reaction based on the use of silyl phosphites is presented below as the method of choice for preparing dinucleotides containing 3′-SP linkages and the overall reaction route is shown in **Fig. 2.2**. The specific example chosen is a thymidine dinucleotide although 3′-thionucleosides have been prepared from both purine and pyrimidine deoxynucleosides (16, 17).

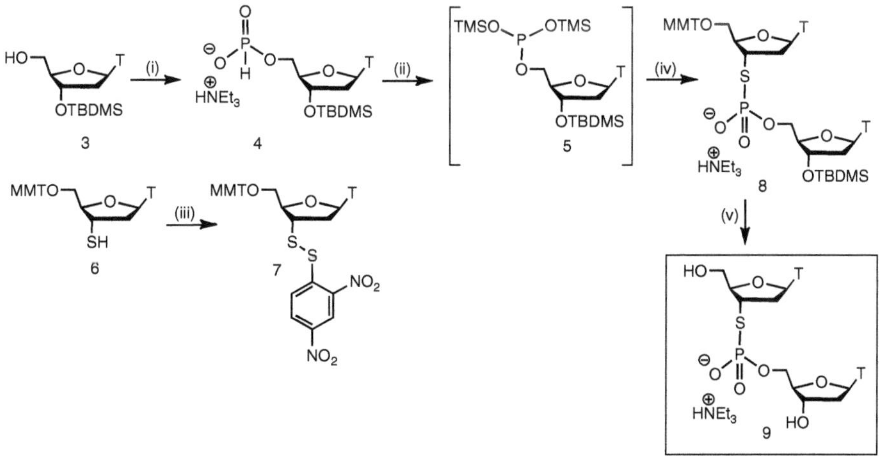

Fig. 2.2. Reagents: (i) phosphorus trichloride, *N*-methylmorpholine, 1,2,4-triazole in dichloromethane; (ii) bis(trimethylsilyl)trifluoroacetamide in dichloromethane; (iii) 2,4-dinitrophenylsulfenyl chloride, pyridine in THF; (iv) compound 7 in acetonitrile; (v) acetic acid:H_2O (4:1) followed by NEt_3.3HF. MMT = monomethoxytrityl, TBDMS = *tertiary*-butyldimethylsilyl.

It should be noted for the purpose of completeness that the phosphoramidite approach has been widely used for the synthesis of oligonucleotides containing up to five 3′-SP linkages (16) and this has been achieved using 3′-*S*-phosphorothioamidites as shown in **Fig. 2.1** (Route B). A detailed protocol for the automated solid-phase synthesis of oligonucleotides incorporating 3′-SP linkages has been published (18), which could certainly be used for the preparation of dinucleotides containing this modification. More recently the amidite method has also been used in a reverse approach, through reaction of a 3′-thionucleoside with a 5′-*O*-phosphoramidite (**Fig. 2.1**, Route C) (17). This route however proved to be relatively inefficient and was most suitable for preparing oligonucleotides that terminated with a 3′-thionucleoside.

2. Materials

2.1. Chemicals

1. 5′-*O*-Monomethoxytrityl-3′-deoxy-3′-thiothymidine (6, **Fig. 2.2**) [synthesised via previously reported protocol using monomethoxytrityl chloride in place of dimethoxytrityl chloride (18)]
2. 3′-*O*-(*tert*-Butyldimethylsilyl)-thymidine (Berry & Associates)
3. Acetic acid
4. Bis(trimethylsilyl)trifluoroacetamide (Sigma-Aldrich)
5. C_{18} reversed-phase silica gel
6. Deuterated chloroform
7. Deuterated water
8. 2,4-Dinitrophenylsulfenyl chloride (Sigma-Aldrich)
9. Dowex 50W-X8 (Hydrogen form) ion exchange resin
10. Magnesium sulphate
11. *N*-methylmorpholine
12. Nitrogen (oxygen-free or white-spot)
13. Phosphorus trichloride
14. Sand (low in iron, 40–100 mesh)
15. Silica gel (particle size 40–63 μm)
16. Sodium bicarbonate aqueous solution (saturated)
17. Sodium sulphate
18. Thin-layer silica gel plated on aluminium backing
19. 1,2,4-Triazole

20. Triethylamine trihydrofluoride
21. Triethylamine

2.2. Reagent Setup

1. Triethylammonium bicarbonate (TEAB) solution (1 M): this is prepared by bubbling carbon dioxide gas (warming solid carbon dioxide) through a solution of triethylamine (139 mL) in distilled water (800 mL) until the required pH is attained (~8.5). The volume of the solution is then adjusted to 1 L using additional distilled water (*see* **Note 1**).
2. TEAB solution (0.1 M): dilute 1 M TEAB (100 mL) into 1 L of distilled water.
3. *p*-Anisaldehyde stain: prepared by mixing *p*-anisaldehyde (6 mL) with sulphuric acid (8 mL), acetic acid (2.4 mL) and ethanol (218 mL).
4. Acetic acid:water (4:1).

2.3. Solvents and Solvent Setup

1. Acetonitrile, anhydrous DNA grade (Link Technologies)
2. Dichloromethane
3. Dichloromethane, anhydrous (Sigma-Aldrich). Can also be distilled from calcium hydride (*see* **Note 2**)
4. Ethanol, absolute
5. Methanol
6. Petroleum ether (40–60)
7. Pyridine, anhydrous (Sigma-Aldrich). Can also be distilled from calcium hydride (*see* **Note 2**)
8. Tetrahydrofuran
9. Tetrahydrofuran, anhydrous (Sigma-Aldrich). Can also be distilled from sodium and benzophenone (*see* **Note 2**)
10. Toluene

2.4. Equipment

1. Disposable 21-gauge hypodermic syringe needles with Luer connection
2. Disposable plastic syringes with Luer connections
3. Reusable, non-sterile 19-gauge metal needles with Luer connection
4. Pasteur pipettes
5. Rubber septum, various sizes
6. Standard organic laboratory glassware

2.5. Equipment/ Procedural Setup

1. Silica gel (*see* **Note 3**) column chromatography: pack the column using the appropriate eluent. Dissolve the impure compound into a minimum volume of eluent and place the resulting crude oily mixture on top of the silica bed using

a Pasteur pipette (*see* **Note 4**). Allow the sample to enter the silica bed. Cover the top of the silica with layer of sand (2.5 cm thick) and run the column using the appropriate eluent system. Use gentle air or nitrogen pressure to aid elution and identify fractions containing pure product by eluting thin-layer silica gel plates on aluminium backing and staining with anisaldehyde (Step 3 of **Section 2.2**). Combine fractions containing pure product and remove all solvents using a rotary evaporator.

2. C_{18} reversed-phase silica gel column chromatography: performed in an analogous way to regular column chromatography above.

3. Methods

Michaelis-Arbuzov chemistry is the best approach for synthesising dinucleotides containing a 3′-*S*-phosphorothiolate linkage. Various disulphides and phosphite triesters have been investigated and the procedure below, as outlined in **Fig. 2.2**, is considered to be the most efficient. The key step in the synthesis is the reaction between a nucleoside silyl phosphite (5, **Fig. 2.2**), which is prepared in situ by silylation of the *H*-phosphonate (4, **Fig. 2.2**), and the nucleoside disulphide (7, **Fig. 2.2**). The resultant 3′,5′-protected dimer (8, **Fig. 2.2**) is then deprotected to give the product dinucleotide (9, **Fig. 2.2**).

3.1. Preparation of 5′-O-Monomethoxytrityl-3′-deoxy-3′-(2,4-dinitrophenyldisulfanyl)-thymidine

See 7, **Fig. 2.2**.

1. Weigh 2,4-dinitrophenylsulfenyl chloride (0.89 g, 3.8 mmol) into a 100-mL round bottomed flask (flask 1), add a Teflon-coated magnetic stir bar and seal the top with a rubber septum.
2. Insert a nitrogen (*see* **Note 5**) inlet and vent (consisting of disposable syringe needles) through the rubber septum and allow the flask to be filled with nitrogen (at least 20 min, *see* **Note 6**).
3. Add anhydrous THF (40 mL) to flask 1, making sure the stock anhydrous THF solution is kept anhydrous by way of a nitrogen inlet.
4. Place flask 1 into an ice bath ensuring the ice bath is maintained at 0–2°C. Leave solution to cool (15 min).
5. Whilst flask 1 is cooling, weigh 5′-*O*-monomethoxytrityl-3′-deoxy-3′-thiothymidine (6, **Fig. 2.2**) (1.0 g, 1.9 mmol) into a separate 50-mL round bottomed flask (flask 2) and add anhydrous pyridine (20 mL). Co-evaporate the pyridine on

a rotary evaporator (to help remove any traces of water from the nucleoside) leaving a thick oil.

6. Seal flask 2 with a rubber septum and insert a nitrogen inlet and vent (consisting of disposable syringe needles) through the rubber septum and allow the flask to be filled with nitrogen (at least 20 min, *see* **Note 7**).
7. Add anhydrous THF (20 mL) to flask 2 and manually agitate to aid dissolution. Transfer the contents of flask 2 into flask 1 using a long metal needle (dropwise) and leave the combined solution to stir at 0–2°C for 30 min.
8. Remove the flask from the ice bath and allow the flask to warm to room temperature. Allow the solution to stir for an additional 1 h.
9. Transfer the contents of the flask to a 250-mL separating funnel containing saturated aqueous $NaHCO_3$ (75 mL) and dichloromethane (75 mL). Shake the flask and discard the aqueous layer (*see* **Note 8**).
10. Wash the dichloromethane layer with two more portions of saturated aqueous $NaHCO_3$ (2 × 75 mL). Dry the combined organic layer by adding magnesium sulphate (generally 4–5 spatulas) until the powdered magnesium sulphate settles slowly ('snowstorm' effect) when the flask is agitated. Filter off the magnesium sulphate under gravity through a fluted filter paper in a glass funnel. Concentrate the filtrate to dryness using a rotary evaporator.
11. Purify the impure compound by column chromatography following the procedure given in **Section 2.5**. Use dichloromethane containing 1% methanol to pack the column, dissolve impure compound using dichloromethane and elute product by slowly increasing the proportion of methanol from 1 to 3%. 5′-*O*-Monomethoxytrityl-3′-deoxy-3′-(2,4-dinitrophenyldisulfanyl)-thymidine will appear as a yellow amorphous solid.

3.2. Preparation of 3′-O-(tert-Butyldimethylsilyl)-thymindin-5′-yl Phosphonate Triethylammmonium Salt

See 4, **Fig. 2.2**.

1. To the side neck of a two-necked flask (flask 1), fit a glass stopper with appropriately sized Keck clip. Insert a Teflon-coated magnetic stir bar to the flask and fit the main neck with a pressure-equalising dropping funnel.
2. Attach a nitrogen bubbler to the top of the pressure-equalising dropping funnel and allow the flask to be filled with nitrogen (at least 20 min, *see* **Note 6**).
3. Remove the stopper from the side neck of flask 1 and add anhydrous dichloromethane (14 mL), making sure the stock anhydrous dichloromethane solution is kept anhydrous by way of a nitrogen inlet. Return the stopper to flask 1.

4. Remove the stopper from the side neck of flask 1 and add phosphorus trichloride (1.2 mL, 14 mmol, *see* **Note 9**) and *N*-methylmorpholine (15.7 mL, 140 mmol). Return the stopper to flask 1 and leave the solution to stir for 5 min to ensure dissolution.
5. To the stirring solution in flask 1, use a spatula to slowly add 1,2,4-triazole (3.2 g, 47 mmol) to flask 1 via the side neck. Leave mixture stirring for 30 min.
6. Whilst flask 1 is stirring, weigh 3′-*O*-(*tert*-butyldimethylsilyl)-thymidine (3, **Fig. 2.2**) (1 g, 2.8 mmol) into a separate 50-mL round bottomed flask (flask 2) and add anhydrous acetonitrile (20 mL). Co-evaporate the acetonitrile on a rotary evaporator to help remove any traces of water from the nucleoside.
7. Seal flask 2 with a rubber septum and insert a nitrogen inlet and vent (consisting of disposable syringe needles) through the rubber septum and allow the flask to be filled with nitrogen (at least 20 min).
8. After flask 1 has been stirred for 30 min, place flask 1 into an ice bath ensuring the ice bath is maintained at 0–2°C. Leave solution to cool (15 min).
9. Whilst flask 1 is cooling, add anhydrous dichloromethane (38 mL) to flask 2 and manually agitate to aid dissolution. Using a long metal needle, transfer the contents of flask 2 to the top of the pressure-equalising dropping funnel on flask 1, ensuring the tap on the funnel is closed.
10. Slowly add the contents of the dropping funnel into the stirring solution in flask 1 over a 20-min period. After addition, allow the mixture to stir for an additional 10 min.
11. To a 500-mL separating funnel, add triethylammonium bicarbonate (TEAB, 1 M, pH 8.5; 110 mL). Transfer the contents of the reaction flask to the separating funnel. Shake the flask and collect the organic layer into a conical flask or another suitable receptacle.
12. To the separating funnel, add a further portion of dichloromethane (200 mL) and wash the aqueous layer. Collect the organic layer and combine with the previous organic layer.
13. Dry the combined organic layer by adding sodium sulphate (generally 4–5 spatulas) until the powdered sodium sulphate settles slowly ('snowstorm' effect) when the conical flask is agitated. Filter off the sodium sulphate under gravity through a fluted filter paper in a glass funnel. Concentrate the filtrate to dryness using a rotary evaporator.

14. Purify the impure compound by column chromatography following the procedure given in **Section 2.5**. Use dichloromethane containing 2% triethylamine and 3% methanol to pack the column, dissolve impure compound using dichloromethane and elute product by slowly increasing the proportion of methanol from 3 to 5% (maintaining 2% triethylamine). 3′-*O*-(*tert*-Butyldimethylsilyl)-thymindin-5′-yl phosphonate triethylammmonium salt will appear as a white amorphous solid (*see* **Note 10**).

3.3. Preparation of 5′-O-Monomethoxytrityl-3′-thiothymidylyl-(3′→5′)-[3′-O-(tert-butyldimethylsilyl)-thymidine] Triethylammonium Salt

See 8, **Fig. 2.2**.

1. Weigh 3′-*O*-(*tert*-butyldimethylsilyl)-thymindin-5′-yl phosphonate triethylammmonium salt (4, **Fig. 2.2**) (0.521 g, 1 mmol) into a 25-mL round bottomed flask (flask 1) and add anhydrous acetonitrile (10 mL). Co-evaporate the acetonitrile on a rotary evaporator to help remove any traces of water from the nucleoside.
2. Add a Teflon-coated magnetic stir bar to flask 1 and seal the top with a rubber septum. Insert a nitrogen inlet and vent (consisting of disposable syringe needles) through the rubber septum and allow the flask to be filled with nitrogen (at least 20 min).
3. Add anhydrous dichloromethane (2 mL), making sure the stock anhydrous dichloromethane solution is kept anhydrous by way of a nitrogen inlet. Stir the solution until dissolution is achieved.
4. Add bis(trimethylsilyl)trifluoroacetamide (0.4 mL, 1.47 mmol) to flask 1 and leave to stir for 15 min. This forms 3′-*O*-(*tert*-butyldimethylsilyl)-thymidin-5′-yl-bis(trimethylsilyl) phosphite (5, **Fig. 2.2**) in situ in almost quantitative yield.
5. Weigh 5′-*O*-monomethoxytrityl-3′-deoxy-3′-(2,4-dinitrophenyldisulfanyl)-thymidine (7, **Fig. 2.2**) (0.698 g, 0.96 mmol) into a 10-mL round bottomed flask (flask 2) and seal the top with a rubber septum.
6. Insert a nitrogen inlet and vent (consisting of disposable syringe needles) through the rubber septum and allow the flask to be filled with nitrogen (at least 10 min)
7. Add anhydrous dichloromethane (1 mL) and manually agitate until dissolution occurs; then transfer the contents of flask 2 to flask 1 dropwise, using a long metal needle. A deep red solution will occur; leave the reaction to stir for 30 min.
8. Quench the reaction with distilled water (0.1 mL) and transfer the contents of the flask to a 50-mL separating funnel containing saturated aqueous $NaHCO_3$ (20 mL) and dichloromethane (20 mL). Shake the separating funnel,

discard the aqueous layer and transfer the organic layer to a conical flask.

9. Dry the organic layer by adding magnesium sulphate (generally 4–5 spatulas) until the powdered magnesium sulphate settles slowly ('snowstorm' effect) when the conical flask is agitated. Filter off the magnesium sulphate under gravity through a fluted filter paper in a glass funnel. Concentrate the filtrate to dryness using a rotary evaporator.
10. Purify the impure fully protected dinucleotide by column chromatography following the procedure given in **Section 2.5**. Use dichloromethane containing 2% triethylamine and 2% methanol to pack the column, dissolve impure fully protected dinucleotide using dichloromethane and elute product by slowly increasing the proportion of methanol from 2 to 5% (maintaining 2% triethylamine). 5′-*O*-Monomethoxytrityl-3′-thiothymidylyl-(3′→5′)-[3′-*O*-(tert-butyldimethylsilyl)thymidine] triethylammonium salt (8, **Fig. 2.2**) will appear as a white amorphous solid.

3.4. Preparation of 3′-Thiothymidylyl-(3′→5′) thymidine Triethylammonium Salt

See 9, **Fig. 2.2**.

1. Weigh out the protected dinucleotide (compound 8, **Fig. 2.2**) (0.25 mmol) into a 50 mL-round bottomed flask and add acetic acid:H_2O (4:1) (25 mL). Add a Teflon-coated magnetic stir bar and leave the solution to stir overnight and evaporate the acetic acid using a rotary evaporator. Wash the residue with ethanol (25 mL) and evaporate.
2. Using petroleum ether (40–60, 25 mL), triturate the residue and decant off the petroleum ether. Add toluene (25 mL) and co-evaporate using a rotary evaporator.
3. Dissolve the residue in THF (4 mL) and add $NEt_3.3HF$ (8.5 mL, 50 mmol). Add a Teflon-coated magnetic stir bar and leave to stir overnight.
4. Add water (1 mL) to quench the reaction and then add triethylamine until the reaction mixture becomes neutral by spotting on pH paper. Evaporate the solution to dryness using a rotary evaporator.
5. Purify the fully deprotected dinucleotide by C_{18} reverse-phase column chromatography following the procedure given in **Section 2.5**. Use methanol to pack the column and wash the column with methanol containing an increasing amount of distilled water (20%, 40%, etc.) until the column is packed in pure water. Dissolve the impure fully deprotected dinucleotide and elute the product by slowly increasing the proportion of methanol from 0 to 5%.

3′-Thiothymidylyl-(3′→5′)-thymidine triethylammonium salt (9, **Fig. 2.2**) will appear as a white amorphous solid.

6. If the sodium salt is required, dissolve the sample into water (5 mL) and pass the solution through a small column of Dowex 50W-X8 (Na^+ form) ion-exchange resin (prepared from the appropriate hydrogen form resin). Evaporate the solution using a rotary evaporator and 3′-thiothymidylyl-(3′→5′) thymidine sodium salt will appear as a glass or a white amorphous solid.

3.5. Yields and Spectral Information

Fast-atom bombardment (FAB) mass spectra were recorded on a VG Analytical 7070E mass spectrometer operating with a PDP 11/250 data system and an Ion Tech FAB ion gun working at 8 kV. High-resolution FAB mass spectra were obtained on a VG ZAB/E spectrometer at the SERC Mass Spectrometry Service Centre (Swansea, UK) and reported masses are accurate to 5 ppm. 3-Nitrobenzyl alcohol was used as a matrix unless stated otherwise. ^1H and ^{13}C NMR spectra were measured on either a Bruker AMX400 or a Bruker AC200 spectrometer, chemical shifts are given in ppm downfield from tetramethylsilane as an internal standard and *J* values are given in Hz. Peaks displaying obvious diastereoisomeric splitting are denoted with an asterisk. ^{31}P NMR spectra are referenced to 85% phosphoric acid.

1. **Compound** 7, 5′-*O*-monomethoxytrityl-3′-deoxy-3′-(2,4-dinitrophenyldisulfanyl)-thymidine): **Yield**, 72%. **Elemental analysis**, Found (C, 59.6; H, 4.5; N, 7.6%); $C_{36}H_{32}N_4O_9S_2$ requires C, 59.3; H, 4.42; N, 7.69%). **δ_H(200 MHz; $CDCl_3$)**, 9.24(1H, s, NH), 9.09(1H, s, ArH), 8.34(2H, s, ArH), 7.61 (1H, s, H6), 7.20–7.48(12H, m, ArH), 6.81(2H, d, *J* 8.8, <u>Ph</u>-OMe), 6.21(1H, t, *J* = 5.2, H1′), 4.09(1H, m, H4′), 3.80(3H, s, OMe), 3.75(1H, m, H3′), 3.65 (1H, m, H5″), 3.39(1H, m, H5′), 2.53(2H, m, H2′) and 1.47(3H, s, 5-Me); **m/z**, (FAB^+) 729 ($M + H^+$).
2. **Compound 4**, 3′-*O*-(*tert*-butyldimethylsilyl)-thymindin-5′-yl phosphonate triethylammmonium salt: **Yield**, 89%. **δ_P(81 MHz; $CDCl_3$)**, 6.3. **δ_H(400 MHz; $CDCl_3$)**, 9.18(1H, br s, NH), 7.74(1H, s, H6), 6.84(1H, d, J_{PH} = 617, HP), 6.34(1H, t, *J* = 7.2, H1′), 4.49(1H, m, H4′), 4.08(1H, m, H3′), 4.02(2H, m, H5′), 3.10(6H, q, *J* = 7.2, $N(\underline{CH_2}CH_3)_3$), 2.18(2H, m, H2′), 1.94(3H, s, 5-Me), 1.32(9 H, t, *J* = 7.2, $N(CH_2\underline{CH_3})_3$), 0.88(9H, s, *t*Bu) and 0.08(6H, s, $SiMe_2$). **m/z**, (FAB^-) 419 ($M - Et_3HN)^-$.
3. **Compound 5**, phosphite triester intermediate: **δ_P(81 MHz; $CDCl_3$)**, 116.0.
4. **Compound 8**, fully protected dinucleotide: **Yield**, 68%. **δ_P(81 MHz; $CDCl_3$)**, 14.1. **δ_H(200 MHz; $CDCl_3$)**,

7.65(1H, s, H6), 7.61(1H, s, H6), 7.19–7.37(12H, m, ArH), 6.78(2H, d, *J* = 8.1, ArH), 6.26(1H, t, H1′), 6.16(1H, t, H1′), 4.38(1H, m), 4.12(1H, m), 3.96(2H, m), 3.86(2H, m), 3.73(3H, s, OMe), 3.43(2H, m), 3.0(6H, q, $N(\underline{CH_2}CH_3)_3$), 2.67(2H, m, H2′), 2.10(2H, m, H2′), 1.84(3H, s, 5-Me), 1.36(3H, s, 5-Me), 1.27(9H, t, $N(CH_2\underline{CH_3})_3$), 0.83 (9H, s, *t*Bu) and 0.01(6H, s, $SiMe_2$); **m/z,** (FAB^-) 947 (M^-).

5. **Compound 9**, 3′-*S*-phosphorothiolate dinucleotide: **δ_P(81 MHz; D_2O)**, 17.82. **δ_H(400 MHz; D_2O)**, 7.82(1H, s, H6), 7.78(1H, s, H6), 6.30(1H, t, *J*=4, H1′a), 6.00(1H, m, H1′b), 4.58(1H, m, H3′a), 4.25(1H, m, H5′a), 4.13(2H, m, H4′a+H5″a), 4.03(2H, m, H4′b+H5′b), 3.95(1H, m, H5″b), 3.55(1H, m, H3′b), 2.65(2H, m, H2′b+H2″b), 2.39(2H, m, H2′a+H2″a), 1.85(6H, m, dd, *J* = 12, 16, 5-Me). **δ_H(50.4 MHz; D_2O)**,168.66(C4), 168.35(C4), 153.97(C2), 153.67(C2), 140.22(C6), 139.77(C6), 113.82(C5), 113.17(C5), 88.60 (C1′), 88.50(C1′), 87.94(C3′b), 87.26(C5′a), 87.17(C4′), 72.11(C4′), 66.74(C5′b), 62.05(C3′a), 61.34(C2′), 49.25 (C2′), 14.18(2C, 5-Me). **m/z,** (FAB^-) 561 (M^-). **HRMS,** Found 561.10426; $C_{20}H_{26}N_4O_{11}PS$ requires 561.10564 (M^-).

4. Notes

1. TEAB solution should be stored in the fridge and generally it can be stored for about 3 months. However, since the buffer is volatile, the pH can increase as carbon dioxide is lost. Test the pH every month and if required, bubble carbon dioxide through the solution to lower the pH (*see* Step 1 of **Section 2.2**).
2. If distillation is the preferred method of obtaining anhydrous solvents, check the COSHH assessments as both calcium hydride and sodium react violently with water. The flasks containing the drying agents and solvents should be quenched appropriately within 6 months of use, as peroxides can build up, which can be exceptionally dangerous.
3. Avoid pouring silica gel dry as silica particles are easily inhaled and prolonged exposure can cause breathing difficulties.
4. If for some reason, the sample does not dissolve in the eluent suggested, then pre-loading the impure material onto silica may be required. To achieve this, dissolve the impure

material into a suitable solvent and form a slurry with a small quantity of silica. Evaporate off the solvent and ensure the silica is dry. Pour the silica bearing the impure sample onto the top of the packed column and run it as usual.

5. Argon could be used instead of nitrogen.
6. A nitrogen inlet is usually in the form of a nitrogen balloon. However, a nitrogen bubbler can also be used, which negates the need for a rubber septum and syringe needles. In the case of a two-necked flask, a bubbler is the preferred route as you can remove the stopper from the side arm to aid nitrogen flow through the flask.
7. It is crucial that the conditions are kept strictly anhydrous whilst making the unsymmetrical disulphide. If 5′-*O*-monomethoxytrityl-3′-deoxy-3′-thiothymidine is maintained in solution under an air atmosphere for a prolonged period of time, the symmetrical disulphide might be formed. One way to avoid this is to place the nitrogen inlet directly into the solution and bubble nitrogen through the solution for approximately 30 min.
8. Take care when doing this extraction, as a by-product to the reaction is HCl; so during the extraction with $NaHCO_3$, carbon dioxide can be produced.
9. Phosphorus trichloride is highly toxic.
10. For some preparations, it might be necessary to remove small amounts of contaminating 1,2,4-triazole. This can be achieved by re-dissolving the sample in dichloromethane and washing with TEAB (0.1 M).

Acknowledgements

JWG was supported by the EPSRC (EP/F011938). The authors would like to thank AP Higson, X Li, GK Scott and JS Vyle who originally worked on the Michaelis–Arbuzov approach.

References

1. Reese, C. B. (2005) Oligo- and polynucleotides: 50 years of chemical synthesis. *Org. Biomol. Chem.* **3**, 3851–3868.
2. Nair, V. (2003) Novel inhibitors of HIV integrase: The discovery of potential anti-HIV therapeutic agents. *Curr. Pharm. Des.* **9**, 2553–2565.
3. Aubert, Y., Chassignol, M., Roig, V., Mbemba, G., Weiss, J., Meudal, H., Mouscadet, J.-F., and Asseline, U. (2009) Synthesis and anti-HIV-1 integrase activity of modified dinucleotides. *Eur. J. Med. Chem.* **44**, 5029–5044.
4. Iyer, R. P., Jin, Y., Roland., A., Morrey, J. D., Mounir, S., and Korba, B. (2004) Phosphorothioate di-and trinucleotides as a novel class of anti-hepatitis B virus agents. *Antimicrob. Agents Chemother.* **48**, 2199–2205.

5. Iyer, R. P., Roland, A., Jin, Y., Mounir, S., Korba, B., Julander, J. G., and Morrey, J. D. (2004) Anti-hepatitis B virus activity of ORI-9020, a novel phosphorothioate dinucleotide, in a transgenic mouse model. *Antimicrob. Agents Chemother.* **48**, 2318–2320.
6. Verma, S., and Eckstein, F. (1998) Modified oligonucleotides: Synthesis and strategy for users. *Annu. Rev. Biochem.* **67**, 99–134.
7. Eckstein, F. (2000) Phosphorothioate oligodeoxynucleotides: What is their origin and what is unique about them? *Antisense Nucleic Acid Drug Dev.* **10**, 117–121.
8. Oka, N., Yamamoto, M., Sato, T., and Wada, T. (2008) Solid-phase synthesis of stereoregular oligodeoxyribonucleoside phosphorothioates using bicyclic oxazaphospholidine derivatives as monomer units. *J. Am. Chem. Soc.* **130**, 16031–16037.
9. Cosstick, R., and Gaynor, J. W. (2008) Synthesis, properties and application of nucleic acids containing phosphorothiolate linkages. *Curr. Org. Chem.* **12**, 291–308.
10. Jahn-Hofmann, K., and Engels, J. W. (2004) Efficient solid phase synthesis of cleavable oligodeoxynucleotides based on a novel strategy for the synthesis of 5′-*S*-(4,4′-dimethoxytrityl)-2′-deoxy-5′-thionucleoside phosphoramidites. *Helv. Chim. Acta* **87**, 2812–2828.
11. Beevers, A. P. G., Witch, E. M., Jones, B., Cosstick, R., Arnold, J. R. P., and Fisher, J. (1999) Conformational analysis of 3′-S-PO3-linked ribo- and deoxyribodinucleoside monophosphates. *Magn. Reson. Chem.* **37**, 814–820.
12. Beevers, A. P. G., Fettes, K. J., O'Neil, I. A., Roberts, S. M., Arnold, J. R. P., Cosstick, R., and Fisher, J. (2002) Probing the effect of a 3′-*S*-phosphorothiolate link on the conformation of a DNA:RNA hybrid; implications for antisense drug design. *Chem. Commun.* 1458–1459.
13. Bentley, J., Brazier, J. A., Fisher, J., and Cosstick, R. (2007) Duplex stability of DNA:DNA and DNA:RNA duplexes containing 3′-*S*-phosphorothiolate linkages. *Org. Biomol. Chem.* **5**, 3698–3702.
14. Li, X., Scott, G. K., Baxter, A. D., Taylor, R. J., Vyle, J. S., and Cosstick, R. (1994) Application of the michaelis-arbusov reaction to the synthesis of internucleoside 3′-*S*-phosphorothiolate linkages. *J. Chem. Soc. Perkin Trans.* **1**, 2123–2129.
15. Higson, A. P., Scott, G. K., Earnshaw, D. J., Baxter, A. D., Taylor, R. A., and Cosstick, R. (1996) Synthesis and structure of *S*-nucleosidyl *S*-aryl disulfides and their reaction with phosphites. *Tetrahedron* **52**, 1027–1034.
16. Sabbagh, G., Fettes, K. J., Gossain, R., O'Neil, I. A., and Cosstick, R. (2004) Synthesis of phosphorothioamidites derived from 3′-thio-3′-deoxythymidine and 3′-thio-2′,3′-dideoxycytidine and the automated synthesis of oligodeoxy-nucleotides containing a 3′-*S*-phosphorothiolate linkage. *Nucleic Acids Res.* **32**, 495–501.
17. Gaynor, J. W., Piperakis, M. W., Fisher, J., and Cosstick, R. (2010) Reverse-direction (5′→3′) synthesis of oligonucleotides containing a3′-*S*-phosphorothiolate linkage and 3′-terminal 3′-thionucleosides. *Org. Biomol. Chem.* **8**, 1463–1470.
18. Gaynor, J. W., Bentley, J., and Cosstick, R. (2007) Synthesis of the 3′-thionucleosides and subsequent automated synthesis of oligodeoxynucleotides containing a 3′-*S*-phosphorothiolate linkage. *Nat. Protoc.* **2**, 3122–3135.

Chapter 3

2′-O,4′-C-Methyleneoxymethylene Bridged Nucleic Acids (2′,4′-BNA^{COC})

Yoshiyuki Hari, Tetsuya Kodama, Takeshi Imanishi, and Satoshi Obika

Abstract

The synthesis of 2′-*O*,4′-*C*-methyleneoxymethylene bridged nucleoside (2′,4′-BNA^{COC}) phosphoramidites and oligonucleotides containing 2′,4′-BNA^{COC} are described. 2′,4′-BNA^{COC} phosphoramidites bearing natural nucleobases, such as thymine, cytosine, 5-methylcytosine, adenine, and guanine were synthesized. Moreover, fully or partially 2′,4′-BNA^{COC}-modified oligonucleotides can be prepared by using a standard protocol except for a prolonged coupling time on an automated DNA synthesizer.

Key words: 2′,4′-BNA^{COC}, BNA, BNA oligonucleotide, bridged nucleic acid, methyleneoxymethylene linkage, nucleoside, nucleotide, oligonucleotide.

1. Introduction

2′-*O*,4′-*C*-Methyleneoxymethylene bridged nucleic acids (2′,4′-BNA^{COC}) possess an *N*-type sugar conformation restricted by an additional carbon-oxygen-carbon (COC) linkage (*see* **Note 1**). The 2′,4′-BNA^{COC}-modified oligonucleotides show several promising features for a variety of nucleic acid-based technologies, such as antisense methodology (1, 2). For example, (i) its enzymatic stability against snake venom phosphodiesterase (SVPDE) is quite excellent and comparable to that of phosphorothioate oligonucleotides (S-oligo) and (ii) it has good duplex-forming ability with single-stranded RNA. The increase in the T_m values per modification (ΔT_m) range from 1.0 to 2.5°C while the ability to form duplexes with single-stranded DNA decreases ($\Delta T_m = -1.1$ to -2.3°C) and (iii) it forms a stable

J. Goodchild (ed.), *Therapeutic Oligonucleotides*, Methods in Molecular Biology 764,
DOI 10.1007/978-1-61779-188-8_3, © Springer Science+Business Media, LLC 2011

triplex nucleic acid with double-stranded DNA. In this chapter, the synthesis of 2′,4′-BNA^{COC} monomers with natural nucleobases and their oligonucleotide derivatives is described. All 2′,4′-BNA^{COC} phosphoramidites are synthesized from the triacetate **2** (**Scheme 3.1**) as a common intermediate. The COC linkage for derivatives bearing pyrimidine nucleobases, namely thymine (T) (**Scheme 3.1**), 5-methylcytosine (^{m}C) (**Scheme 3.2**), and cytosine (C) (**Scheme 3.3**), is constructed by formation of a

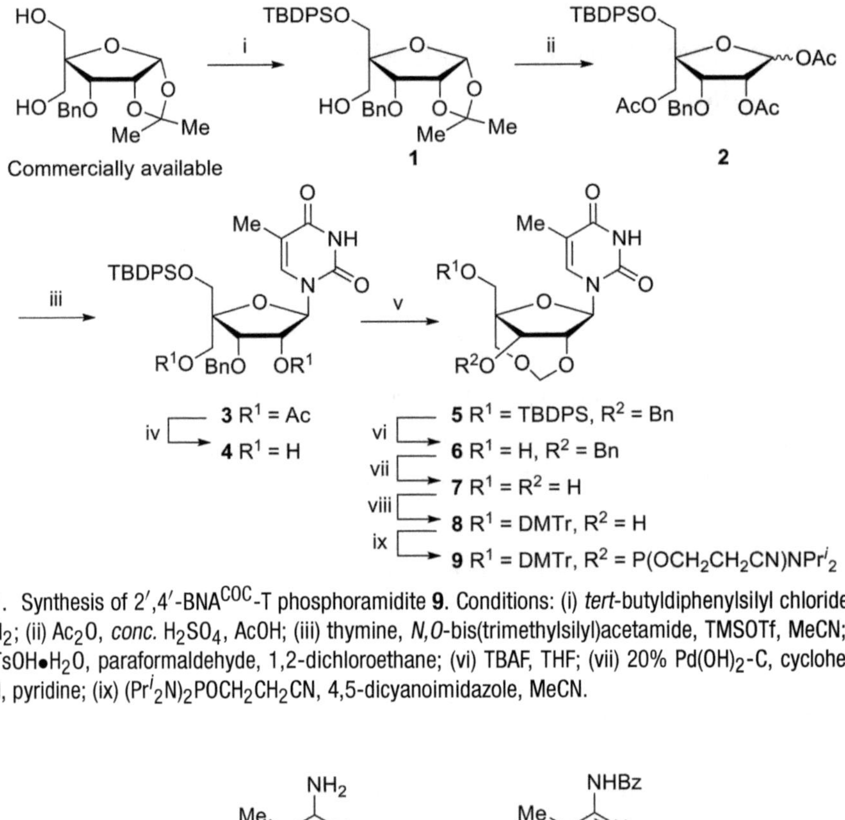

Scheme 3.1. Synthesis of 2′,4′-BNA^{COC}-T phosphoramidite **9**. Conditions: (i) *tert*-butyldiphenylsilyl chloride (TBDPSCl), Et_3N, CH_2Cl_2; (ii) Ac_2O, *conc.* H_2SO_4, AcOH; (iii) thymine, *N,O*-bis(trimethylsilyl)acetamide, TMSOTf, MeCN; (iv) K_2CO_3, MeOH; (v) $TsOH{\bullet}H_2O$, paraformaldehyde, 1,2-dichloroethane; (vi) TBAF, THF; (vii) 20% $Pd(OH)_2$-C, cyclohexene, EtOH; (viii) DMTrCl, pyridine; (ix) $(Pr^i{}_2N)_2POCH_2CH_2CN$, 4,5-dicyanoimidazole, MeCN.

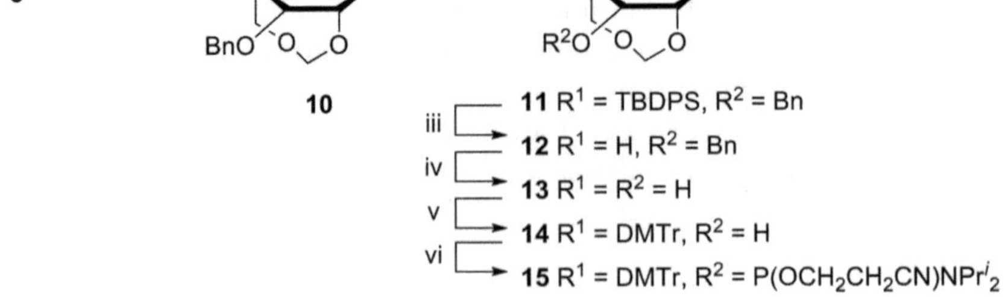

Scheme 3.2. Synthesis of 2′,4′-BNA^{COC}-^{m}C phosphoramidite **15**. Conditions: (i) 1,2,4-1*H*-triazole, $POCl_3$, Et_3N, MeCN, then 28% NH_3 aq., 1,4-dioxane; (ii) BzCl, pyridine, then 28% NH_3 aq.; (iii) TBAF, THF; (iv) 20% $Pd(OH)_2$-C, cyclohexene, THF; (v) DMTrCl, pyridine; (vi) $(Pr^i{}_2N)_2POCH_2CH_2CN$, 5-ethylthio-1*H*-tetrazole, MeCN.

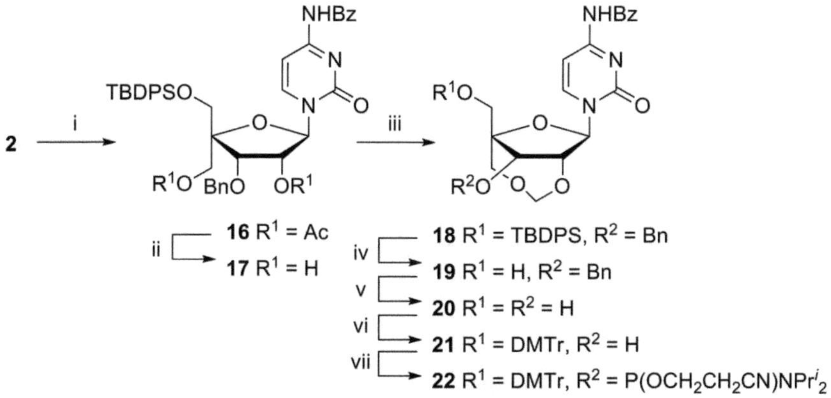

Scheme 3.3. Synthesis of 2′,4′-BNA^{COC}-C phosphoramidite **22**. Conditions: (i) N^4-benzoylcytosine, *N,O*-bis(trimethylsilyl)acetamide, TMSOTf, MeCN; (ii) LiOH•H_2O, THF-H_2O; (iii) TsOH•H_2O, paraformaldehyde, 1,2-dichloroethane; (iv) TBAF, THF; (v) 20% $Pd(OH)_2$-C, cyclohexene, MeOH; (vi) DMTrCl, pyridine; (vii) $(Pr^i_2N)_2POCH_2CH_2CN$, 5-ethylthio-1*H*-tetrazole, MeCN.

methylene acetal by reaction of diols **4** (**Scheme 3.1**) and **17** (**Scheme 3.3**) with paraformaldehyde under acidic conditions. For an adenine (A) analog, the desired 2′,4′-BNA^{COC} skeleton can be constructed by treatment of diol **25** (**Scheme 3.4**) with *N*-bromosuccinimide and DMSO (3), although the same reaction conditions to those for pyrimidine analogs give a complex mixture. This reaction using *N*-bromosuccinimide and DMSO is also applicable to a guanine (G) analog **33** (**Scheme 3.5**). The fully or partially 2′,4′-BNA^{COC}-modified oligonucleotides are prepared by conventional phosphoramidite chemistry on an automated DNA synthesizer (*see* **Note 2**). A prolonged coupling time and more reactive activator for introduction of the 2′,4′-BNA^{COC} unit are required due to steric hindrance of the bulky bridged structure. Eventually, a coupling time of 20 min and 5-ethylthio-1*H*-tetrazole (4) as an activator were found to increase the coupling yield of 2′,4′-BNA^{COC} phosphoramidites to over 95%, which is comparable to that of natural phosphoramidites.

2. Materials

1. Melting points are measured with a Yanagimoto micro melting point apparatus and are uncorrected.
2. ^{1}H NMR (270 or 300 MHz) spectra are recorded on JEOL EX-270 and JEOL-AL-300 spectrometers. Tetramethylsilane is used as an internal standard.

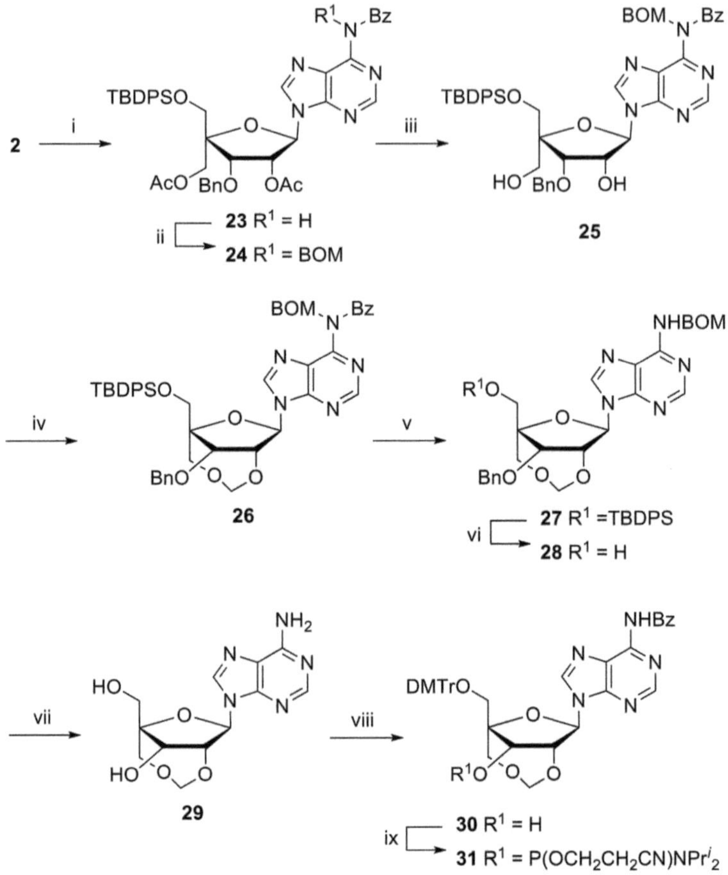

Scheme 3.4. Synthesis of 2′,4′-BNA^{COC}-A phosphoramidite **31**. Conditions: (i) silylated N^6-benzoyladenine, TMSOTf, 1,2-dichloroethane; (ii) benzyloxymethyl chloride (BOMCl), DBU, DMF; (iii) K_2CO_3, MeOH; (iv) *N*-bromosuccinimide, DMSO; (v) 28% NH_3 aq., THF; (vi) TBAF, THF; (vii) 20% $Pd(OH)_2$-C, ammonium formate, EtOH-AcOH; (viii) DMTrOTf then BzCl, CH_2Cl_2-pyridine, then 1 M LiOH aq., THF; (ix) $(Pr^i_2N)_2POCH_2CH_2CN$, 4,5-dicyanoimidazole, MeCN.

3. ^{31}P NMR (202.4 MHz) spectra are recorded on JEOL GX-500 spectrometer. 85% H_3PO_4 is used as an external standard.
4. Mass spectra are measured on JEOL JMS-600 or JMS-700 mass spectrometers.

2.1. Synthesis of 2′,4′-BNA^{COC} Phosphoramidites

Reagents should be purified before use or the purity should be guaranteed by providers. The following list omits solvents and common reagents found in most laboratories.

1. 1,2,4-1*H*-Triazole
2. 1,8-Diazabicyclo[5.4.0]undec-7-ene (DBU)
3. 20% $Pd(OH)_2$-on carbon

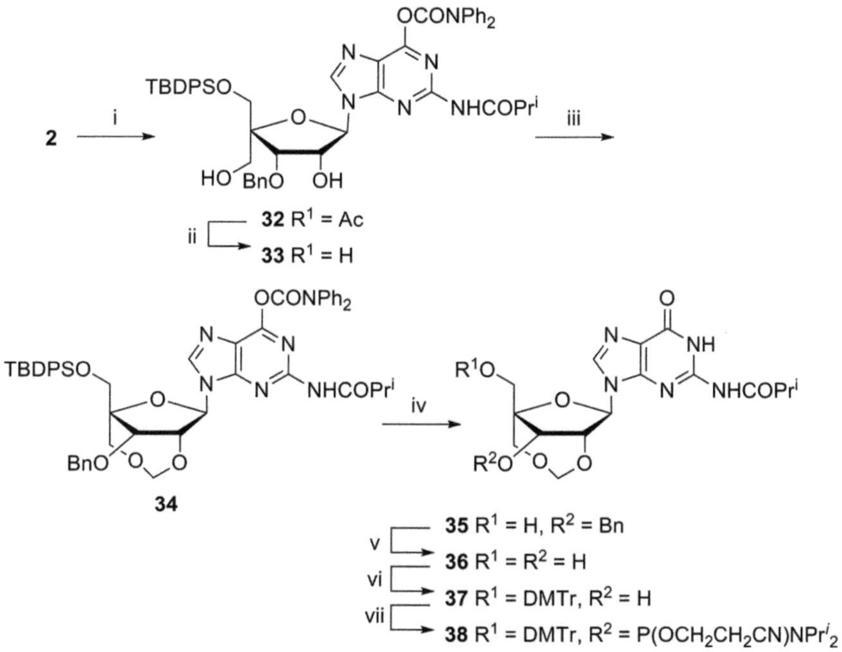

Scheme 3.5. Synthesis of 2′,4′-BNA^{COC}-G phosphoramidite **38**. Conditions: (i) silylated O^6-(diphenylcarbamoyl)-N^2-isobutyrylguanine, TMSOTf, toluene; (ii) K_2CO_3, MeOH; (iii) *N*-bromosuccinimide, DMSO; (iv) $NaNO_2$, DMSO, then TBAF, THF; (v) H_2, 20% $Pd(OH)_2$-C, MeOH; (vi) DMTrCl, pyridine; (vii) $(Pr^i_2N)_2POCH_2CH_2CN$, 4,5-dicyanoimidazole, MeCN-THF.

4. 2-Cyanoethyl N,N,N′,N′-tetraisopropylphosphorodiamidate [$(Pr^i_2N)_2POCH_2CH_2CN$]
5. 3-*O*-Benzyl-4-*C*-hydroxymethyl-1,2-*O*-isopropylidene-α-D-*erythro*-pentofuranose (Chemical Soft R&D Inc.)
6. 4,4′-Dimethoxytrityl chloride (DMTrCl)
7. 4,5-Dicyanoimidazole: 0.25 M in acetonitrile
8. 5-Ethylthio-1*H*-tetrazole
9. Benzoyl chloride (BzCl)
10. Benzyloxymethyl chloride (BOMCl)
11. Diphenylcarbamoyl chloride
12. Guanine
13. Hexamethyldisilazane
14. Isobutyric anhydride
15. *N,N*-Diisopropylethylamine
16. *N,N*-Dimethylacetamide
17. *N,O*-Bis(trimethylsily)acetamide
18. N^4-Benzoylcytosine

19. N^6-Benzoyladenine
20. *N*-Bromosuccinimide: recrystallize from boiling water just before use
21. Phosphorus oxychloride
22. Silver triflate
23. *tert*-Butyldiphenylsilyl chloride (TBDPSCl)
24. Tetra-*n*-butylammonium fluoride (TBAF): 1 M solution in tetrahydrofuran (THF)
25. Thymine
26. Trimethylsilyl chloride (TMSCl)
27. Trimethylsilyl triflate (trimethylsilyl triflate)

2.2. Synthesis of 2′,4′-BNA^{COC}-Modified Oligonucleotides

1. Automated DNA synthesizer: Expedite® 8909 (Applied Biosystems) (*see* **Note 3**).
2. Oxidizing solution: 0.02 M iodine in THF/pyridine/H_2O (Glen Research, VA).
3. Deblocking solution: 3% trichloroacetic acid in dichloromethane.
4. Capping solution A: THF/Ac_2O (9:1, v/v).
5. Capping solution B: 10% 1-methylimidazole in THF/pyridine (8:1, v/v).
6. Activator: 0.25 M 5-ethylthio-1*H*-tetrazole in anhydrous MeCN.
7. Support for oligonucleotide synthesis: 500 Å controlled pore glass derivatized with the standard long chain alkylamino linker.
8. Triethylammonium acetate: 0.1 M, pH 7.0.
9. Prepacked reversed-phase column: Sep-Pak® Plus C18 cartridges (Waters).
10. MALDI-TOF-MS: Autoflex II TOF/TOF instruments (Bruker Daltonics).
11. Ion exchange resin: AG 50W-X8 Resin, biotechnology grade 100–200 mesh, hydrogen form (Bio-Rad).
12. Matrix solution: equal volumes of 3-hydroxypicolinic acid (10 mg/mL in H_2O) and diammonium hydrogen citrate (1 mg/mL in H_2O) mixed just before use.

3. Methods

3.1. Synthesis of 2′,4′-BNA^{COC} Phosphoramidite Bearing T (2′,4′-BNA^{COC}-T Phosphoramidite)

3.1.1. 3-O-Benzyl-5-O-tert-butyldiphenylsilyl-4-C-hydroxymethyl-1,2-O-isopropylidene-α-D-erythro-pentofuranose, 1

1. Under N_2 atmosphere, add triethylamine (2.69 g, 26.6 mmol) and *tert*-butyldiphenylsilyl chloride (TBDPSCl) (7.09 g, 25.8 mmol) to a stirred solution of 3-*O*-benzyl-4-*C*-hydroxymethyl-1,2-*O*-isopropylidene-α-*D*-*erythro*-pentofuranose (2.50 g, 8.1 mmol) in dichloromethane (50 mL) at room temperature, and stir the resulting mixture for 11 h at the same temperature.
2. After addition of saturated aqueous sodium bicarbonate solution, extract the mixture with chloroform twice. Combine the organic layers and dry over sodium sulfate. Filter insoluble and concentrate the obtained filtrate under reduced pressure.
3. Purify the residue by silica gel column chromatography [hexane/ethyl acetate (4:1 to 3:1, v/v)] to give **1** (2.97 g, 67%) (5). White solid; mp 98–99°C (hexane). ^{1}H NMR ($CDCl_3$) δ 1.13 (9H, s), 1.50 (3H, s), 1.78 (3H, s), 2.56 (1H, t, *J* = 7 Hz), 3.82, 3.92 (2H, ABq, *J* = 11 Hz), 3.94 (2H, m), 4.57 (1H, d, *J* = 5 Hz), 4.64, 4.95 (2H, ABq, *J* = 12 Hz), 4.83 (1H, dd, *J* = 4, 5 Hz), 5.95 (1H, d, *J* = 4 Hz), 7.44–7.55 (11H, m), 7.72–7.78 (4H, m). Anal. calcd for $C_{32}H_{41}O_6Si$: C, 70.04; H, 7.35. Found: C, 70.19; H, 7.38. Mass (FAB): *m/z* 549 (MH^+).

3.1.2. 4-C-Acetoxymethyl-1,2-di-O-acetyl-3-O-benzyl-5-O-tert-butyldiphenylsilyl-D-erythro-pentofuranose, 2

1. Add acetic anhydride (1 mL) and concentrated sulfuric acid (0.1 mL) to a solution of compound **1** (5) (900 mg, 1.64 mmol) in acetic acid (4 mL) at room temperature and stir the mixture at room temperature for 1 h.
2. After addition of the reaction mixture to saturated aqueous sodium bicarbonate solution, extract the mixture with ethyl acetate. Wash the organic extracts with water and brine, dry over sodium sulfate, and concentrate under reduced pressure. Purify the residue by flash silica gel column chromatography [hexane/ethyl acetate (3:1, v/v)] to give compound **2** (1.03 g, 99%). Colorless oil. ^{1}H NMR ($CDCl_3$) δ 1.07 (9H, s), 1.84 (3H, s), 1.95 (3H, s), 2.10 (3H, s), 3.66, 3.88 (2H, AB, *J* = 11 Hz), 4.38, 4.44 (2H, AB, *J* = 12 Hz), 4.50 (1H, d, *J* = 5 Hz), 4.55, 4.61 (2H, AB, *J* = 11 Hz), 5.40 (1H, d, *J* = 5 Hz), 6.19 (1H, s), 7.23–7.45 (11H, m), 7.63–7.66 (4H, m). Mass (FAB): *m/z* 657 (MNa^+).

*3.1.3. 4′-C-Acetoxymethyl-2′-O-acetyl-3′-O-benzyl-5′-O-tert-butyldiphenylsilyl-5-methyluridine, **3***

1. Under N_2 atmosphere, add thymine (296 mg, 2.35 mmol) and *N,O*-bis(trimethylsilyl)acetamide (*see* **Note 4**) (0.78 mL, 5.48 mmol) to a solution of compound **2** (1.00 g, 1.57 mmol) in anhydrous acetonitrile (10 mL) at room temperature and reflux the mixture for 1 h. Add trimethylsilyl triflate (0.14 mL, 0.78 mmol) to the reaction mixture at 0°C and reflux the mixture for 1 h.
2. After addition of saturated aqueous sodium bicarbonate solution, extract the mixture with ethyl acetate. Wash the organic extracts with water and brine, dry over sodium sulfate, and concentrate under reduced pressure. Purify the residue by flash silica gel column chromatography [hexane/ethyl acetate (3:2, v/v)] to give compound **3** (1.07 g, 93%). White powder; mp 68–70°C. ^1H NMR ($CDCl_3$) δ 1.11 (9H, s), 1.63 (3H, s), 1.93 (3H, s), 2.09 (3H, s), 3.71, 3.89 (2H, AB, *J* = 11 Hz), 4.10, 4.40 (2H, AB, *J* = 12 Hz), 4.47, 4.59 (2H, AB, *J* = 11 Hz), 4.52 (1H, d, *J* = 6 Hz), 5.42 (1H, dd, *J* = 6, 6 Hz), 6.22 (1H, d, *J* = 6 Hz), 7.27–7.48 (12H, m), 7.63–7.67 (4H, m), 9.37 (1H, s). Mass (FAB): *m/z* 701 (MH^+).

*3.1.4. 3′-O-Benzyl-5′-O-tert-butyldiphenylsilyl-4′-C-hydroxymethyl-5-methyluridine, **4***

1. Add potassium carbonate (355 mg, 2.56 mmol) to a solution of compound **3** (360 mg, 0.52 mmol) in methanol (10 mL) at room temperature and stir the mixture at room temparature for 4 h.
2. After neutralization with diluted aqueous HCl solution, extract the mixture with ethyl acetate. Wash the organic extracts with water and brine, dry over sodium sulfate, and concentrate under reduced pressure. Purify the residue by flash silica gel column chromatography [hexane/ethyl acetate (2:1, v/v)] to give compound **4** (288 mg, 90%). White powder; mp 63–65°C. ^1H NMR ($CDCl_3$) δ 1.06 (9H, s), 1.50 (3H, s), 3,42 (1H, brs), 3.55, 3.77 (2H, AB, *J* = 12 Hz), 3.64, 3.80 (2H, AB, *J* = 11 Hz), 4.31 (1H, d, *J* = 6 Hz), 4.38 (1H, dd, *J* = 5,6 Hz), 4.52, 4.87 (2H, AB, *J* = 12 Hz), 4.96 (1H, brd), 6.14 (1H, d, *J* = 5 Hz), 7.30–7.44 (12H, m), 7.57–7.60 (4H, m), 10.1 (1H, brs). Mass (FAB): *m/z* 617 (MH^+).

*3.1.5. 3′-O-Benzyl-5′-O-tert-butyldiphenylsily-5-methyl-2′-O,4′-C-(methyleneoxymethylene) uridine, **5***

1. Under N_2 atmosphere, add paraformaldehyde (1.20 g) and *p*-toluenesulfonic acid monohydrate (180 mg, 0.94 mmol) to a solution of compound **4** (360 mg, 0.52 mmol) in anhydrous 1,2-dichloroethane (30 mL) at room temperature and reflux the mixture for 3 h (*see* **Note 5**).
2. After addition of saturated aqueous sodium bicarbonate solution at room temperature, extract the mixture with ethyl

acetate. Wash the organic extracts with water and brine, dry over sodium sulfate, and concentrate under reduced pressure. Purify the residue by flash silica gel column chromatography [hexane/ethyl acetate (3:1, v/v)] to give compound **5** (664 mg, 81%). White powder; mp 59–62°C. ^{1}H NMR ($CDCl_3$) δ 1.09 (9H, s), 1.57 (3H, s), 3.67, 3.80 (2H, AB, J = 12 Hz), 3.74, 3.92 (2H, AB, J = 12 Hz), 4.57 (1H, d, J = 6 Hz), 4.59, 4.83 (2H, AB, J = 11 Hz), 4.61 (1H, d, J = 6 Hz), 5.25, 5.33 (2H, AB, J = 6 Hz), 6.22 (1H, s), 7.28–7.43 (12H, m), 7.59–7.67 (4H, m), 9.78 (1H, brs). Mass (FAB): m/z 629 (MH^+).

*3.1.6. 3′-O-Benzyl-5-methyl-2′-O,4′-C-(methyleneoxymethylene)uridine, **6***

1. Under N_2 atmosphere, add tetra-*n*-butylammonium fluoride (TBAF) (1 M solution in THF, 4.0 mL, 4.00 mmol) to a solution of compound **5** (2.20 g, 3.50 mmol) in anhydrous THF (50 mL) at room temperature and stir the mixture at room temperature for 4 h.
2. After removal of the solvent under reduced pressure, purify the residue by flash silica gel column chromatography [chloroform/methanol (20:1, v/v)] to give compound **6** (1.16 g, 85%). White powder; mp 103–105°C. ^{1}H NMR ($CDCl_3$) δ 1.86 (3H, s), 2.99 (1H, brs), 3.65–3.81 (4H, m), 4.55 (1H, d, J = 6 Hz), 4.63, 4.77 (2H, AB, J = 10 Hz), 4.68 (1H, d, J = 6 Hz), 5.19, 5.36 (2H, AB, J = 6 Hz), 6.02 (1H, s), 7.31–7.39 (6H, m), 9.54 (1H, brs). Mass (FAB): m/z 391 (MH^+).

*3.1.7. 5-Methyl-2′-O,4′-C-(methyleneoxymethylene)uridine, **7***

1. Add 20% $Pd(OH)_2$-C (25 mg) and cyclohexene (0.38 mL) to a solution of compound **6** (36 mg, 75 μmol) in ethanol (2 mL) at room temperature and reflux the mixture for 3 h.
2. Filter the mixture through a paper filter. After addition of silica (0.2 g) to the filtrate, concentrate the mixture under reduced pressure. Purify the residue by flash silica gel column chromatography [chloroform/methanol (12:1, v/v)] to give compound 7 (25 mg, 89%). Colorless crystals; mp 294–295°C (methanol). ^{1}H NMR (CD_3OD) δ 1.86 (3H, d, J = 1 Hz), 3.65, 3.71 (2H, AB, J = 12 Hz), 3.70 (2H, s), 4.18 (1H, d, J = 6 Hz), 4.49 (1H, d, J = 6 Hz), 5.07, 5.32 (2H, AB, J = 6 Hz), 6.05 (1H, s), 7.99 (1H, q, J = 1 Hz). Mass (FAB): m/z 301 (MH^+).

*3.1.8. 5′-O-(4,4′-Dimethoxytrityl)-5-methyl-2′-O,4′-C-(methyleneoxymethylene)uridine, **8***

1. Under N_2 atmosphere, add 4,4′-dimethoxytrityl chloride (DMTrCl) (266 mg, 0.79 mmol) to a solution of compound 7 (157 mg, 0.52 mmol) in anhydrous pyridine (3 mL) at room temperature and stir the mixture at room temperature for 3 h.

2. After addition of saturated aqueous sodium bicarbonate solution, extract the mixture with ethyl acetate. Wash the organic extracts with water and brine, dry over sodium sulfate, and concentrate under reduced pressure. Purify the residue by flash silica gel column chromatography [chloroform/methanol (50:1, v/v)] to give compound **8** (315 mg, quant). White powder; mp 189–194°C. ^{1}H NMR (acetone-d_6) δ 1.41 (3H, d, J = 1 Hz), 3.33, 3.40 (2H, AB, J = 11 Hz), 3.68, 3.85 (2H, AB, J = 12 Hz), 3.79 (6H, s), 4.34 (1H, d, J = 6 Hz), 4.91 (1H, dd, J = 6, 6 Hz), 5.07, 5.30 (2H, AB, J = 6 Hz), 5.32 (1H, d, J = 6 Hz), 6.15 (1H, s), 6.89 (4H, d, J = 8 Hz), 7.22–7.39 (7H, m), 7.50 (2H, d, J = 10 Hz), 7.61 (1H, s), 10.04 (1H, s). Mass (FAB): *m/z* 625 [MNa^+].

*3.1.9. 3′-O-[2-Cyanoethoxy(diisopropylamino)phosphino]-5′-O-(4,4′-dimethoxytrityl)-5-methyl-2′-O,4′-C-(methyleneoxymethylene)uridine, **9***

1. Under N_2 atmosphere, add $(Pr^i{}_2N)_2POCH_2CH_2CN$ (0.33 mL, 1.04 mmol) to a solution of compound **8** (250 mg, 0.42 mmol) and 4,5-dicyanoimidazole (98 mg, 0.83 mmol) in anhydrous acetonitrile (20 mL) at room temperature and stir the mixture at room temperature for 2 h. Then, add $(Pr^i{}_2N)_2POCH_2CH_2CN$ (0.33 mL, 1.04 mmol) and stir the mixture for 10 h.
2. If the reaction is not completed yet, add $(Pr^i{}_2N)_2POCH_2CH_2CN$ (0.33 mL, 1.04 mmol) and stir the mixture for 10 h (*see* **Note 6**).
3. After addition of saturated aqueous sodium bicarbonate solution, extract the mixture with ethyl acetate. Wash the organic extracts with water and brine, dry over sodium sulfate, and concentrate under reduced pressure. Purify the residue by flash silica gel column chromatography [hexane/ethyl acetate (1:1, v/v)] and by the reprecipitation (*see* **Note 7**) from hexane/ethyl acetate to give compound **9** (328 mg, 94%). White powder; mp 81–87°C. ^{31}P NMR (acetone-d_6) δ 150.7, 151.1. Mass (FAB): *m/z* 803 (MH^+).

3.2. Synthesis of 2′,4′-BNA^{COC} Phosphoramidite Bearing ^{m}C (2′,4′-BNA^{COC}-^{m}C Phosphoramidite)

*3.2.1. 3′-O-Benzyl-5′-O-tert-butyldiphenylsilyl-5-methyl-2′-O,4′-C-(methyleneoxymethylene)cytidine, **10***

1. Under N_2 atmosphere, add phosphorus oxychloride (1.09 g, 7.15 mmol) dropwise to a suspension of 1,2,4-1*H*-triazole (1.98 g, 28.6 mmol) in anhydrous acetonitrile (20 mL) at 0°C and vigorously stir the mixture at 0°C for 10 min. Add triethylamine (2.89 g, 26.6 mmol) dropwise and further stir the mixture at 0°C for 35 min. Then, add a solution of compound **5** (300 mg, 0.48 mmol) in anhydrous acetonitrile (10 mL) dropwise and stir the mixture at room temperature for 5.5 h (*see* **Note 8**).
2. After addition of saturated aqueous sodium bicarbonate solution, extract the mixture with ethyl acetate. Wash the

organic extracts with water and brine, dry over sodium sulfate, and concentrate under reduced pressure.

3. Add 28% aq. ammonia/1,4-dioxane (1:6, v/v, 11 ml) to the residue and stir the mixture at room temperature for 2 h.
4. After removal of the solvent under reduced pressure, purify the residue by flash silica gel column chromatography [ethyl acetate/methanol (10:1, v/v)] and by reprecipitation (*see* **Note 7**) from hexane/ethyl acetate to give compound **10** (246 mg, 82%). White powder; mp 63–66°C. ^{1}H NMR ($CDCl_3$) δ 1.09 (9H, s), 1.50 (3H, s), 3.67, 3.76 (2H, AB, *J* = 12 Hz), 3.76, 3.97 (2H, AB, *J* = 11 Hz), 4.46 (1H, d, *J* = 6 Hz), 4.52, 4.80 (2H, AB, *J* = 11 Hz), 4.56 (1H, d, *J* = 6 Hz), 5.26, 5.32 (2H, AB, *J* = 6 Hz), 6.22 (1H, s), 7.27–7.47 (10H, m), 7.59–7.68 (5H, m), 8.17 (1H, s). MS (FAB): *m/z* 628 (MH^+).

3.2.2. 4-N-Benzoyl-3′-O-benzyl-5′-O-tert-butyldiphenylsily-5-methyl-2′-O,4′-C-(methyleneoxymethylene) cytidine, ***11***

1. Under N_2 atmosphere, add benzoyl chloride (1.59 g, 11.3 mmol) to a solution of compound **10** (2.37 g, 3.77 mmol) in anhydrous pyridine (20 mL) at room temperature and stir the mixture at room temperature for 3.5 h. After addition of 28% aqueous ammonia solution, further stir the mixture at room temperature for 1 h.
2. After removal of the solvent under reduced pressure, extract the mixture with ethyl acetate. Wash the organic extracts with water and brine, dry over sodium sulfate, and concentrate under reduced pressure. Purify the residue by flash silica gel column chromatography [hexane/ethyl acetate (3:1, v/v)] to give compound **11** (2.40 g, 87%). White powder; mp 56–58°C. ^{1}H NMR ($CDCl_3$) δ 1.11 (9H, s), 1.69 (3H, s), 3.67, 3.78 (2H, AB, *J* = 12 Hz), 3.76, 3.96 (2H, AB, *J* = 12 Hz), 4.53 (1H, d, *J* = 6 Hz), 4.57 (1H, d, *J* = 6 Hz), 4.58, 4.83 (2H, AB, *J* = 11 Hz), 5.25, 5.33 (2H, AB, *J* = 6 Hz), 6.23 (1H, s), 7.30–7.48 (14H, m), 7.60–7.68 (5H, m), 8.27–8.30 (2H, m). MS (FAB): *m/z* 732 (MH^+).

3.2.3. 4-N-Benzoyl-3′-O-benzyl-5-methyl-2′-O,4′-C-(methyleneoxy-methylene)cytidine, ***12***

1. Under N_2 atmosphere, add TBAF (1 M solution in THF, 1.9 mL, 1.90 mmol) to a solution of compound **11** (1.00 g, 1.37 mmol) in anhydrous THF (20 mL) at room temperature and stir the mixture at room temperature for 10 h.
2. After removal of the solvent under reduced pressure, purify the residue by flash silica gel column chromatography [chloroform/methanol (60:1, v/v)] to give compound **12** (631 mg, 97%). White powder; mp 255–257°C. ^{1}H NMR (DMSO-d_6) δ 2.02 (3H,s), 3.58, 3.70 (2H, AB, *J* = 12 Hz), 3.59, 3.70 (2H, AB, *J* = 12 Hz), 4.42 (1H, d, *J* = 5 Hz), 4.61 (1H, d, *J* = 5 Hz), 4.64, 4.71 (2H, AB, *J* = 12 Hz),

5.11, 5.20 (2H, AB, $J = 6$ Hz), 5.67 (1H, brs), 5.97 (1H, s), 7.28–7.36 (5H, m), 7.47–7.62 (3H, m), 8.18 (2H, d, $J = 8$ Hz), 8.27 (1H, s), 12.90 (1H, brs). MS (FAB): *m/z* 494 (MH^+).

3.2.4. 4-N-Benzoyl-5-methyl-2′-O,4′-C-(methyleneoxymethylene) cytidine, ***13***

1. Add 20% $Pd(OH)_2$-C (800 mg) and cyclohexene (4 mL) to a solution of compound **12** (216 mg, 0.44 mmol) in anhydrous THF (20 mL) at room temperature and reflux the mixture for 4 h.
2. After filtration of the mixture, concentrate the filtrate under reduced pressure. Purify the residue by flash silica gel column chromatography [chloroform/THF (8:1, v/v)] to give compound **13** (121 mg, 68%). Colorless crystals; mp 204–206°C. ^1H NMR (C_5D_5N) δ 1.93 (3H, s), 3.98, 4.07 (2H, AB, $J = 12$ Hz), 4.09, 4.18 (2H, ABX, $J = 4$, 12 Hz), 4.86 (1H, dd, $J = 5$, 6 Hz), 5.29 (1H, brs), 5.44, 5.82 (2H, AB, $J = 6$ Hz), 6.69 (1H, d, $J = 3$ Hz), 7.43–7.61 (4H, m), 8.46 (1H, brs), 8.55 (2H, dd, $J = 1$, 8 Hz), 8.66 (1H, s), 13.6 (1H, brs). MS (FAB): *m/z* 404 (MH^+).

3.2.5. 4-N-Benzoyl-5′-O-(4,4′-dimethoxytrityl)-5-methyl-2′-O,4′-C-(methyleneoxymethylene) cytidine, ***14***

1. Under N_2 atmosphere, add DMTrCl (55 mg, 0.16 mmol) to a solution of compound **13** (50 mg, 0.12 mmol) in anhydrous pyridine (3 mL) at room temperature and stir the mixture at room temperature for 12 h.
2. After addition of saturated aqueous sodium bicarbonate solution, extract the mixture with ethyl acetate. Wash the organic extracts with water and brine, dry over sodium sulfate, and concentrate under reduced pressure. Purify the residue by flash silica gel column chromatography [chloroform/methanol (50:1, v/v)] to give compound **14** (81 mg, 93%). White powder; mp 123–126°C. ^1H NMR (acetone-d_6) δ 1.62 (3H, d, $J = 1$ Hz), 3.38, 3.46 (2H, AB, $J = 11$ Hz), 3.71, 3.85 (2H, AB, $J = 12$ Hz), 3.79 (6H,s), 4.45 (1H, d, $J = 6$ Hz), 4.96 (1H, dd, $J = 5$, 6 Hz), 5.09, 5.31 (2H, AB, $J = 6$ Hz), 5.41 (1H, d, $J = 5$ Hz), 6.16 (1H, s), 6.89–6.93 (4H, m), 7.23–7.55 (12H, m), 7.95 (1H, d, $J = 1$ Hz), 8.26–8.30 (2H, m), 8.56 (1H, brs). MS (FAB): *m/z* 706 (MH^+).

3.2.6. 4-N-Benzoyl-3′-O-[2-cyanoethoxy (diisopropylamino) phosphino]-5′-O-(4,4′-dimethoxytrityl)-5-methyl-2′-O,4′-C-(methyleneoxymethylene) cytidine, ***15***

1. Under N_2 atmosphere, add $(Pr^i{}_2N)_2POCH_2CH_2CN$ (45 μL, 0.14 mmol) to a solution of compound **14** (50 mg, 71 μmol) and 5-ethylthio-1*H*-tetrazole (14 mg, 0.11 mmol) in anhydrous acetonitrile (10 mL) at room temperature and stir the mixture at room temperature for 6 h. Then, add $(Pr^i{}_2N)_2POCH_2CH_2CN$ (23 μL, 71 μmol) and further stir the mixture for 6 h.

2. If the reaction is not completed yet, add $(Pr^i{}_2N)_2POCH_2CH_2CN$ (23 μL, 71 μmol) and further stir the mixture for 6 h (*see* **Note 6**).
3. After addition of saturated aqueous sodium bicarbonate solution, extract the mixture with ethyl acetate. Wash the organic extracts with water and brine, dry over sodium sulfate, and concentrate under reduced pressure.
4. Purify the residue by flash silica gel column chromatography [hexane/ethyl acetate (3:1, v/v)] and by reprecipitation (*see* **Note 7**) from hexane/ethyl acetate to give compound **15** (61 mg, 95%). White powder; mp 65–68°C. ^{31}P NMR (acetone-d_6) ∂ 149.5, 150.9. MS (FAB): *m/z* 906 (MH^+).

3.3. Synthesis of 2′,4′-BNA^{COC} Phosphoramidite Bearing C (2′,4′-BNA^{COC}-C Phosphoramidite)

3.3.1. 4′-C-Acetoxymethyl-2′-O-acetyl-4-N-benzoyl-3′-O-benzyl-5′-O-tert-butyldiphenylsilylcytidine, 16

1. Under N_2 atmosphere, add N^4-benzoylcytosine (8.13 g, 37.8 mmol) and *N,O*-bis(trimethylsilyl)acetamide (16.3 mL, 0.11 mol) to a solution of compound **2** (16.0 g, 25.2 mmol) in anhydrous acetonitrile (75 mL) at room temperature and reflux the mixture for 3 h. Add trimethylsilyl triflate (11.4 mL, 53.0 mmol) to the reaction mixture at 0°C and reflux the mixture for 9 h (*see* **Note 4**).
2. After addition of saturated aqueous sodium bicarbonate solution at 0°C, extract the mixture with ethyl acetate. Wash the organic extracts with water and brine, dry over sodium sulfate, and concentrate under reduced pressure. Purify the residue by flash silica gel column chromatography [hexane/ethyl acetate (3:2, v/v)] to give compound **16** (15.1 g, 76%). White powder; mp 113–115°C. ^{1}H NMR ($CDCl_3$) δ 1.12 (9H, s), 1.94 (3H, s), 2.12 (3H, s), 3.78, 4.04 (2H, AB, $J = 11$ Hz), 4.12, 4.53 (2H, AB, $J = 13$ Hz), 4.40, 4.61 (2H, AB, $J = 11$ Hz), 4.54 (1H, d, $J = 6$ Hz), 5.53 (1H, dd, $J = 5, 6$ Hz), 6.26 (1H, d, $J = 5$ Hz), 7.21–7.53 (14H, m), 7.58–7.67 (5H, m), 7.89 (2H, d, $J = 7$ Hz), 8.13 (1H, d, $J = 7$ Hz), 8.67 (1H, brs). MS (FAB): *m/z* 790 (MH^+).

3.3.2. 4-N-Benzoyl-3′-O-benzyl-5′-O-tert-butyldiphenylsilyl-4′-C-(hydroxymethyl)cytidine, 17

1. Add lithium hydroxide monohydrate(1.60 g, 8.23 mmol) to a solution of compound **16** (6.50 g, 8.23 mmol) in THF/H_2O (1:1, v/v, 90 mL) at room temperature and stir the mixture at room temperature for 2 h (*see* **Note 9**).
2. Extract the mixture with ethyl acetate. Wash the organic extracts with water and brine, dry over sodium sulfate, and concentrate under reduced pressure. Purify the residue by recrystallization from ethyl acetate to give compound **17** (5.24 g, 84%). Colorless crystals; mp 144–145°C (ethyl acetate). ^{1}H NMR ($CDCl_3$) δ 1.06 (9H, s), 3.57, 3.84 (2H, AB, $J = 12$ Hz), 3.68, 3.87 (2H, AB, $J = 12$ Hz), 4.36 (1H,

d, $J = 6$ Hz), 4.38 (1H, d, $J = 6$ Hz), 4.42, 4.78 (2H, AB, $J = 11$ Hz), 4.49 (1H, brs), 5.75 (1H, brs), 6.23 (1H, s), 7.24–7.62 (19H, m), 7.96 (2H, d, $J = 7$ Hz), 8.25 (1H, d, $J = 7$ Hz), 9.46 (1H, brs). MS (FAB): *m/z* 706 (MH^+).

3.3.3. 4-N-Benzoyl-3′-O-benzyl-5′-O-tert-butyldiphenylsily-2′-O,4′-C-(methyleneoxymethylene) cytidine, ***18***

1. Under N_2 atmosphere, add *p*-toluenesulfonic acid monohydrate (130 mg, 0.68 mmol) to a solution of compound **17** (324 mg, 0.46 mmol) in anhydrous 1,2-dichloroethane (30 mL) at room temperature and add paraformaldehyde (41 mg) at 60°C. Stir the mixture at 60°C for 1.5 h (*see* **Note 5**).
2. After addition of saturated aqueous sodium bicarbonate solution at 0°C, extract the mixture with ethyl acetate. Wash the organic extracts with water and brine, dry over sodium sulfate, and concentrate under reduced pressure. Purify the residue by flash silica gel column chromatography [hexane/ethyl acetate (2:3, v/v)] to give compound ***18*** (230 mg, 70%). White powder; mp 210–212°C. ^1H NMR ($CDCl_3$) δ 1.05 (9H, s), 3.56, 3.63 (2H, AB, $J = 12$ Hz), 3.64, 3.86 (2H, AB, $J = 12$ Hz), 4.38 (1H, d, $J = 6$ Hz), 4.45, 4.73 (2H, AB, $J = 11$ Hz), 4.48 (1H, d, $J = 6$ Hz), 5.19, 5.22 (2H, AB, $J = 6$ Hz), 6.21 (1H, s), 7.15–7.59 (19H, m), 7.81 (2H, d, $J = 7$ Hz), 8.33 (1H, d, $J = 7$ Hz), 8.63 (1H, brs). MS (FAB): *m/z* 718 (MH^+).

3.3.4. 4-N-Benzoyl-3′-O-benzyl-2′-O,4′-C-(methyleneoxymethylene) cytidine, ***19***

1. Under N_2 atmosphere, add TBAF (1 M solution in THF, 5.0 mL, 5.00 mmol) to a solution of compound **18** (2.37 g, 3.30 mmol) in anhydrous THF (50 mL) at room temperature and stir the mixture at room temperature for 18 h.
2. After removal of the solvent under reduced pressure, purify the residue by flash silica gel column chromatography [hexane/ethyl acetate (1:3, v/v)] to give compound **19** (1.45 g, 92%). White powder; mp 242–243°C. ^1H NMR ($CDCl_3$) δ 3.23 (1H, brs), 3.72, 3.80 (2H, AB, $J = 12$ Hz), 3.72, 3.90 (2H, AB, $J = 12$ Hz), 4.50 (1H, d, $J = 6$ Hz), 4.55 (1H, d, $J = 6$ Hz), 4.57, 4.77 (2H, AB, $J = 11$ Hz), 5.22, 5.36 (2H, AB, $J = 6$ Hz), 6.17 (1H, s), 7.31–7.59 (9H, m), 7.80 (2H, d, $J = 8$ Hz), 8.36 (1H, d, $J = 7$ Hz), 8.93 (1H, brs). MS (FAB): *m/z* 480 (MH^+).

3.3.5. 4-N-Benzoyl-2′-O,4′-C-(methyleneoxymethylene) cytidine, ***20***

1. Add 20% $Pd(OH)_2$-C (50 mg) and cyclohexene (2 ml) to a solution of compound **19** (50 mg, 0.10 mmol) in methanol (20 ml) at room temperature and reflux the mixture for 1 h. After filtration of the reaction mixture, add 20% $Pd(OH)_2$-C (50 mg) to the filtrate and further reflux the mixture for 1 h.

2. Filter the mixture through a paper filter and after addition of silica (0.2 g) to the filtrate, concentrate the mixture under reduced pressure. Purify the residue by flash silica gel column chromatography [chloroform/methanol (15:1, v/v)] to give compound **20** (29 mg, 71%). White powder; mp 244–246°C. ^{1}H NMR (C_5D_5N) δ 4.03, 4.10 (2H, AB, *J* = 12 Hz), 4.13, 4.21 (2H, AB, *J* = 12 Hz), 4.90 (1H, d, *J* = 6 Hz), 5.29 (1H, d, *J* = 6 Hz), 5.47, 5.85 (2H, AB, *J* = 6 Hz), 6.87 (1H, s), 7.42–7.67 (4H, m), 8.18 (2H, d, *J* = 7 Hz), 8.35 (1H, brs), 9.11 (1H, d, *J* = 8 Hz), 12.19 (1H, brs). MS (FAB): *m/z* 390 (MH^+).

3.3.6. 4-N-Benzoyl-5′-O-(4,4′-dimethoxytrityl)-2′-O,4′-C-(methyleneoxymethylene)cytidine, 21

1. Under N_2 atmosphere, add DMTrCl (79 mg, 0.23 mmol) to a solution of compound **20** (70 mg, 0.18 mmol) in anhydrous pyridine (3 mL) at room temperature and stir the mixture at room temperature for 5 h.
2. After addition of saturated aqueous sodium bicarbonate solution, extract the mixture with ethyl acetate. Wash the organic extracts with water and brine, dry over sodium sulfate, and concentrate under reduced pressure. Purify the residue by flash silica gel column chromatography [chloroform/methanol (20:1, v/v)] to give compound **21** (110 mg, 88%). White powder; mp 155–161°C. ^{1}H NMR (acetone-d_6) δ 3.38, 3.49 (2H, AB, *J* = 11 Hz), 3.71, 3.80 (2H, AB, *J* = 12 Hz), 3.82 (6H, s), 4.35 (1H, d, *J* = 6 Hz), 4.87 (1H, brd, *J* = 6 Hz), 5.08, 5.31 (2H, AB, *J* = 6 Hz), 5.25 (1H, brs). 6.12 (1H, s), 6.92 (4H, d, *J* = 9 Hz), 7.15 (1H, brs), 7.25–7.41 (7H, m), 7.49–7.68 (5H, m), 8.15 (2H, d, *J* = 7 Hz), 8.44 (1H, d, *J* = 8 Hz), 9.72 (1H, brs). MS (FAB): *m/z* 692 (MH^+).

3.3.7. 4-N-Benzoyl-3′-O-[2-cyanoethoxy(diisopropylamino)phosphino]-5′-O-(4,4′-dimethoxytrityl)-2′-O,4′-C-(methyleneoxymethylene)cytidine, 22

1. Under N_2 atmosphere, add $(Pr^i{}_2N)_2POCH_2CH_2CN$ (0.28 mL, 0.87 mmol) to a solution of compound **21** (304 mg, 0.43 mmol) and 5-ethylthio-1*H*-tetrazole (86 mg, 0.66 mmol) in anhydrous acetonitrile (20 mL) at room temperature and stir the mixture at room temperature for 20 h (*see* **Note 6**).
2. If the reaction is not completed, add $(Pr^i{}_2N)_2POCH_2CH_2CN$ (0.14 mL, 0.43 mmol) and further stir the mixture for 6 h.
3. After addition of saturated aqueous sodium bicarbonate solution, extract the mixture with ethyl acetate. Wash the organic extracts with water and brine, dry over sodium sulfate, and concentrate under reduced pressure.
4. Purify the residue by flash silica gel column chromatography [hexane/ethyl acetate (1:2, v/v)] and by the reprecipitation

(*see* **Note 7**) from hexane/ethyl acetate to give compound **22** (319 mg, 81%). White powder; mp 97–101°C. ^{31}P NMR (acetone-d_6) δ 151.58, 151.65. MS (FAB): *m/z* 892 (MH^+).

3.4. Synthesis of 2′,4′-BNA^{COC} Phosphoramidite Bearing A (2′,4′-BNA^{COC}-A Phosphoramidite)

3.4.1. 4′-C-Acetoxymethyl-2′-O-acetyl-6-N-benzoyl-3′-O-benzyl-5′-O-tert-butyldiphenylsilyladenosine, 23

1. Under N_2 atmosphere, add TMSCl (0.5 mL) to a suspension of *N*^{6}-benzoyladenine (2.17 g, 9.07 mmol) in anhydrous hexamethyldisilazane (35 mL) at room temperature and reflux the mixture for 24 h. Evaporate the resulting clear solution under reduced pressure (*see* **Note 10**).
2. Add a solution of compound **2** (4.80 g, 7.56 mmol) in anhydrous 1,2-dichloroethane (25 mL) and trimethylsilyl triflate (0.14 mL, 0.756 mmol) to the residue and reflux the mixture for 11 h.
3. After addition of saturated aqueous sodium bicarbonate solution at 0°C, extract the mixture with ethyl acetate. Wash the organic extracts with water and brine, dry over sodium sulfate, and concentrate under reduced pressure. Purify the residue by silica gel column chromatography [hexane/ethyl acetate (3:4→1:2, v/v)] to give compound **23** (4.68 g, 76%). White powder; mp 71–73°C. ^{1}H NMR ($CDCl_3$): δ 1.06 (9H, s), 1.97 (3H, s), 2.05 (3H, s), 3.81(1H, d, *J* = 11 Hz), 3.96 (1H, d, *J* = 11 Hz), 4.30 (1H, d, *J* = 12 Hz), 4.66 (1H, d, *J* = 12 Hz), 4.62 (2H, s), 4.91 (1H, d, *J* = 6 Hz), 6.08 (1H, t, *J* = 6 Hz), 6.25 (1H, d, *J* = 5 Hz), 7.31–7.66 (18H, m), 8.01 (2H, d, *J* = 7 Hz), 8.09 (1H, s), 8.59 (1H, s), 9.04 (1H, brs). Mass (FAB): *m/z* 814 (MH^+).

3.4.2. 4′-C-Acetoxymethyl-2′-O-acetyl-6-N-benzoyl-3′-O-benzyl-6-N-benzyloxymethyl-5′-O-tert-butyldiphenylsilyladenosine, 24

1. Under N_2 atmosphere, add DBU (0.39 mL, 2.58 mmol) to a solution of compound **23** (1.05 g, 1.29 mmol) in anhydrous DMF (6.4 mL) at 0°C and stir the mixture for 10 min. Add benzyloxymethyl chloride (BOMCl) (0.27 mL, 1.93 mmol) to the reaction mixture at 0°C and stir the mixture at 0°C for 1 h.
2. After addition of water at 0°C, extract the mixture with Et_2O. Wash the organic extracts with 0.5 M aqueous potassium hydrogen sulfate solution, water and brine, dry over magnesium sulfate, and concentrate under reduced pressure. Purify the residue by silica gel column chromatography [hexane/ethyl acetate (2:1, v/v)] to give compound **24** (769 mg, 64%). White foam. ^{1}H NMR ($CDCl_3$): δ 1.04 (9H, s), 1.96 (3H, s), 2.03 (3H, s), 3.79 (1H, d, *J* = 11 Hz), 3.92 (1H, d, *J* = 11 Hz), 4.28 (1H, d, *J* = 12 Hz), 4.62 (1H, d, *J* = 12 Hz), 4.59 (2H, s), 4.76 (2H, s), 4.87 (1H, d, *J* = 6 Hz), 5.87 (2H, s), 5.98 (1H, t, *J* = 6 Hz), 6.18 (1H, d, *J* = 5 Hz), 7.14–7.46 (19H, m), 7.50 (2H, d, *J* = 7 Hz), 7.60–7.64 (4H, m), 7.99 (1H, s), 8.38 (1H, s). Mass (FAB): *m/z* 934 (MH^+).

3.4.3. 6-N-Benzoyl-3′-O-benzyl-6-N-benzyloxymethyl-5′-O-tert-butyldiphenylsilyl-4′-C-(hydroxymethyl) adenosine, ***25***

1. Add potassium carbonate (1.61 g, 11.6 mmol) to a solution of compound **24** (3.62 g, 3.88 mmol) in methanol (80 mL) at 0°C and stir the mixture for 50 min.
2. After neutralization with a 10% aqueous HCl solution (4.3 mL), extract the mixture with ethyl acetate. Wash the organic extracts with water and brine, dry over sodium sulfate, and concentrate under reduced pressure. Purify the residue by silica gel column chromatography [hexane/ethyl acetate (3:2 to 1:1, v/v)] to give compound **25** (3.21 g, 97%). White powder; mp 54–57°C. ^{1}H NMR ($CDCl_3$): δ 0.99 (9H, s), 2.51 (1H, dd, J = 5, 7 Hz), 3.70 (2H, s), 3.80 (1H, dd, J = 7, 12 Hz), 3.97 (1H, dd, J = 5, 12 Hz), 4.47 (1H, d, J = 9 Hz), 4.58 (1H, d, J = 6 Hz), 4.63 (1H, d, J = 11 Hz), 4.80 (1H, d, J = 11 Hz), 4.77 (1H, m), 4.78 (2H, s), 5.88 (2H, s), 6.04 (1H, d, J = 4 Hz), 7.10–7.44 (19H, m), 7.48–7.56 (6H, m), 8.00 (1H, s), 8.43 (1H, s). Mass (FAB): m/z 850 (MH^+).

3.4.4. 6-N-Benzoyl-3′-O-benzyl-6-N-benzyloxymethyl-5′-O-tert-butyldiphenylsilyl-2′-O,4′-C-(methyleneoxymethylene) adenosine, ***26***

1. Under N_2 atmosphere, add *N*-bromosuccinimide (312 mg, 1.75 mmol) to a solution of compound **25** (373 mg, 0.438 mmol) in anhydrous DMSO (5.5 mL) at room temperature and stir the mixture at 60°C for 1.5 h (*see* **Note 5**).
2. After addition of saturated aqueous sodium bicarbonate solution at 0°C, extract the mixture with ethyl acetate. Wash the organic extracts with water and brine, dry over sodium sulfate, and concentrate under reduced pressure. Purify the residue by silica gel column chromatography [hexane/ethyl acetate (3:1, v/v)] to give compound **26** (281 mg, 74%). White foam. ^{1}H NMR ($CDCl_3$): δ 0.97 (9H, s), 3.68 (1H, d, J = 12 Hz), 3.81 (1H, d, J = 12 Hz), 3.72 (1H, d, J = 12 Hz), 3.88 (1H, d, J = 12 Hz), 4.62 (1H, d, J = 11 Hz), 4.81 (1H, d, J = 11 Hz), 4.79 (2H, s), 4.83 (1H, d, J = 6 Hz), 4.99 (1H, d, J = 6 Hz), 5.25 (1H, d, J = 6 Hz), 5.40 (1H, d, J = 6 Hz), 5.91 (2H, s), 6.42 (1H, s), 7.10–7.42 (19H, m), 7.50 (2H, d, J = 7 Hz), 7.54–7.59 (4H, m), 8.14 (1H, s), 8.50 (1H, s). Mass (FAB): m/z 862 (MH^+).

3.4.5. 3′-O-Benzyl-6-N-benzyloxymethyl-5′-O-tert-butyldiphenylsilyl-2′-O,4′-C-(methylenoxy-methylene)adenosine, ***27***

1. Add 28% aqueous ammonia solution (1.2 mL) to a solution of compound **26** (209 mg, 0.242 mmol) in THF (2.4 mL) at room temperature. Seal the flask and stir the mixture at 50°C for 48 h.
2. Extract the mixture with ethyl acetate. Wash the organic extracts with water and brine, dry over sodium sulfate, and concentrate under reduced pressure. Purify the residue by flash silica gel column chromatography [hexane/ethyl

acetate (2:1, v/v)] to give compound **27** (186 mg, quant). White foam. ^{1}H NMR ($CDCl_3$): δ 0.99 (9H, s), 3.69 (1H, d, J= 11 Hz), 3.73 (1H, d, J= 12 Hz), 3.83 (1H, d, J= 11 Hz), 3.90 (1H, d, J= 12 Hz), 4.62 (1H, d, J= 11 Hz), 4.66 (2H, s), 4.82 (1H, d, J= 11 Hz), 4.92 (1H, d, J= 6 Hz), 5.04 (1H, d, J= 6 Hz), 5.27 (1H, d, J= 6 Hz), 5.28 (2H, brd, J= 7 Hz), 5.41 (1H, d, J= 6 Hz), 6.43 (1H, s), 6.52 (1H, brs), 7.21–7.43 (16H, m), 7.57–7.60 (4H, m), 8.03 (1H, s), 8.38 (1H, s). Mass (FAB): m/z 758 (MH^+).

3.4.6. 3′-O-Benzyl-6-N-benzyloxymethyl-2′-O,4′-C-(methyleneoxy-methylene)adenosine, ***28***

1. Add TBAF (1.0 M solution in THF, 0.19 mL, 0.19 mmol) to a solution of compound **27** (128 mg, 0.169 mmol) in THF (1.7 mL) at room temperature and stir the mixture at room temperature for 3 h.
2. After addition of hexane/ethyl acetate (1:1, v/v, 2.0 mL) at room temperature, evaporate the mixture under reduced pressure to 1/10th of its volume. Purify the residue by flash silica gel column chromatography [hexane/ethyl acetate (2:5, v/v)] to give compound **28** (80 mg, 91%). White foam. ^{1}H NMR ($CDCl_3$): δ 2.21 (1H, brs), 3.69 (1H, d, J= 12 Hz), 3.82 (1H, d, J= 12 Hz), 3.79 (2H, s), 4.52 (1H, d, J= 6 Hz), 4.65 (2H, s), 4.70 (1H, d, J= 11 Hz), 4.78 (1H, d, J= 11 Hz), 5.25 (1H, d, J= 6 Hz), 5.26 (2H, brs), 5.26 (1H, d, J= 6 Hz), 5.43 (1H, d, J= 6 Hz), 6.35 (1H, s), 7.14 (1H, brs), 7.22–7.38 (10H, m), 7.86 (1H, s), 8.35 (1H, s). Mass (FAB): m/z 520 (MH^+).

3.4.7. 2′-O,4′-C-(Methyleneoxymethylene) adenosine, ***29***

1. Add 20% $Pd(OH)_2$-C (750 mg) and ammonium formate (3.00 g, 47.43 mmol) to a solution of compound **28** (493 mg, 0.949 mmol) in ethanol/acetic acid (100:3, v/v, 24 mL) at room temperature and reflux the mixture for 3.5 h.
2. Filter the hot solution through a celite pad and wash with boiling methanol (200 mL) (*see* **Note 11**).
3. After addition of silica (3.0 g) to the filtrate, concentrate the mixture under reduced pressure. Purify the residue by silica gel column chromatography [chloroform/methanol (15:1→12:1, v/v)] to give compound **29** (162 mg, 55%). Colorless crystals; mp 255–257°C (methanol). ^{1}H NMR (DMSO-d_6): δ 3.53 (1H, dd, J = 6, 12 Hz), 3.61 (1H, dd, J= 6, 12 Hz), 3.65 (1H, d, J= 12 Hz), 3.80 (1H, d, J= 12 Hz), 4.41 (1H, d, J= 6 Hz), 4.85 (1H, t, J= 6 Hz), 5.09 (1H, d, J= 6 Hz), 5.30 (1H, d, J= 6 Hz), 5.14 (1H, t, J= 6 Hz), 6.09 (1H, brd, J= 4 Hz), 6.23 (1H, s), 7.27 (2H, s), 8.13 (1H, s), 8.32 (1H, s). Mass (FAB): m/z 310 (MH^+).

*3.4.8. 6-N-Benzoyl-5′-O-(4,4′-dimethoxytrityl)-2′-O,4′-C-(methyleneoxymethylene) adenosine, **30***

1. Under N_2 atmosphere, add DMTrCl (1.40 g, 4.13 mmol) in dichloromethane (7.0 mL) dropwise to a suspension of silver triflate (1.06 g, 4.13 mmol) in dichloromethane (3.0 mL) at room temperature and stir the mixture for 1 h (*see* **Note 12**).
2. Add the supernatant fluid (3.0 mL, 3.0 equiv) to a solution of compound **29** (128 mg, 0.413 mmol) in dichloromethane/pyridine (1:2, v/v, 8.4 mL) at room temperature and stir the mixture for 1.5 h.
3. Add benzoyl chloride (0.24 mL, 2.07 mmol) to the mixture at room temperature and stir the mixture for 24 h (*see* **Note 13**).
4. After addition of saturated aqueous sodium bicarbonate solution at 0°C, extract the mixture with ethyl acetate. Wash the organic extracts with saturated aqueous sodium bicarbonate solution twice, followed by wash with water and brine, then dry over sodium sulfate, and concentrate under reduced pressure.
5. Dissolve the residue in THF (5.6 mL) and add 1 M aqueous lithium hydroxide solution (2.8 mL, 2.8 mmol). Stir the mixture for 5 h at room temperature (*see* **Note 9**).
6. Extract the mixture with ethyl acetate. Wash the organic extracts with saturated aqueous sodium bicarbonate solution twice, followed by wash with water and brine, then dry over sodium sulfate, and concentrate under reduced pressure. Purify the residue by silica gel column chromatography [hexane/ethyl acetate (1:2, v/v)] to give compound **30** (272 mg, 92%). White powder; mp 125–127°C. ^{1}H NMR ($CDCl_3$): δ 3.37 (1H, d, *J* = 10 Hz), 3.42 (1H, d, *J* = 10 Hz), 3.39 (1H, d, *J* = 9 Hz), 3.77 (6H, s), 4.01 (1H, d, *J* = 13 Hz), 4.06 (1H, d, *J* = 13 Hz), 4.92 (1H, d, *J* = 6 Hz), 4.99 (1H, dd, *J* = 6, 9 Hz), 5.23 (1H, d, *J* = 7 Hz), 5.33 (1H, d, *J* = 7 Hz), 6.39 (1H, s), 6.78 (2H, d, *J* = 9 Hz), 6.79 (2H, d, *J* = 9 Hz), 7.19–7.27 (7H, m), 7.34–7.38 (2H, m), 7.48–7.64 (3H, m), 8.01 (2H, d, *J* = 7 Hz), 8.10 (1H, s), 8.74 (1H, s), 9.07 (1H, s). Mass (FAB): *m/z* 716 (MH^+).

*3.4.9. 6-N-Benzoyl-3′-O-[2-cyanoethoxy (diisopropylamino) phosphino]-5′-O-(4,4′-dimethoxytrityl)-2′-O,4′-C-(methyleneoxy-methylene)adenosine, **31***

1. Under N_2 atmosphere, add 4,5-dicyanoimidazole (0.25 M in acetonitrile, 1.9 mL, 0.48 mmol) and $(Pr^i{}_2N)_2POCH_2CH_2CN$ (0.19 mL, 0.59 mmol) to a solution of compound **30** (283 mg, 0.396 mmol) in acetonitrile (3.0 mL) at room temperature and stir the mixture at room temperature for 16 h (*see* **Note 6**).

2. After removal of the solvent under reduced pressure, dilute the residue with ethyl acetate. Wash the mixture with saturated aqueous sodium bicarbonate solution, water and brine, dry over sodium sulfate, and concentrate under reduced pressure.
3. Purify the residue by silica gel column chromatography [hexane/ethyl acetate (2:3, v/v)] followed by the precipitation (*see* **Note 7**) from hexane/ethyl acetate to give compound **31** (304 mg, 84%). White powder; mp 99–102°C. ^{31}P NMR ($CDCl_3$): δ 150.7, 151.2. MS (FAB): *m/z* 916 (MH^+).

3.5. Synthesis of 2′,4′-BNA^{COC} Phosphoramidite Bearing G (2′,4′-BNA^{COC}-G Phosphoramidite)

3.5.1. 2-N-Isobutyrylguanine, 6

1. Stir a mixture of guanine (7.56 g, 50.0 mmol) and isobutyric anhydride (22.4 mL, 135 mmol) in dried *N,N*-dimethylacetamide (100 mL) at 150°C for 3 h.
2. Evaporate the resulting mixture under reduced pressure to 1/10th of its volume.
3. Collect the resulting solid.
4. Recrystallize the solid from a boiling mixture of water and ethanol (1:1, ca. 1500 mL) to give 2-*N*-isobutyrylguanine (6) (9.83 g, 82%). ^{1}H NMR (DMSO-d_6): δ 1.10 (6H, d, $J = 7$ Hz), 2.75 (1H, sept, $J = 7$ Hz), 8.10 (1H, brs), 11.50 (1H, brs), 12.06 (1H, brs), 13.15 (1H, brs).

3.5.2. 6-O-(Diphenylcarbamoyl)-2-N-isobutyrylguanine, 7

1. Stir a mixture of **6** (9.83 g, 41.1 mmol) and acetic anhydride (10.0 mL, 107 mmol) in dried *N,N*-dimethylformamide (50 mL) at 100°C to give a clear solution.
2. Evaporate the resulting mixture under reduced pressure.
3. Wash the solid well with ethanol (ca. 20 mL), collect by vacuum filtration, and dry the solid in vacuo.
4. Stir the solid in dried pyridine (190 mL) and add *N,N*-diisopropylethylamine (13.1 mL, 75.4 mmol) and diphenylcarbamoyl chloride (9.60 g, 41.5 mmol). Continue stirring the mixture at room temperature for 3 h.
5. Add H_2O (15 mL) and further stir the mixture for 10 min.
6. Evaporate the solvent then coevaporate the residue with toluene (40 mL × 3). Suspend the residue in 50% aqueous ethanol (235 mL) and reflux for 2 h (*see* **Note 13**).
7. After cooling, collect the solid by vacuum filtration and wash with ethanol until the washings are colorless to give 6-*O*-(diphenylcarbamoyl)-2-*N*-isobutyrylguanine as a white powder (7) (12.29 g, 72%). ^{1}H NMR (DMSO-d_6): δ 1.08 (6H, d, $J = 7$ Hz), 2.79 (1H, sept, $J = 7$ Hz), 7.29–7.47 (10H, m), 8.45 (1H, s), 11.57 (1H, brs), 13.54 (1H, brs).

3.5.3. 4′-C-Acetoxymethyl-2′-O-acetyl-3′-O-benzyl-5′-O-tert-butyldiphenylsilyl-6-O-diphenylcarbamoyl-2-N-isobutyrylguanosine, ***32***

1. Under N_2 atmosphere, add *N,O*-bis(trimethylsily)acetamide (0.17 mL, 0.678 mmol) to a suspension of compound **7** (121 mg, 0.291 mmol) in 1,2-dichloroethane (1.9 mL) at room temperature and stir the mixture at 80°C for 30 min. Evaporate the resulting clear solution under reduced pressure.
2. Add a solution of compound **2** (123 mg, 0.194 mmol) in anhydrous toluene (1.9 mL) and trimethylsilyl triflate (0.04 mL, 0.233 mmol) to the residue and stir the mixture at 80°C for 2 h.
3. After addition of saturated aqueous sodium bicarbonate solution at 0°C, extract the mixture with ethyl acetate. Wash the organic extracts with water and brine, dry over sodium sulfate, and concentrate under reduced pressure. Purify the residue by silica gel column chromatography [hexane/ethyl acetate (2:1 to 3:2, v/v)] to give compound **32** (161 mg, 84%). White powder; mp 86–88°C. ^{1}H NMR ($CDCl_3$): δ 1.01 (9H, s), 1.15 (3H, d, *J* = 7 Hz), 1.19 (3H, d, *J* = 7 Hz), 1.94 (3H, s), 2.06 (3H, s), 2.77 (1H, m), 3.85 (2H, s), 4.31 (1H, d, *J* = 12 Hz), 4.66 (1H, d, *J* = 12 Hz), 4.65 (1H, d, *J* = 11 Hz), 4.74 (1H, d, *J* = 11 Hz), 5.23 (1H, d, *J* = 6 Hz), 5.88 (1H, dd, *J* = 4, 6 Hz), 6.09 (1H, d, *J* = 4 Hz), 7.18–7.45 (21H, m), 7.54–7.61 (4H, m), 7.82 (1H, brs), 7.98 (1H, s). Mass (FAB): *m/z* 991 (MH^+).

3.5.4. 3′-O-Benzyl-5′-O-tert-butyldiphenylsilyl-6-O-diphenylcarbamoyl-4′-C-hydroxymethyl-2-N-isobutyrylguanosine, ***33***

1. Stir a mixture of **32** (100 mg, 0.10 mmol) and potassium carbonate (42 mg, 0.30 mmol) in methanol (1.4 mL) at 0°C for 40 min.
2. After neutralization with a 10% aqueous HCl solution (0.11 mL), extract the mixture with ethyl acetate. Wash the organic extracts with water and brine, dry over sodium sulfate, and concentrate under reduced pressure. Purify the residue by silica gel column chromatography [hexane/ethyl acetate (3:2 to 1:1, v/v)] to give compound **33** (88 mg, 96%). White powder; mp 94–96°C. ^{1}H NMR ($CDCl_3$): δ 0.91 (9H, s), 1.24 (3H, d, *J* = 7 Hz), 1.25 (3H, d, *J* = 7 Hz), 2.57 (1H, m), 3.55 (1H, d, *J* = 11 Hz), 3.67 (1H, d, *J* = 11 Hz), 3.80 (1H, d, *J* = 12 Hz), 3.96 (1H, d, *J* = 12 Hz), 4.66 (1H, d, *J* = 5 Hz), 4.74 (1H, d, *J* = 12 Hz), 5.16 (1H, d, *J* = 12 Hz), 4.80 (1H, t, *J* = 5 Hz), 6.02 (1H, d, *J* = 5 Hz), 7.18–7.49 (25H, m), 8.05 (1H, brs), 8.11 (1H, s). Mass (FAB): *m/z* 907 (MH^+).

*3.5.5. 3′-O-Benzyl-5′-O-tert-butyldiphenylsilyl-6-O-diphenylcarbamoyl-2-N-isobutyryl-2′-O,4′-C-(methyleneoxymethylene) guanosine, **34***

1. Under N_2 atmosphere, add *N*-bromosuccinimide (145 mg, 0.816 mmol) to a solution of **33** (123 mg, 0.136 mmol) in anhydrous DMSO (1.7 mL) at room temperature and stir the mixture at 60°C for 40 min.
2. After addition of saturated aqueous sodium bicarbonate solution at 0°C, extract the mixture with ethyl acetate. Wash the organic extracts with water and brine, dry over sodium sulfate, and concentrate under reduced pressure. Purify the residue by silica gel column chromatography [hexane/ethyl acetate (3:1, v/v)] to give compound **34** (56 mg, 45%). White powder; mp 97–99°C. ^{1}H NMR ($CDCl_3$): δ 0.97 (9H, s), 1.11 (3H, d, *J* = 7 Hz), 1.17 (3H, d, *J* = 7 Hz), 2.69 (1H, m), 3.80 (1H, d, *J* = 12 Hz), 4.12 (1H, d, *J* = 12 Hz), 3.93 (2H, s), 4.79 (1H, d, *J* = 6 Hz), 4.82 (2H, s), 5.23 (1H, d, *J* = 6 Hz), 5.47 (1H, d, *J* = 6 Hz), 5.41 (1H, d, *J* = 6 Hz), 6.32 (1H, s), 7.05–7.10 (2H, m), 7.20–7.59 (23H, m), 7.84 (1H, brs), 8.07 (1H, s). Mass (FAB): *m/z* 919 (MH^+).

*3.5.6. 3′-O-Benzyl-2-N-isobutyryl-2′-O,4′-C-(methyleneoxymethylene) guanosine, **35***

1. Add sodium nitrite (1.20 g, 17.42 mmol) to a solution of **34** (801 mg, 0.871 mmol) in anhydrous DMSO (7.2 mL) at room temperature and stir the mixture at 70°C for 19 h.
2. After addition of saturated aqueous sodium bicarbonate solution at 0°C, extract the mixture with ethyl acetate. Wash the organic extracts with water and brine, dry over sodium sulfate, and concentrate under reduced pressure.
3. Dissolve the residue in THF (8.7 mL) and add TBAF (1.0 M in THF, 1.05 mL, 1.05 mmol). Stir the mixture for 15 h at room temperature.
4. After removal of the solvent under reduced pressure, purify the residue by silica gel column chromatography [hexane/ethyl acetate (1:2, v/v) and ethyl acetate/ethanol (50:1, v/v)] to give compound **35** (264 mg, 62% over two steps). Pale brown powder; mp 144–146°C. ^{1}H NMR ($CDCl_3$): δ 1.25 (6H, d, *J* = 5 Hz), 2.72 (1H, m), 3.75 (1H, d, *J* = 12 Hz), 3.97 (1H, d, *J* = 12 Hz), 3.83 (1H, d, *J* = 12 Hz), 3.92 (1H, d, *J* = 12 Hz), 4.44 (1H, d, *J* = 5 Hz), 4.62 (1H, d, *J* = 11 Hz), 4.72 (1H, d, *J* = 11 Hz), 4.94 (1H, d, *J* = 5 Hz), 5.16 (1H, d, *J* = 6 Hz), 5.36 (1H, d, *J* = 6 Hz), 6.18 (1H, s), 7.24–7.33 (5H, m), 8.04 (1H, s), 9.92 (1H, brs), 12.09 (1H, brs). Mass (FAB): *m/z* 486 (MH^+).

*3.5.7. 2-N-Isobutyryl-2′-O,4′-C-(methyleneoxy-methylene)guanosine, **36***

1. Add 20% $Pd(OH)_2$-C (170 mg) to a solution of compound **35** (171 mg, 0.351 mmol) in methanol (7.0 mL) at room temperature and stir the mixture under H_2 atmosphere for 40 h.

2. After filtration of the mixture, concentrate the filtrate under reduced pressure. Purify the residue by silica gel column chromatography [chloroform/methanol (15:1 to 10:1, v/v)] to give compound **36** (118 mg, 85%). White powder; mp 182–184°C. ^{1}H NMR (CD_3OD): δ 1.22 (6H, d, *J* = 7 Hz), 2.71 (1H, sept, *J* = 7 Hz), 3.67 (1H, d, *J* = 12 Hz), 3.74 (1H, d, *J* = 12 Hz), 3.75 (2H, s), 4.39 (1H, d, *J* = 6 Hz), 4.73 (1H, d, *J* = 6 Hz), 5.07 (1H, d, *J* = 7 Hz), 5.35 (1H, d, *J* = 7 Hz), 6.23 (1H, s), 8.44 (1H, brs). Mass (FAB): *m/z* 396 (MH^+).

*3.5.8. 5′-O-(4,4′-Dimethoxytrityl)-2-N-isobutyryl-2′-O,4′-C-(methyleneoxymethylene) guanosine, **37***

1. Under N_2 atmosphere, add DMTrCl (454 mg, 1.340 mmol) to a solution of **36** (265 mg, 0.670 mmol) in anhydrous pyridine (5.2 mL) at room temperature and stir the mixture for 9 h.
2. After addition of saturated aqueous sodium bicarbonate solution at 0°C, extract the mixture with ethyl acetate. Wash the organic extracts with water and brine, dry over sodium sulfate, and concentrate under reduced pressure. Purify the residue by silica gel column chromatography [chloroform/methanol (50:1 to 20:1, v/v)] to give compound **37** (442 mg, 95%). White powder; mp 173–175°C. ^{1}H NMR ($CDCl_3$): δ 1.23 (3H, d, *J* = 7 Hz), 1.27 (3H, d, *J* = 7 Hz), 2.75 (1H, sept, *J* = 7 Hz), 3.37 (1H, d, *J* = 10 Hz), 3.56 (1H, d, *J* = 10 Hz), 3.70 (6H, s), 3.90 (1H, d, *J* = 13 Hz), 4.19 (1H, d, *J* = 13 Hz), 4.57 (1H, d, *J* = 5 Hz), 5.10 (1H, brd, *J* = 5 Hz), 5.14 (1H, brs), 5.17 (1H, d, *J* = 7 Hz), 5.37 (1H, d, *J* = 7 Hz), 6.19 (1H, s), 6.70 (4H, d, *J* = 9 Hz), 7.11–7.18 (3H, m), 7.24 (4H, d, *J* = 9 Hz), 7.32–7.36 (2H, m), 7.67 (1H, s), 9.58 (1H, brs), 12.34 (1H, brs). Mass (FAB): *m/z* 698 (MH^+).

*3.5.9. 3′-O-[2-Cyanoethoxy(diisopropylamino) phosphino]-5′-O-(4,4′-dimethoxytrityl)-2-N-isobutyryl-2′-O,4′-C-(methyleneoxymethylene) guanosine, **38***

1. Under N_2 atmosphere, add 4,5-dicyanoimidazole (0.25 M in acetonitrile, 3.0 mL, 0.75 mmol) and $(Pr^i{}_2N)_2POCH_2CH_2CN$ (0.30 mL, 0.945 mmol) to a solution of **37** (440 mg, 0.630 mmol) in acetonitrile/THF (1:1, v/v, 5.0 mL) at room temperature and stir the mixture for 19 h (*see* **Note 6**).
2. If the reaction is not completed yet, add 4,5-dicyanoimidazole (0.25 M in acetonitrile, 2.6 mL, 0.65 mmol) and $(Pr^i{}_2N)_2POCH_2CH_2CN$ (0.30 mL, 0.945 mmol) and further stir the mixture at room temperature for 6 h.
3. After removal of the solvent under reduced pressure, dilute the residue with ethyl acetate. Wash the mixture with saturated aqueous sodium bicarbonate solution, water, and

brine, dry over sodium sulfate, and concentrate under reduced pressure.

4. Purify the residue by silica gel column chromatography [hexane/ethyl acetate (2:3, v/v)] followed by precipitation (*see* **Note 6**) from hexane/ethyl acetate to give compound **38** (360 mg, 64%). White powder; mp 124–127°C. ^{31}P NMR ($CDCl_3$):δ 150.5, 150.6. MS (FAB): *m/z* 898 (MH^+).

3.6. Synthesis of 2′,4′-BNA^{COC}-Modified Oligonucleotides

1. The synthesis is performed on an automated DNA synthesizer. The synthesis scale is 0.2 μmol, and the 1.0 μmol synthesis also is practicable.
2. Each 2′,4′-BNA^{COC} phosphoramidite is dissolved in anhydrous acetonitrile to prepare the optimum 0.067 M solutions suitable for the Expedite® 8909 synthesizer. Other synthesizers may require different concentrations.
3. Follow the conventional phosphoramidite protocol except that the coupling time for 2′,4′-BNA^{COC} phosphoramidites increases to 20 min (*see* **Notes 2, 14**). Use conventional reagents for oxidizing step, capping step, and deblocking step (*see* **Section 2.2**). Use 5-ethylthio-1*H*-tetrazole as an activator for every coupling step. The synthesis is performed using either DMTr-ON or DMTr-OFF mode. Use the universal CPG support when the 3′-terminal nucleotide of desired oligonucleotide is a 2′,4′-BNA^{COC} analog.
4. Following synthesis, the 2′,4′-BNA^{COC}-modified oligonucleotide is cleaved from the CPG support by treatment with 28% aqueous ammonia solution at room temperature for 1.5 h. Cleavage from the universal CPG support is performed by treatment with 2 M methanolic ammonia solution at room temperature for 1.5 h. Perform an additional treatment with 28% aqueous ammonia solution at 55°C for 15 h to remove the protecting groups.
5. Purify the crude oligonucleotide with prepacked reversed-phase column followed by reversed-phase HPLC [0.1 M triethylammonium acetate buffer (pH 7.0)/acetonitrile] (for DMTr-ON mode) or prepacked gel filtration column followed by reversed-phase HPLC [0.1 M triethylammonium acetate buffer (pH 7.0)/acetonitrile] (for DMTr-OFF mode).
6. The composition of the 2′,4′-BNA^{COC}-modified oligonucleotide is confirmed by MALDI-TOF-MS analysis. Mix a solution of oligonucleotide (15–40 μM, 1 μL) with ion exchange resin and make it stand for 30 s for desalting. Pipette the resulting oligonucleotide, mix it with 1 μL of matrix solution on plate, and dry it. Perform the MALDI-TOF-MS analysis under negative mode (*see* **Note 15**).

3.7. Reprecipitation and Storage of 2′,4′-BNA^{COC} Phosphoramidite (see Note 7)

1. Dissolve silica gel purified 2′,4′-BNA^{COC} phosphoramidite in the minimum volume of ethyl acetate and dilute by adding no more than 1 volume of ethyl acetate.
2. Slowly add the resulting solution into vigorously stirred, ice-cooled hexane (100 times the volume of ethyl acetate used) to give the phosphoramidite as a powder.
3. Collect the powder by vacuum filtration and dry it under reduced pressure.
4. Store the powdered phosphoramidite under an inert gas atmosphere in a freezer.

4. Notes

1. Many bridged nucleic acids with restricted *N*-type sugar conformations have been reported, *e.g.*, 2′,4′-BNA/LNA (8–10), ENA (11, 12), 2′,4′-BNA^{NC} (13–15), PrNA (12), and so on.
2. On an automated DNA synthesizer (Expedite® 8909), 96 s of coupling time, 0.067 M phosphoramidite in anhydrous acetonitrile, and 0.25 M 4,5-dicyanoimidazole as an activator are used conventionally for natural base phosphoramidites.
3. This series of automated DNA synthesizer from Applied Biosystems has been discontinued.
4. *N*,*O*-Bis(trimethylsilyl)acetamide (BSA) is recommended for the silylation of nucleobases in glycosylation reactions. Hexamethyldisilazane (HMDS) can be used as a silylation reagent, but it CANNOT tolerate the one-pot synthesis. HMDS should be removed before mixing with sugar moiety.
5. The bridge construction on pyrimidine nucleoside derivatives can be performed using formaldehyde and *p*-toluenesulfonic acid. On the other hand, purine nucleoside derivatives should be subjected to a mixture of *N*-bromosuccinimide (NBS) and DMSO to prevent the elimination of purine bases occurring under harsh acidic conditions. DMSO activated by NBS acts as methylene source.
6. Ordinarily several activators, such as 1*H*-tetrazole, diisopropylammoniun tetrazolide, 4,5-dicyanoimidazole and 5-ethylthio-1*H*-tetrazole, are used for phosphitylation using 2-cyanoethyl *N*,*N*,*N*′,*N*′-tetraisopropylphosphorodiamidite. But for the phosphitylation of 2′,4′-BNA^{COC},

4,5-dicyanoimidazole and 5-ethylthio-1*H*-tetrazole are recommended. 1*H*-Tetrazole and diisopropylammoniun tetrazolide, which are less reactive activators, lead to quite slow phosphitylation because of steric hindrances around bridged structure.

7. Reprecipitation of phosphoramidite compound is sometimes helpful to remove a trace of impurity such as residue of phosphitylation reagent, which improves the reproducibility of coupling yields on automated DNA synthesizer. The method is described in **Section 3.7**.
8. Stirring "vigorously" is extremely important for formation of the triazolide.
9. Lithium hydroxide (lithium hydroxide) is sometimes effective for deacylation of hydroxyl group in presence of acylated nucleobase. Potassium carbonate (potassium carbonate)/methanol may remove acyl groups on nucleobase.
10. *N,O*-Bis(trimethylsilyl)acetamide (BSA) is used for the silylation of nucleobases in glycosylation reaction.
11. Because this compound is easily absorbed to activated carbon, careful washing during the filtration is necessary to recover it. (Caution! Combustible.)
12. 4,4′-Dimethoxytrityl chloride (DMTrCl), which is a typical reagent for 4,4′-dimethoxytritylation of nucleoside, may not be suitable for compound **13** because selective 4,4′-dimethoxytritylation at 5′-hydroxyl group does NOT proceed reproducibly. Prepare this DMTrOTf reagent just before you use it because of its chemical instability.
13. It is difficult to demonstrate that that the reaction is complete by TLC.
14. The coupling efficiency of 2′,4′-BNA^{COC} phosphoramidites is decreased by using a commonly used activator, 1*H*-tetrazole or 4,5-dicyanoimidazole, due to steric hindrance of the seven-membered ring structure with a methylenoxymethylene linkage.
15. Oligothymidylate can be used as standard compound for calibration.

References

1. Hari, Y., Obika, S., Ohnishi, R., Eguchi, K., Osaki, T., Ohishi, H., and Imanishi, T. (2006) Synthesis and properties of 2'-*O*,4'-*C*-methyleneoxymethylene bridged nucleic acid. *Bioorg. Med. Chem.* **14**, 1029–1038.
2. Mitsuoka, Y., Kodama, T., Ohnishi, R., Hari, Y., Imanishi, T., and Obika, S. (2009) A bridged nucleic acid, 2',4'-BNA^{COC}: Synthesis of fully modified oligonucleotides bearing thymine, 5-methylcytosine, adenine and guanine 2',4'-BNA^{COC} monomers and RNA-selective nucleic-acid recognition. *Nucleic Acids Res.* **37**, 1225–1238.

3. Hanessian, S., Yang-Chung, G., Lavallee, P., and Pernet, A. G. (1972) Chemistry of α-alkoxy sulfoxides. Formation of methylene acetals from dimethyl sulfoxide and alcohols. *J. Am. Chem. Soc.* **94**, 8929–8931.
4. Sproat, B., Colonna, F., Mullah, B., Tsou, D., Andrus, A., Hampel, A., and Vinayak, R. (1995) An efficient method for the isolation and purification of oligoribonucleotides. *Nucleos. Nucleot.* **14**, 255–73.
5. Obika, S., Morio, K., Nanbu, D., Hari, Y., Itoh, H., and Imanishi, T. (2002) Synthesis and conformation of 3',4'-BNA monomers, 3',4'-methyleneribonucleosides. *Tetrahedron* **58**, 3039–3049.
6. Jenny, T. F., Schneider, K. C., and Benner, S. A. (1992) *N*2-Isobutyryl-*O*6-[2-(*p*-nitrophenyl)ethyl]guanine: A new building block for the efficient synthesis of carbocyclic guanosine analogs. *Nucleos. Nucleot.* **11**, 1257–1261.
7. Robins, M. J., Zou, R., Guo, Z., and Wnuk, S. F. (1996) Nucleic acid related compounds. 93. A solution for the historic problem of regioselective sugar-base coupling to produce 9-glycosylguanines or 7-glycosylguanines. *J. Org. Chem.* **61**, 9207–9212.
8. Obika, S., Nanbu, D., Hari, Y., Morio, K., In, Y., Ishida, T., and Imanishi, T. (1997) Synthesis of 2'-*O*,4'-*C*-methyleneuridine and -cytidine. Novel bicyclic nucleosides having a fixed C3'-*endo* sugar puckering. *Tetrahedron Lett.* **38**, 8735–8738.
9. Obika, S., Nanbu, D., Hari, Y., Andoh, J., Morio, K., Doi, T., and Imanishi, T. (1998) Stability and structural features of the duplexes containing nucleoside analogs with a fixed N-type conformation, 2'-*O*,4'-*C*-methyleneribonucleosides. *Tetrahedron Lett.* **39**, 5401–5404.
10. Singh, S. K., Nielsen, P., Koshkin, A. A., and Wengel, J. (1998) LNA (locked nucleic acids): Synthesis and high-affinity nucleic acid recognition. *Chem. Commun.* 455–456.
11. Morita, K., Hasegawa, C., Kaneko, M., Tsutsumi, S., Sone, J., Ishikawa, T., Imanishi, T., and Koizumi, M. (2002) 2'-*O*,4'-*C*-Ethylene-bridged nucleic acids (ENA): Highly nuclease-resistant and thermodynamically stable oligonucleotides for antisense drug. *Bioorg. Med. Chem. Lett.* **12**, 73–76.
12. Morita, K., Takagi, M., Hasegawa, C., Kaneko, M., Tsutsumi, S., Sone, J., Ishikawa, T., Imanishi, T., and Koizumi, M. (2003) Synthesis and properties of 2'-*O*,4'-*C*-ethylene-bridged nucleic acids (ENA) as effective antisense oligonucleotides. *Bioorg. Med. Chem.* **11**, 2211–2226.
13. Rahman, S. M. A., Seki, S., Obika, S., Haitani, S., Miyashita, K., and Imanishi, T. (2007) Highly stable pyrimidine-motif triplex formation at physiological pH values by a bridged nucleic acid analog. *Angew. Chem. Int. Ed.* **46**, 4306–4309.
14. Miyashita, K., Rahman, S. M. A., Seki, S., Obika, S., and Imanishi, T. (2007) *N*-Methyl substituted 2',4'-BNA^{NC}: A highly nuclease-resistant nucleic acid analog with high-affinity RNA selective hybridization. *Chem. Commun.* 3765–3767.
15. Rahman, S. M. A., Seki, S., Obika, S., Yoshikawa, H., Miyashita, K., and Imanishi, T. (2008) Design, synthesis, and properties of 2',4'-BNA^{NC}: A bridged nucleic acid analogue. *J. Am. Chem. Soc.* **130**, 4886–4896.

Chapter 4

A Non-covalent Peptide-Based Strategy for Ex Vivo and In Vivo Oligonucleotide Delivery

Laurence Crombez, May C. Morris, Frederic Heitz, and Gilles Divita

Abstract

The dramatic acceleration in identification of new nucleic acid-based therapeutic molecules such as short interfering RNA (siRNA) and peptide–nucleic acid (PNA) analogues has provided new perspectives for therapeutic targeting of specific genes responsible for pathological disorders. However, the poor cellular uptake of nucleic acids together with the low permeability of the cell membrane to negatively charged molecules remain major obstacles to their clinical development. Several non-viral strategies have been proposed to improve the delivery of synthetic short oligonucleotides both in cultured cells and in vivo. Cell-penetrating peptides constitute very promising tools for non-invasive cellular import of oligonucleotides and analogs. We recently described a non-covalent strategy based on short amphiphatic peptides (MPG8/PEP3) that have been successfully applied ex vivo and in vivo for the delivery of therapeutic siRNA and PNA molecules. PEP3 and MPG8 form stable nanoparticles with PNA analogues and siRNA, respectively, and promote their efficient cellular uptake, independently of the endosomal pathway, into a wide variety of cell lines, including primary and suspension lines, without any associated cytotoxicity. This chapter describes easy-to-handle protocols for the use of MPG-8 or PEP-3-nanoparticle technologies for PNA and siRNA delivery into adherent and suspension cell lines as well as in vivo into cancer mouse models.

Key words: Cell-penetrating peptide, peptide-based non-covalent strategy, amphiphatic peptide, non-endosomal pathway, peptide acid nucleic, sirna, drug delivery, cancer mouse models.

1. Introduction

The design of potent systems for the delivery of charged and non-charged molecules that target genes of interest remains a major challenge in therapeutics (1, 2). Short interfering RNAs (siRNA) constitute powerful biomedical tools to specifically control protein activation and/or gene expression

J. Goodchild (ed.), *Therapeutic Oligonucleotides*, Methods in Molecular Biology 764,
DOI 10.1007/978-1-61779-188-8_4, © Springer Science+Business Media, LLC 2011

post-transcriptionally and has provided great hope for therapeutic targeting of specific genes responsible for pathological disorders (3–5). Among antisense DNA mimics, peptide nucleic acids (PNAs) and their derivatives are very promising tools for antisense therapy in both eukaryotic and prokaryotic cells (6–8). They present several advantages including specific gene targeting, high stability, resistance to nucleases and proteases, and high affinity of RNA or DNA targets (6–8). However, the major obstacle to clinical application of most antisense or nucleic acid-based strategies remains their low cellular uptake associated with poor ability to reach their intracellular target (9–11). Therefore, numerous non-viral strategies have been proposed to improve the delivery of synthetic small oligonucleotides both in cultured cells and in vivo, including lipids, cationic polymers, antibody-protamines, RNA-aptamers, nanoparticles, and cell-penetrating peptides (12–14). However, so far there is no universal efficient method for oligonucleotide delivery. Cell-penetrating peptides (CPP) or protein transduction domain (PTD) constitute very promising tools for non-invasive cellular import of cargo and have been successfully applied for ex vivo and in vivo delivery of therapeutic molecules (13–15). PTD/CPP can be grouped into two major classes, the first requiring chemical linkage with the drug for cellular internalization and the second involving the formation of stable, non-covalent complexes with drugs (13). CPPs including synthetic and natural cell-permeable peptides, protein transduction domains (PTDs), and membrane-translocating sequences have been successfully used to improve the delivery of covalently linked antisense molecules into cells (16–19). Several chemical modifications based on covalently linked cell-penetrating peptides have been successfully used to improve PNA and PNA analog delivery into cultured cells as well as PNA bioavailability and activity in vivo (17, 18). In contrast, the delivery of charged oligonucleotides and siRNA is more challenging as multiple anionic charges of the nucleic acid interact with the CPP moiety and inhibit uptake by steric hindrance. Recently, CPPs have also been optimized for siRNA delivery, and the non-covalent strategy has been clearly shown to be more appropriate for siRNA delivery (13, 14, 19). The primary amphiphatic MPG peptide has been reported to improve siRNA delivery ex vivo into a large panel of cell lines and in vivo (20–22). The non-covalent approach for siRNA delivery has been extended to other well-known CPPs including polyarginine, Penetratin and TAT-derived peptides (13, 14).

We have recently described a new peptide-based strategy for oligonucleotide delivery based on primary amphiphatic peptides: PEP3 (23) and MPG8 (22). PEP3 is a short 15-residue peptide (ac-KWFETWFTEWPKKRK-cysteamide) derived from PEP1 (24) and MPG8 is a 21-residue (β AFLGWLGAWGTMGWSP-

KKKRK-cysteamide) optimized from MPG sequence (20). Both peptides form stable complexes with either siRNA for MPG8 or PNA and negatively charged PNA-like DNA mimic, HypNA-*p*PNA, for PEP3 (22, 23). HypNA-*p*PNA consists of a phosphonate analog of a PNA and a PNA-like monomer based on a *trans*-4-hydroxyl-L-proline (25). We have demonstrated that MPG8 and PEP3 form stable complexes with their respective cargoes, through non-covalent interactions, thereby increasing their stability and improving their delivery into a wide variety of cell lines, including suspension and primary cell lines. MPG8/PEP3-mediated oligonucleotide cellular uptake mechanism is independent of the major endocytosis pathways and controlled by both PEP3/MPG8 structural versatility and their ability to interact with phospholipids (26).

This chapter will describe easy-to-handle protocols for the use of the non-covalent MPG/PEP technologies for the delivery of PNA/HypNA-*p*PNA and siRNA into mammalian adherent, "hard-to-transfect" suspension cell lines and in vivo using xenografted mouse tumor models. It will also highlight different critical points in the peptide/nucleic acid complex preparation and transfection protocols.

2. Materials

2.1. Preparation of Peptide/Nucleic Acid Complexes

1. Stock solution of siRNA targeting *gapdh* mRNA (5′-CAUCAUCCCUGCCUCUACUTT-3′ for the sense strand) (Eurogentec): 5 μM in 50 mM Tris, 0.5 mM EDTA buffer, or RNase-free water.
2. Stock solution of siRNA targeting *Cyclin B1* mRNA (5′-GGCGAAGAUCAACAUGGCATT-3′ for the sense strand) (Eurogentec): 5 μM in 50 mM Tris, 0.5 mM EDTA buffer, or RNase free water.
3. Stock solution of antisense PNA (5′-TGC CAT CGG GCT TGG AGG-3′) targeting the first codons of the open reading frame of the *cyclin B1* gene (Applied Biosystems): 10 or 100 μM in water, PBS, or other suitable buffer.
4. Stock solution of antisense HypNA-*p*PNA (GripNA) (5′-TGC CAT CGG GCT TGG AGG-3′) targeting the first codons of the open reading frame of the *cyclin B1* gene (Active Motif Inc.): 10 or 100 μM in water or PBS.
5. Phosphate buffered saline (PBS): 1.1 mM KH_2PO_4, 155 mM NaCl, 3 mM Na_2HPO_4, pH 7.4.
6. MPG8: 2 mg/mL (780 μM) in ultra pure RNAse, Dnase free water, 2% DMSO.

7. PEP3: 2 mg/mL (902 μM) in ultra pure RNAse, Dnase free water, 2% DMSO.
8. Reverse-phase C^{18} HPLC column: UP5 WOD/25M Uptispere 300 5 ODB, 250 × 21.2 mm (Interchrom).

2.2. MPG8/PEP3-Mediated Oligonucleotide Transfection into Adherent Cell Lines

1. Dulbecco's Modified Eagle's Medium (DMEM)
2. Complete growth medium: DMEM; 2 mM glutamine; 1% antibiotics (streptomycin 10,000 μg/mL, penicillin, 10,000 IU/mL); 10% (w/v) fetal calf serum (FCS)
3. Growth medium: DMEM containing GlutaMAX and 10% FCS
4. Complete growth medium with 16% FCS
5. PBS (*see* **Section 2.1**)
6. QuantiGene® 2.0 Reagent System (Affimetrix/Panomics Inc.)
7. Anti-GAPDH (6C5) monoclonal mouse primary antibody (Santa Cruz Biotechnology)
8. Anti-Cyclin B1 (SC-245) monoclonal mouse primary antibody (Santa Cruz Biotechnology)
9. Rabbit anti-actin antibody (Sigma-Aldrich)
10. Sheep secondary anti-mouse HRP-linked whole antibody (Amersham)

2.3. MPG8/PEP3-Mediated Oligonucleotide Transfection into Suspension Cell Lines

1. RPMI 1640
2. RPMI 1640 supplemented with 10% FCS
3. PBS (*see* **Section 2.1**)

2.4. MPG8/PEP3-Mediated Oligonucleotide Delivery In Vivo

1. PEP3 and MPG8 conjugated with polyethylene glycol (PEG) or a cholesterol (Chol) moiety at their *N*-terminus.

3. Methods

The protocols described below outline (1) the formation and storage of PEP3/PNA or MPG8/siRNA complexes, (2) optimized protocols for PNA and siRNA transfections in mammalian adherent cells, (3) in suspension "hard-to-transfect" cell lines, and (4) in vivo in tumor mouse models. The different procedures were performed using a siRNA or a PNA targeting *gapdh* and *Cyclin B1* genes and modified according to *refs* (22 and 23).

3.1. Preparation of Peptide/Nucleic Acid Complexes

The procedure for complex formation constitutes a major factor in the success and efficiency of PEP and MPG technologies and should be followed carefully (**Notes 1–3**).

MPG8 (21-residue: βAFLGWLGAWGTMGWSPKKKRK-cya; MW: 2567 Da) and PEP3 (15-residue: ac-KWFETWFTEWPKKRK-cya; MW: 2217 Da) were synthesized by solid-phase peptide synthesis using AEDI-expensin resin with (fluorenylmethoxy)-carbonyl (Fmoc) on a Pioneer Peptide Synthesizer starting from Fmoc-PAL-PEG-PS resin on a 0.2-mmol scale, as described previously (20, 22, 23). The coupling reactions were performed with 0.5 M of HATU in the presence of 1 M DIEA. Protecting group removal and final cleavage from the resin were carried out with TFA/phenol/H_2O/thioanisol/ethanedithiol (82.5/5/5/5/2.5%) for 3.5 h. All peptides were *N*-acetylated and bear a cysteamide group at their carboxy-terminus (-NH-CH_2-CH_2-SH), both of which are essential for MPG/PEP stability, cellular uptake, and formation of peptide/nucleic acid particles (22, 23). The crude peptides were purified by semi-preparative reverse-phase high-performance liquid chromatography (RP-HPLC) on a C^{18} column and identified by electrospray mass spectrometry and amino acid analysis (22, 23). Peptides are stable for at least 1 year when stored at −20°C in lyophilized form. Mono-PEGylated-Pep-3 and Cholesterol-MPG-8 conjugations were performed at the primary amino group of the *N*-terminal residues, then PEGylated-PEP3 or Cholesterol-MPG8 were further purified by RP-HPLC and analyzed by electrospray ionization mass spectroscopy (20–24).

3.1.1. Stock Solutions of Vector-Peptides and Nucleic Acids

1. Take the vial containing the peptide powder out of the freezer and equilibrate for 30 min at room temperature without opening the vial. Resuspend MPG8 or PEP3 at a final concentration of 2 mg/mL (MPG-8: 780 μM; PEP-3: 902 μM) in ultra pure water (RNAse, Dnase free) containing 2% DMSO. Peptide powder should be first solubilized directly in DMSO; then add the calculated volume of water to reach the 2 mg/mL final carrier peptide concentration and 2% DMSO.
2. Mix gently by tapping the tube.
3. Sonicate the carrier peptide solution for 10 min in a water bath sonicator. Sonication is essential to prevent aggregation and finalize the solubilization of MPG8 and PEP3.
4. For siRNA or PNA/HypNA-*p*PNA transfection experiments, dilute the MPG8 or PEP3 solution at 100 μM in ultra pure water (RNAse, Dnase free). Repeated freeze/thaw cycles can induce peptide aggregation; therefore, it is recommended

to aliquot the peptide carrier stock solution into tubes containing the amount you expect to use in a typical experiment prior to freezing. The MPG8 or PEP3 stock solution is stable for about 4 months when stored at –20°C.

5. Prepare a stock solution of siRNA in water at 5 or 100 μM. Usually the concentration of commercially available solution of annealed siRNA is at 100 μM and should be diluted to 5 μM in water or in buffer containing 50 nM Tris, pH 7.5, and 2 mM EDTA.
6. Prepare a stock solution of HypNA-*p*PNA/PNA in water or PBS at 10 or 100 μM. Depending on the sequence of PNA and its modifications, solubility of the PNA can be a problem and different buffers should be screened in order to improve the solubility. In contrast, HypNA-*p*PNAs are soluble in water and PBS.

3.1.2. MPG8/siRNA and PEP3/HypNA-pPNA Complexes for Transfection

1. If needed defrost on ice the carrier peptide solutions at 100 μM. At this stage, sonication of the peptide solution is recommended, to limit aggregation, for 5 min in a water bath sonicator. Alternatively a probe sonicator can also be used: place the tube in cold water and sonicate for 1 min at amplitude of 30%. If needed, defrost siRNA or PNAs solution at 5 and 10 μM. Vortex siRNA and PNA/HypNA-*p*PNA before use. Do not vortex carrier peptides and their complexes with nucleic acid. Although MPG8/siRNA and PEP3/HypNA-*p*PNA complexes are stable 2 weeks at 4°C, we suggested, for a high efficiency, to prepare them freshly for each experiment (**Notes 4, 6**).
2. Depending on the cell line, the biological response expected, and the target gene, siRNA can be used at concentrations varying from 5 to 200 nM. 50 nM of siRNA are sufficient for a gene expression knockdown greater than 80%. Accordingly, the protocols described are for a 35-mm culture plate using a final concentration of 40 nM of siRNA complexed with 800 nM of MPG8, corresponding to a peptide/siRNA molar ratio of 20/1, respectively. Dilute 5 μM siRNA (12.8 μL) in PBS (87.2 μL) and 100 μM MPG8 (12.8 μL) in water (87.2 μL) into two separate tubes.
3. PNA/HypNA-*p*PNA are used at concentrations ranging from 20 nM to 1 μM. Using PEP3, we demonstrated than 100 nM of PNA/HypNA-*p*PNA is sufficient for an antisense response greater than 80%. Accordingly, the protocols described are for a 35-mm culture plate using a final concentration of 100 nM of PNA/HypNA-*p*PNA complexed with 2000 nM of PEP3, corresponding to a peptide/PNA PNA/HypNA-*p*PNA molar ratio of 20/1, respec-

tively. Dilute 10 μM PNA PNA/HypNA-*p*PNA (15.6 μL) in PBS (84.4 μL) and 100 μM PEP3 (32 μL) in water (68 μL) into two separate tubes.

4. Add 100 μL of diluted carrier peptide solution to 100 μL siRNA or PNA/HypNA-*p*PNA solutions. Mix gently by tapping the tube.
5. Incubate at 37°C for 20–30 min to allow the carrier peptide/nucleic acid complexes to form, and then proceed immediately to the transfection experiments. For lower concentrations of nucleic acid, dilute the peptide/nucleic acid complex in PBS (0.5×) or in water (**Notes 3–5**) using serial dilutions, to reach the needed concentration. For multiple assays, a mix of 6–12 reactions can be used at a molar ratio of 20/1. Do not exceed the volume required for 12 reactions and the 20/1 peptide/nucleic acid ratio, as this may induce aggregation.

3.2. MPG8/PEP3-Mediated Oligonucleotide Transfection into Adherent Cell Lines

The protocol is described for Hela, U2OS, PC3, and HUVEC cell lines cultured in 6-well plates, using a siRNA or a PNA/HypNA-*p*PNA targeting the *Cyclin B1* gene (**Fig. 4.1**). A common protocol for MPG8 and PEP3 has been established and optimized. Cells were maintained as monolayer cultures in Dulbecco's modified Eagle's medium (DMEM) supplemented with 2 mM glutamine, 1% antibiotics (streptomycin 10,000 μg/mL, penicillin, 10,000 IU/mL), and 10% (w/v) fetal calf serum (FCS), at 37°C in a humidified atmosphere containing 5% CO_2.

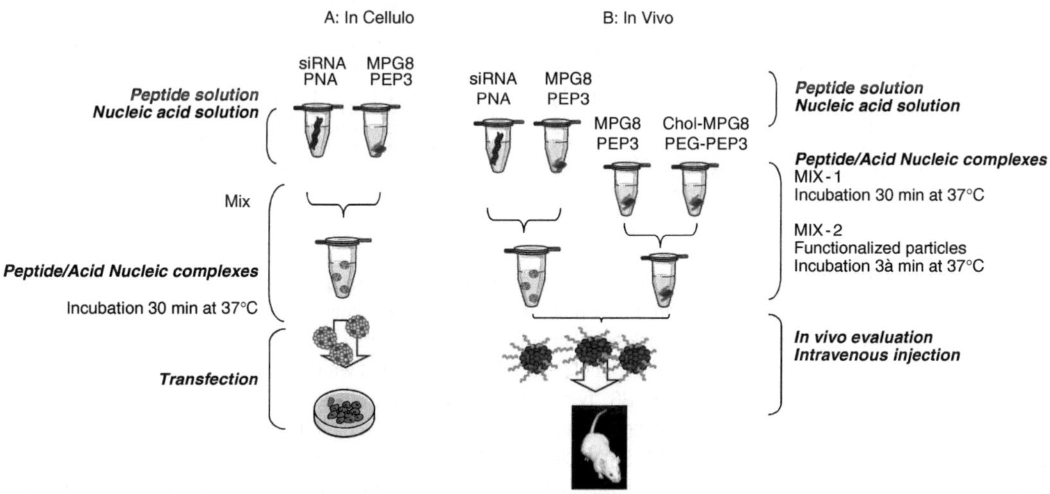

Fig. 4.1. Schematic diagram of the protocol for PEP3/MPG8-mediated nucleic acid delivery. (**a**) The protocol involves the formation and handling of PEP/HypNA-*p*PNA and MPG8/siRNA complexes and then nucleic acid transfection into either adherent or suspension "hard-to-transfect" cell lines. (**b**) The protocol involves the formation and handling of PEG-PEP3/PEP3/HypNA-*p*PNA and Chol-MPG8/MPG8/siRNA particles and then nucleic acid administration by systemic intravenous injection.

The siRNA-associated silencing responses were followed at both the mRNA (**Fig. 4.2A**) and protein levels (**Fig. 4.2a**) using Quantigen technology and Western blot analysis, respectively. The antisense response linked to PNA/HypNA-*p*PNA was monitored at the protein level using Western blot analysis (**Fig. 4.2b**). The amount of siRNA, PNA/HypNA-*p*PNA, MPG8, and PEP3, and transfection volume and number of cells should be adjusted according to the size of the culture plate used (**Notes 1, 3, 6, 7**).

1. Trypsinize and count the cells on the day before transfection; then split cells in 6-well plates at a density of 1.3×10^4 cells per well with 2 mL of preheated complete growth medium. It is recommended to pass the cell the day before treatment for a better response following transfection. Incubate cells overnight at 37°C in a humidified atmosphere containing 5% CO_2 until the cells are 50–70% confluent. It is important (1) not to add antibiotics to the media during transfection, (2) to minimize trypsinization treatment, and (3) to use prewarmed trypsin solution (at least 15 min at room temperature) to limit cell cycle arrest or/and cell death.
2. Preheat growth medium (DMEM containing GlutaMAX and 10% FBS) at 37°C in a CO_2 incubator for at least 30 min before use.

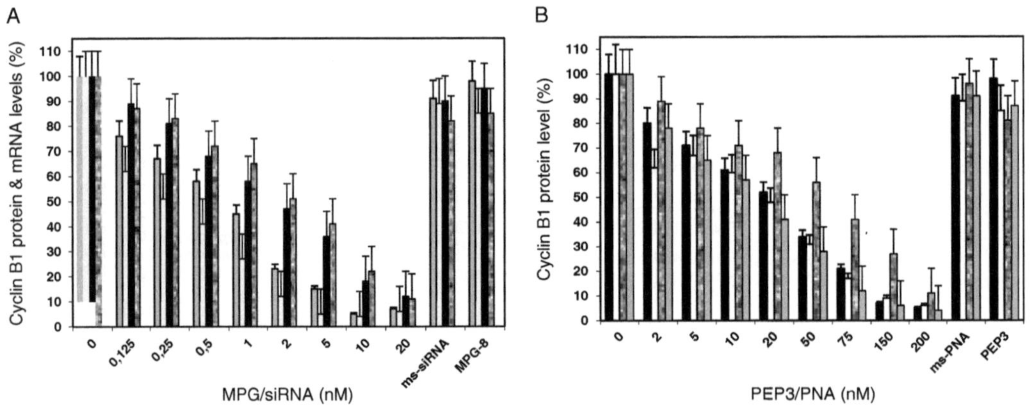

Fig. 4.2. PEP/MPG8-mediated nucleic acid delivery into adherent and challenging cell lines. (**a**) Stock solutions of MPG8/siRNA (100 nM) particles were prepared at a molar ratio of 1/20, and lower concentrations (from 50 nM to 0.3 nM) were obtained by serial dilution of the stock solution in PBS. HUVEC (in *grey* for mRNA and in *white* for protein) and JURKAT (*black* for mRNA and in *dashed* for protein) cells were then transfected with varying concentrations (0.3–40 nM) of human Cyclin B1 siRNA complexed with MPG8. Twenty-four hours post transfection, cells were lysed and *cyclinB1* mRNA and protein levels were quantified by Quantigen technology and Western Blotting, respectively. Mismatched siRNA associated with MPG8 (50 nM) and empty MPG8 particles (5 μM) were used as a control. Changes in *Cyclin B1* mRNA and Cyclin B1 protein levels were normalized to cyclophillin B gene expression and non-transfected cells, respectively. (**b**) Stock solutions of PEP3/PNA (100 nM) particles were prepared at a molar ratio of 1/10. PC3 (*in white*), HUVEC (*in grey*), Hela (*in black*) and JURKAT (*in dashed*) cells were then transfected with varying concentrations (1–200 nM) of human Cyclin B1 antisense PNA complexed with PEP3. Twenty-four hours post transfection, cells were lysed and Cyclin B1 protein level was quantified by Western Blotting. Mismatched PNA associated with PEP3 (200 nM) and empty PEP3 particles (5 μM) were used as a control.

3. Remove growth medium from the cells by aspiration and rinse the cells twice with PBS. It is important to remove the entire growth medium, as serum will lower the transfection efficiency of the peptide/nucleic acid complex.
4. Add the 200 μL of peptide/nucleic complex directly onto the cells and incubated for 3–5 min at 37°C. Do not exceed 5 min to avoid drying the cells.
5. Add 400 μL of the appropriate medium without serum (DMEM or others) to achieve a final volume of 600 μL for a 35-mm plate. At that stage, 5% serum can be added to the medium for sensitive cell lines.
6. Incubate at 37°C in a humidified atmosphere containing 5% CO_2 for 30 min.
7. Add 1 mL of complete growth medium with 16% FBS to obtain a final concentration of 10% FBS. Do not remove the peptide/nucleic acid complex.
8. Incubate at 37°C in a humidified atmosphere containing 5% CO_2 for 24–48 h, depending on the cellular response expected and on the analysis approaches (**Note 2**). The nucleic acids are fully released in the cells after 1 h.
9. Process the cells for observation or detection assays. Cyclin B1 protein and mRNA levels were determined by Western blot and quantigenTM, respectively. mRNA quantification was performed using QuantiGene R_ 2.0 Reagent System directly on cell lysates without mRNA purification or amplification according to the manufacturer's instructions. Cyclin B1 protein level was analyzed by Western blot as described in (22), using monoclonal mouse primary antibodies anti-GAPDH (6C5) and anti-Cyclin B1 (SC-245) and rabbit anti-actin antibody (as a loading control) and sheep secondary anti-mouse HRPlinked whole antibody from.

3.3. MPG8/PEP3-Mediated Oligonucleotide Transfection into Suspension Cell Lines

A common protocol for PEP3 and MPG8 has been optimized on Jurkat T cells, using a siRNA or antisense PNA HypNA-*p*PNA, targeting the *cyclin B1 or GAPDH* genes. Jurkat suspension cell line was cultured in RPMI-1640 medium supplemented with 10% FCS and 0.05 mM 2-mercaptoethanol. The siRNA- and PNA/HypNA-*p*PNA-associated responses were followed at the mRNA (**Fig. 4.2a**) and protein (**Fig. 4.2b**) levels as described in **Section 3.2**.

1. The same number of cells recommended for seeding adherent cells is recommended for suspension cells (confluency between 50 and 70%). Cells are cultured in appropriate medium (RPMI 1640 supplemented with 10% FSC) in 35-mm dishes or 6-well plates (*see* **Notes 5, 6**).

2. The peptide/nucleic acid complexes at molar ratio are formed as described for adherent cells (Steps 1–5, **Section 3.1.2**).
3. Collect the suspension cells by centrifugation at 400×*g* for 5 min. Remove the supernatant and wash the cells twice with PBS.
4. Centrifuge at 400×*g* for 5 min to pellet the cells. Remove the supernatant.
5. Add 400 μL serum-free medium to the peptide/nucleic acid complex solution (200 μL) in order to achieve a final transduction volume of 600 μL, then solubilize the cell pellet in the 600 μL of peptide/nucleic acid complex solution.
6. Incubate at 37°C in a humidified atmosphere containing 5% CO_2 for 30 min to 1 h depending on the cell lines.
7. Add complete growth medium to the cells and adjust serum levels according to culture requirements. Do not remove the peptide/nucleic acid complex. Continue to incubate at 37°C in a humidified atmosphere containing 5% CO_2 for 24–48 h depending on the expected cellular response. As described for adherent cells, nucleic acids are fully released into cells 1 h later.
8. Process the cells for observation or detection assays as reported in Step 9, **Section 3.2**.

3.4. MPG8/PEP3-Mediated Oligonucleotide Delivery In Vivo

Protocols for in vivo systemic intravenous administration of PEP3/HypNA-*p*PNA and MPG8/siRNA nanoparticles have been established using a mouse model for prostate cancer and targeting cyclin B1. Experiments were performed on 5–10 animals per group and the silencing or antisense responses were quantified on the growth of the tumor and on cyclin B1 protein level within the tumor (**Fig. 4.3a, b**). Experiments were performed according to national regulations and approved by the local animal experimentation ethical committee.

1. To improve the stability of Pep-3/HypNA-*p*PNA and of MPG8/siRNA particles in vivo, the sequences of PEP3 and of MPG8 have been modified by attachment of a polyethylene glycol (PEG) or a cholesterol (Chol) moiety at their *N*-terminus, respectively. Protocols were established to form peptide-based nanoparticle containing 15% of functionalized peptides, PEG-PEP3, or cholesterol-MPG8, at their surface.
2. Make a stock solution of PEG-PEP3 or Chol-MPG8 at a final concentration of 0.6 mM in ultra pure water (RNAse, Dnase free) and 2% DMSO. The Chol-MPG8 or PEG-PEP3 stock solution is stable for about 3 weeks when stored at –20°C.

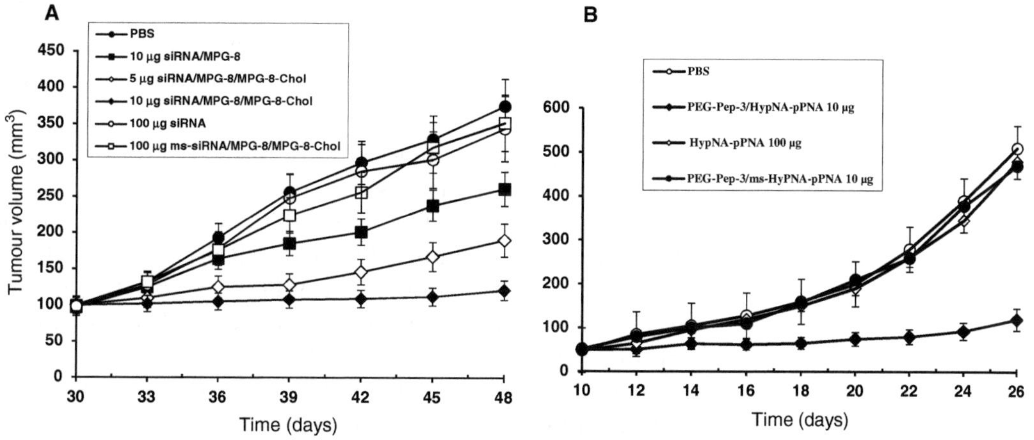

Fig. 4.3. MPG8/PEP3-mediated cyclin B1 siRNA or antisense HypNA-*p*PNA delivery blocks tumor growth in vivo. (**a**) *Inhibition of PC3 tumor growth upon intravenous injection of MPG8/siRNA.* Swiss nude nice (a cohort of *N*=6 animals) were injected subcutaneously with 10^6 PC3 cells and tumour diameter was measured in two directions at regular intervals using a digital calliper. Animals were treated by intravenous tail vein injection, every 3 days, with a solution of 0.1 ml of either PBS (•), free Cyc-B1 siRNA (100 µg: ○), Cyc-B1 siRNA (10 µg) complexed with MPG-8 (□), control siRNA Cyc-B3 (100 µg: ○, or Cyc-B1 siRNA (5 µg: ◇ and 10 µg: ♦) complexed with MPG-8/chol-MPG-8 at a 1/20 molar ratio. Curves show the mean value of tumour size in a *group of six animals.* (**b**) *Tumour growth inhibition by antisense HypNA-pPNA/PEP3 via intravenous administration.* Xenografted mice were treated with 100 µg of naked HypNA-pPNA (○), 10 µg of antisense HyPNA-*p*PNA containing two mutations associated with PEG-PEP3 (•), or 10 µg of antisense associated with PEP3 (◇), PEP2 (□), and PEG-PEP3 (♦) at a 20:1 molar ratio. Treatment started after 7 days, when tumour size reached about 50–100 mm^3. The tumour diameter was measured in two directions at regular intervals using a digital calliper. Curves represent the mean value of tumour size in a group of five animals.

3. Formulations containing 15% Chol-MPG8 or PEG-PEP3 were prepared in a stepwise fashion (**Fig. 4.1**). Protocols are described for four injections of 10 µg of nucleic acid and similar protocol can be easy adapted for different quantities of nucleic acid as well as for a mix of 6 up to 24 injections. However, the homogeneity of the particle solution should be checked when using larger volume and higher concentration of nucleic acid (**Note 3**).
4. First PEP3/HyPNA-*p*PNA precomplexes were formed at molar ratio of 10/1. For four injections of 100 µL of 10 µg (10 µg: 0.5 mg/Kg) of HyPNA-*p*PNA associated to PEP3/PEG-PEP3 mix particles. Dilute 100 µM HypNA-*p*PNA (141.6 µL) in water (58.4 µL) and 600 µM PEP3 (18 µL) in water (82 µL) into two separate tubes. Add 100 µL of carrier peptide solution to 200 µL PNA/HypNA-*p*PNA solutions. Mix gently by tapping the tube. Incubate at 37°C for 20 min to allow the carrier peptide/nucleic acid complexes to be formed.
5. Dilute 600 µM PEG-PEP3 (8 µL) into 600 µM PEP3 (8 µL) and make up to 100 µL with water (84 µL). Add the 100 µL of the mixed peptide solution to the

PEP3/HypNA-*p*PNA. Mix gently by tapping the tube. Incubate at 37°C for 20 min. Before intravenous injection, adjust salt condition to 0.9% NaCl; then proceed to intravenous injection.

6. MPG8/siRNA precomplexes were formed at molar ratio of 20/1. For four injections of 100 μL of 10 μg (10 μg: 0.5 mg/Kg) of siRNA associated to MPG8/Chol-MPG8 mixed particles. Dilute 100 μM siRNA (141.6 μL) in water (58.4 μL) and 600 μM MPG8 (37.6 μL) in water (62.4 μL) into two separate tubes. Add 100 μL of carrier peptide solution to 200 μL siRNA solution. Mix gently by tapping the tube. Incubate at 37°C for 20 min to allow the carrier peptide/nucleic acid complexes to be formed.
7. Dilute 600 μM Chol-MPG8 (12.5 μL) into 600 μM MPG8 (12.5 μL) and make up to 100 μL with water (65 μL). Add the 100 μL of the mixed peptide solution to the MPG8/siRNA. Mix gently by tapping the tube. Incubate at 37°C for 20 min. Before intravenous injection adjust salt condition to 0.9% NaCl; then proceed to intravenous injection.
8. Athymic female nude mice (6–8 weeks of age) are subcutaneously inoculated in the flank with 1×10^6 PC3 cells in 100 μL PBS.
9. Two to three weeks after tumor implant, when tumor size reached about 100 mm^3, animals are treated by intravenous injection, every 3 days, with a solution of 0.1 mL of different PEG-PEP3/PEP3/HypNA-*p*PNA and Chol-MPG8/MPG8/siRNA formulations.
10. The tumor diameter was measured in two directions at regular intervals using a digital calliper and volumes were calculated using the formula length × width × height × 0.52 (27). At the end of the experiments, tumors are removed and Cyclin B1 protein levels are evaluated by Western blotting.

4. Notes

1. This technology is not dependent on the siRNA or PNA sequence and so is usable for targeting any gene without further optimization. A large variety of siRNA/PNA/HypNA-*p*PNA have been successfully applied using MPG8/PEP3 technology for silencing activity in different cell lines (13, 19, 22, 23).
2. The incubation time of the peptide/nucleic acid complex onto transfected cells is an important point and is directly

related to the target protein half-life. Hence the RNA interference or HypNA-*p*PNA antisense effect must be long enough to allow existing protein to be degraded and permit detection of the knock-down effect.

3. Homogeneity, size, and global charge of the MPG8/siRNA and PEP3/HypNA-*p*PNA particles are major parameters that need to be controlled for in vivo application. Size of the particles should be lower than 300 nm diameter and particle charge should not exceed a zeta potential of 20 volts. Characterization of complexes can be performed by fluorescence polarization titration experiments (22, 23). Mean particle size and charge distributions can be evaluated by dynamic light scattering and zeta potential. For Chol-MPG8/MPG8/siRNA and PEG-PEP3/PEP3/HypNA-*p*PNA particles ideal sizes are 180 ± 45 nm and 120 ± 15 nm with zeta potential of 14 ± 2 volts and 12 ± 5 volts, respectively.
4. The advantages of MPG/PEP technologies are directly associated with lack of toxicity and their cellular uptake mechanism, which is independent of the endosomal pathway, favors a rapid release of the nucleic acid within the cytoplasm, significantly limits degradation, and preserves the biological activity of internalized cargoes for prolonged time periods. Moreover, both peptides are not toxic and do not induce any immune response in vivo (22, 23).
5. It is essential to perform complex formation between MPG-8/siRNA and PEP3/HypNA-pPNA in the absence of serum to limit degradation of siRNA and interactions with serum proteins. However, the transfection process itself is not affected by the presence of serum, which is a considerable advantage for most biological applications (22, 23). For in vivo systemic delivery, the peptide/nucleic acid complexes can be formed in water and then adjusted to physiological conditions using NaCl (0.9%) or glucose (15%).
6. Although this protocol was tested on several cells lines, conditions for efficient siRNA delivery should be optimized for every new cell line, including reagent concentration, cell number, and exposure time of cells to the MPG8/siRNA or PEP3/HypNA-*p*PNA complexes. A well-characterized siRNA should always be used as a positive control of transfection.
7. Low efficiency may be associated with several parameters: *(a) Cell confluency:* For adherent cells the optimal confluence is about 50–60%; higher confluence (90%) dramatically reduces the transduction efficiency. Cells must be in the exponential growth stage at the time of transfection; thus,

confluency of 40–60% is recommended. *(b) Formation of Peptide/nucleic acid complexes:* Conditions for the formation of MPG8/siRNA and PEP3/HypNA-*p*PNA complexes are critical and should be respected. Special attention should be paid to the recommended volumes, incubation times for the formation of the complexes, and time of exposure of these complexes to cells.

Acknowledgments

This work was supported in part by the Centre National de la Recherche Scientifique (CNRS), by the Agence Nationale de la Recherche (ANR, ANR-06-BLAN-0071-Pepvec4Ther), and by a grant from Panomics Inc. LC was supported by a grant from the Ligue de Recherche contre le Cancer (LNCC). We thank all members of the laboratory for fruitful discussions.

References

1. Opalinska, J. B., and Gewirtz, A. M. (2002) Nucleic acid therapeutics: basis principles and recent application. *Nat. Rev. Drug Discov.* **1**, 503–514.
2. Sazani, P., Vacek, M. M., and Kole, R. (2002) Short-term and long-term modulation of gene expression by antisense therapeutics. *Curr. Opin. Biotechnol.* **13**, 468–472.
3. Fire, A., Xu, S., Montgomery, M. K., Kostas, S. A., Driver, S. E., and Mello, C. C. (1988) Potent and specific genetic interference by double-stranded RNA in Caenorhabditis elegans. *Nature* **391**, 806–811.
4. Dorsett, Y., and Tuschl, T. (2004) siRNAs: applications in functional genomics and potential as therapeutics. *Nat. Rev. Drug Discov.* **3**, 318–329.
5. Castanotto, D., and Rossi, J. J. (2009) The promises and pitfalls of RNA-interference-based therapeutics. *Nature* **457**, 426–433.
6. Nielsen, P. E., and Egholm, M. (1999) Introduction to peptide nucleic acid. *Curr. Issues Mol. Biol.* **1**, 89–104.
7. Nielsen, P. E. (2006) Addressing the challenges of cellular delivery and bioavailability of peptide nucleic acids (PNA). *Q. Rev. Biophys.* **38**, 345–350.
8. Egholm, M., Buchardt, O., Christensen, L., Behrens, C., Freier, S. M., Driver, D. A., Berg, R. H., Kim, S. K., Norden, B., and Nielsen, P. E. (1993) PNA hybridizes to complementary oligonucleotides obeying the Watson-Crick hydrogen-bonding rules. *Nature* **365**, 566–568.
9. De Fougerolles, A., Vornlocher, H. -P., Maraganore, J., and Lieberman, J. (2007) Interfering with disease: a progress report on siRNA-based therapeutics. *Nat. Rev. Drug Discov.* **6**, 443–453.
10. Whitehead, K. A., Langer, R., and Anderson, D. G. (2009) Knocking down barriers: advances in siRNA delivery. *Nat. Rev. Drug Discov.* **8**, 129–138.
11. Sibley, C. R., Seow, Y., and Wood, M. J. (2010) Novel RNA-based strategies for therapeutic gene silencing. *Mol. Ther.* **18**, 466–476.
12. Juliano, R., Alam, M. R., Dixit, V., and Kang, H. (2008) Mechanisms and strategies for effective delivery of antisense and siRNA oligonucleotides. *Nucleic Acids Res.* **36**, 4158–4171.
13. Heitz, F., Morris, M. C., and Divita, G. (2009) Twenty years of cell-penetrating peptides: from molecular mechanisms to therapeutics. *Br. J. Pharmacol.* **157**, 195–206.
14. Eguchi, A., and Dowdy, S. F. (2009) siRNA delivery using peptide transduction domains. *Trends Pharmacol. Sci.* **30**, 341–345.
15. El-Andaloussi, S., Holm, T., and Langel, U. (2005) Cell-penetrating peptides: mech-

anisms and applications. *Curr. Pharm. Des.* **11**, 3597–3611.

16. Pooga, M., Soomets, U., Hallbrink, M., Valkna, A., Saar, K., Rezaei, K., Kahl, U., Hao, J. X., Xu, X. J., Wiesenfeld-Hallin, Z., Hokfelt, T., Bartfai, T., and Langel, U. (1998) Cell penetrating PNA constructs regulate galanin receptor levels and modify pain transmission in vivo. *Nat. Biotechnol.* **16**, 857–861.
17. Lebleu, B., Moulton, H. M., Abes, R., Ivanova, G. D., Abes, S., Stein, C. A., Iversen, P. L., Arzumanov, A. A., and Gait, M. J. (2008) Cell penetrating peptide conjugates of steric block oligonucleotides. *Adv. Drug Deliv. Rev.* **60**, 517–529.
18. Said Hassane, F., Saleh, A. F., Abes, R., Gait, M. J., and Lebleu, B. (2010) Cell penetrating peptides: overview and applications to the delivery of oligonucleotides. *Cell Mol. Life Sci.* **67**, 715–26.
19. Crombez, L., Morris, M. C., Deshayes, S., and Divita, G. (2008) Peptide-based Nanoparticles for ex vivo and in vivo drug delivery. *Curr. Pharm. Des.* **14**, 3656–3665.
20. Simeoni, F., Morris, M. C., Heitz, F., and Divita, G. (2003) Insight into the mechanism of the peptide-based gene delivery system MPG: implications for delivery of siRNA into mammalian cells. *Nucleic Acids Res.* **31**, 2717–2724.
21. Zeineddine, D., Papadimou, E., Chebli, K., Gineste, M., Liu, J., Grey, C., Thurig, S., Behfar, A., Wallace, V. A., Skerjanc, I. S., and Puceat, M. (2006) Oct-3/4 dose dependently regulates specification of embryonic stem cells toward a cardiac lineage and early heart development. *Dev. Cell* **11**, 535–546.
22. Crombez, L., Morris, M. C., Dufort, S., Aldrian-Herrada, G., Nguyen, Q., Mc Master, G., Coll, J. L., Heitz, F., and Divita, G. (2009) Targeting cyclin B1 through peptide-based delivery of siRNA prevents tumour growth. *Nucleic Acids Res.* **37**, 4559–4569.
23. Morris, M. C., Gros, E., Aldrian-Herrada, G., Choob, M., Archdeacon, J., Heitz, F., and Divita, G. (2007) A non-covalent peptide-based carrier for in vivo delivery of DNA mimics. *Nucleic Acids Res.* **35**, e49–e59.
24. Morris, M. C., Depollier, J., Mery, J., Heitz, F., and Divita, G. (2001) A peptide carrier for the delivery of biologically active proteins into mammalian cells. *Nat. Biotechnol.* **19**, 1173–1176.
25. Efimov, V., Choob, M., Buryakova, A., Phelan, D., and Chakhmakhcheva, O. (2001) PNA-related oligonucleotide mimics and their evaluation for nucleic acid hybridization studies and analysis. *Nucleosides Nucleotides Nucleic Acids* **20**, 419–428.
26. Deshayes, S., Morris, M., Heitz, F., and Divita, G. (2008) Delivery of proteins and nucleic acids using a non-covalent peptide-based strategy. *Adv. Drug Deliv. Rev.* **60**, 537–547.
27. Umekita, Y., Hiipakka., R. A., Kokontis, J. M., and Liao, S. (1996) Human prostate tumor growth in athymic mice: inhibition by androgens and stimulation by finasteride. *Proc. Natl. Acad. Sci. USA* **93**, 11802–11807.

Chapter 5

Cell-Penetrating Peptides-Based Strategies for the Delivery of Splice Redirecting Antisense Oligonucleotides

Samir El Andaloussi, Fatouma Said Hassane, Prisca Boisguerin, Rannar Sillard, Ülo Langel, and Bernard Lebleu

Abstract

Progress in our understanding of the molecular pathogenesis of human malignancies has provided therapeutic targets amenable to oligonucleotide (ON)-based strategies. Antisense ON-mediated splicing regulation in particular offers promising prospects since the majority of human genes undergo alternative splicing and since splicing defects have been found in many diseases. However, their implementation has been hampered so far by the poor bioavailability of nucleic acids-based drugs. Cell-penetrating peptides (CPPs) now appear as promising non-viral delivery vector for non-permeant biomolecules. We describe here new CPPs allowing the delivery of splice redirecting steric-block ON using either chemical conjugation or non-covalent complexation. We also describe a convenient and robust splice redirecting assay which allows the quantitative assessment of ON nuclear delivery.

Key words: Antisense oligonucleotides, delivery, splicing redirection, cell-penetrating peptides.

1. Introduction

The clinical implementation of oligonucleotide (ON)-based strategies to regulate gene expression has been hampered by the poor bioavailability of nucleic acids (1). Although numerous ON analogs have been developed with improved pharmacological properties (e.g. with increased metabolic stability or affinity for their RNA, DNA or protein target), no chemical modifications have significantly improved cellular uptake and subsequent escape from endocytotic vesicles. Neutral ON mimics such as peptide nucleic acids (PNA) (2) or phosphodiamidate morpholino oligomers (PMO) (3) are not taken up more efficiently than

J. Goodchild (ed.), *Therapeutic Oligonucleotides*, Methods in Molecular Biology 764,
DOI 10.1007/978-1-61779-188-8_5,

negatively charged ONs since diffusion across lipid bilayers does not occur in this range of molecular masses. It is significant in this respect that some of the most encouraging data in pre-clinical studies or clinical trials have been reported in applications that do not require ON cellular uptake. As an example, a VEGF-specific aptamer has been FDA-approved for the treatment of wet age-related macular degeneration with the advantage of having to be administered topically and not requiring cellular internalization (4). As far as systemic administration is concerned, many of the reported antitumoural and antiviral activities of ON analogues (and in particular of phosphorothioate (PS) derivatives) may, for a large part, be accounted for by the activation of innate immune responses after binding to cognate Toll-like receptors (5). Finally, encouraging data reported recently with free steric-block ON analogs (2′OMe-PS, PNA or PMO) promoting exon skipping as a potential therapeutic strategy for Duchenne muscular dystrophy (DMD) have benefited from facilitated uptake across damaged muscle membranes (6).

Limitations in ON bioavailability have prompted many groups to search for appropriate delivery strategies including conjugation to cholesterol, complexation with cationic lipids to form lipoplexes or association/encapsulation to/into nanoparticles (7).

Our group became interested very early in conjugation/complexation to cationic peptides. We initially capitalized on poly-L-lysine (PLL) conjugation as initially proposed by H. Ryser et al. for the delivery of anti-cancer drugs. We and other groups did demonstrate improved and, importantly, sequence-specific responses of PLL-conjugated antisense ON using several in vitro biological models. However, PLL-based delivery systems had little future due to their cytotoxicity for some cell types and to complement activation. Interest for cationic peptides-based delivery strategies was revived with the characterization of cell-penetrating peptides (CPPs) also named protein transduction domains (PTD). Unexpectedly, purified proteins such as the Drosophila Antennapedia transcription factor or the HIV-1 Tat transactivating protein were shown to exert their activity upon incubation with cells in culture thus implying their ability to cross (several) biological barriers before reaching their nuclear targets. Short, basic amino acids-rich peptides responsible for cellular uptake (named penetratin and Tat 48-60, respectively) were rapidly identified (8, 9). Many natural or synthetic CPPs have now been described such as transportan (10) or oligo-arginine (11). Moreover and importantly for biotechnological applications, CPP chemical conjugation to various non-permeant cargoes (ranging from low molecular weight drugs to nanoparticles for cell imaging) was shown to improve their cellular delivery in many cell types including primary cells (12, 13). Although largely used for the delivery of proteins or peptides, Tat- and penetratin-based delivery vectors were poorly efficient for ON delivery. A major limitation turned out to be entrapment into endocytotic vesicles since biological activity was dramatically improved upon treatment by endosomolytic agents such as chloroquine (14). Second generation CPPs, with a largely improved potential for the delivery of antisense ONs or of siRNAs, have now been designed and optimized (15).

Their efficiency can be routinely evaluated in the splicing redirection assay introduced by Kole and colleagues (**Fig. 5.1**) (16). This is used to quantitate the nuclear delivery of CPP-conjugated or CPP-complexed splice redirecting ONs. The assay is based on genetically modified HeLa cells named HeLa pLuc705 cells. These cells are stably transfected with a luciferase-encoding sequence interrupted by the insertion of intron 2 from β-globin pre-mRNA carrying an aberrant splice site. Unless this aberrant splice site is masked by a splice redirecting ON, this cryptic splice site is activated and leads to improperly processed luciferase pre-mRNA. Thus, measuring luciferase expression after treatment with ONs conjugated to or complexed with CPPs allows the quantitative assessment of the potency of CPPs since increase in luciferase expression directly correlates to the amount of active ON in cell nuclei. A similar assay with EGFP (enhanced green fluorescent protein) reporter gene has also been described allowing assessment of CPP-ON biodistribution in vivo in transgenic mice. These assays are advantageous in providing a positive readout over a low background and a large dynamic response range. Protocols described here should be applicable for the nuclear delivery of other ONs as well.

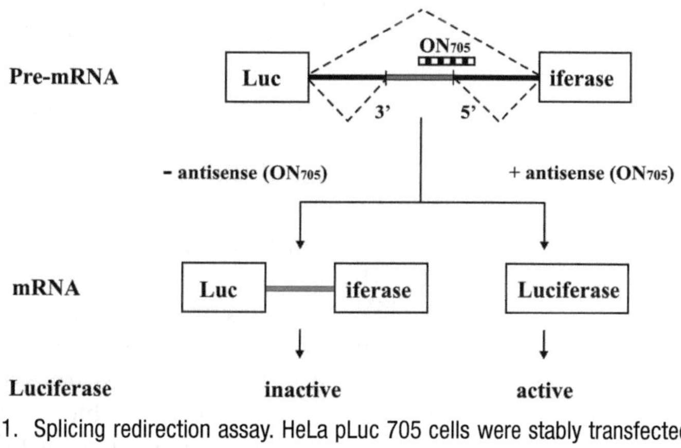

Fig. 5.1. Splicing redirection assay. HeLa pLuc 705 cells were stably transfected with a construction in which the coding sequence of the luciferase gene is interrupted by a mutated intron 2 of the human β-globin gene. This mutation creates a 5′splice site and activates a 3′splice site. Masking of the 5′splice site by a RNase H-incompetent antisense ON (705) restores the production of functional luciferase mRNA and protein.

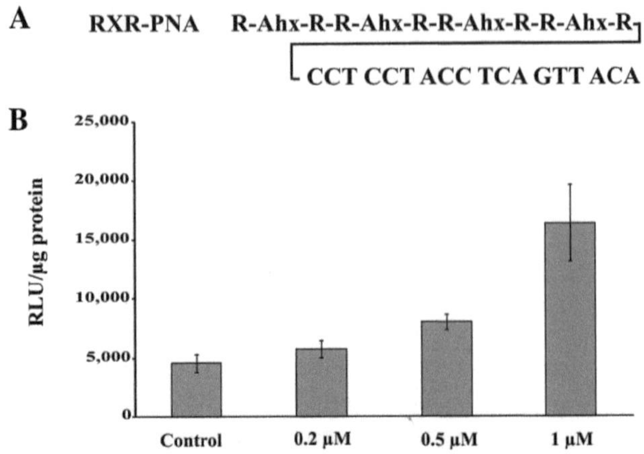

Fig. 5.2. Splice redirection specificity of RXR-PNA. **(a)** Sequence of RXR-PNA (with R = arginine and X = Ahx or aminohexanoic acid). (**b**) 1.75×10^5 HeLa pLuc705 cells were incubated for 4 h in OptiMEM in the absence (control) or presence of RXR-PNA splice redirecting conjugate at the indicated concentrations. Luciferase expression was quantified 20 h later and was expressed as relative luminescence units (RLU) per microgram protein. Each experiment was made in triplicate and error bars (standard deviations) are indicated.

Antisense-mediated splicing redirection indeed represents a promising new strategy to regulate gene expression since more than 75% of human genes undergo alternative splicing and since splicing defects have been found in many human diseases.

We first describe a family of arginine-rich CPPs (whose prototype is $(\text{R-Ahx-R})_n$) (*see* **Fig. 5.2a** for structure), which proved efficient in vitro and in vivo for the delivery of conjugated steric-block ON derivatives with a neutral backbone (such as PNA or PMO) (13). As an example, $(\text{R-Ahx-R})_4$–PNA conjugates (with Ahx for an aminohexanoic acid) were able to promote splicing redirection at submicromolar concentration in the in vitro splicing redirection assay described above and in **Fig. 5.2b** (17). Importantly, PMOs formulated with closely related CPPs were much more efficient than free ONs when administered systemically in a murine mdx model of DMD (18, 19).

We also describe a series of stearylated CPP derivatives (*see* **Figs. 5.3a** and **5.4a** for structures), which allow splicing redirection in the same in vitro assay when non-covalently complexed with charged steric-block ON derivatives (such as 2′O-methyl phosphorothioate RNA (2′O-Me-PS RNA) as described in **Figs. 5.3** and **5.4**. (10, 20). Non-covalent CPP strategies will obviously be advantageous in several applications. Importantly, some of these stearylated CPP derivatives also allow the efficient delivery of siRNAs (unpublished observations).

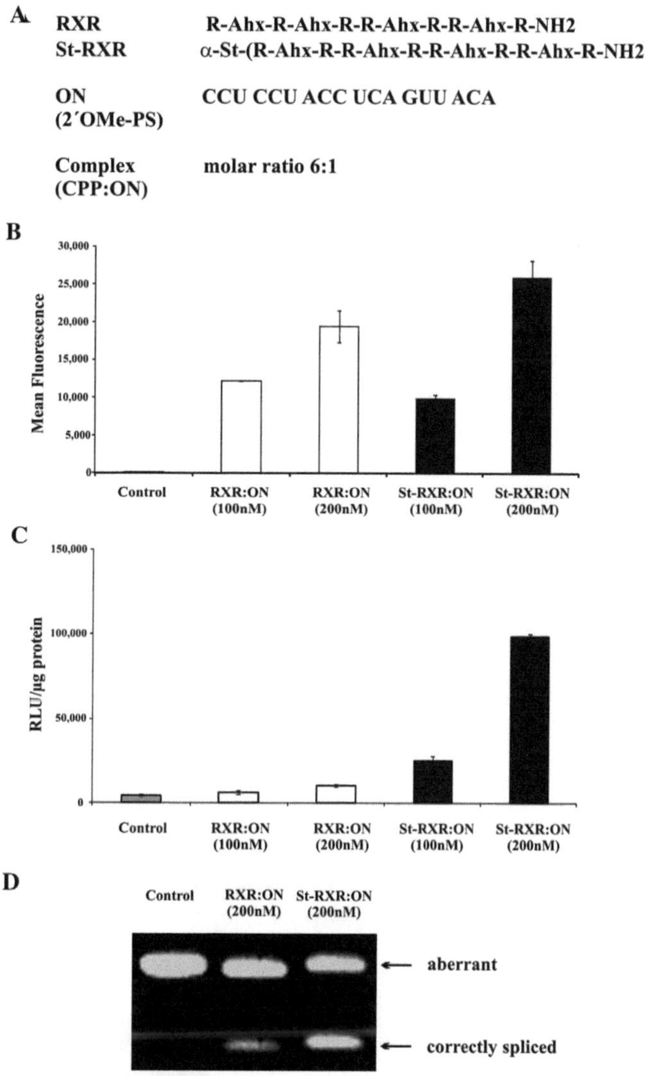

Fig. 5.3. Effect of stearylation of RXR on cellular uptake and splice redirection. (**a**) Sequences of CPPs RXR, stearylated RXR (St-RXR) and 2′OMe-PS ON (with PS = phosphorothioate). (**b**) 2×10^5 HeLa pLuc 705 cells were seeded 24 h prior to experiment in 24-well plates. Cell uptake was monitored by FACS analysis Cells were incubated for 2 h in OptiMEM in the absence (control) or presence of correcting CPP:ON complexes (6:1). (**c**) 5×10^4 HeLa pLuc 705 cells were seeded 24 h prior to experiment in 24-well plates. Cells were incubated for 4 h in OptiMEM in the absence (control) or presence of correcting CPP:ON complexes (6:1). Luciferase expression was quantified 20 h later and was expressed as RLU per microgram protein. (**d**) RT-PCR analysis of splice redirection. Total RNA was extracted from the same cellular lysates as for luciferase assay and amplified by RT-PCR. PCR products from aberrant and correctly spliced luciferase pre-mRNA were analysed on a 2% agarose gel.

A

TP10	AGYLLGKINLKALAALAKKIL-NH2
St-TP10	α-St- AGYLLGKINLKALAALAKKIL-NH2
ON (2′OMe-PS)	CCU CCU ACC UCA GUU ACA
Complex (CPP:ON)	molar ratio 10:1

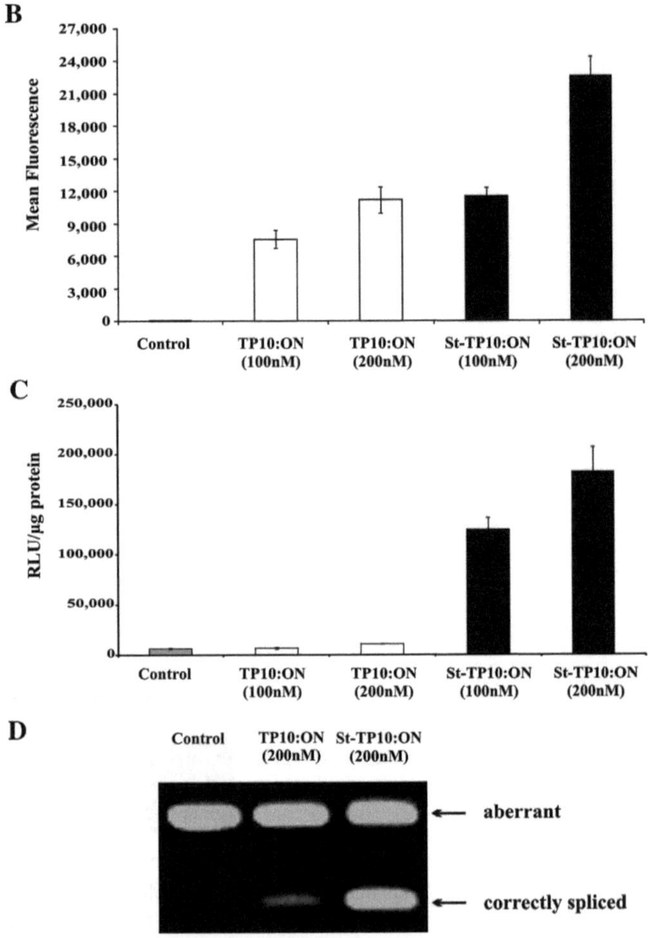

Fig. 5.4. Effect of stearylation of TP-10 on cellular uptake and splice redirection. (**a**) Sequences of TP-10 CPP, stearylated tP-10 (St-TP10) and ON. (**b**) 2×10^5 HeLa pLuc 705 cells were seeded 24 h prior to experiment in 24-well plates. Cells were incubated for 2 h in OptiMEM in the absence (control) or presence of correcting CPP:ON complexes (10:1). Cell uptake was monitored by FACS analysis (**c**) 5×10^4 HeLa pLuc 705 cells were seeded 24 h prior to experiment in 24-well plates. Cells were incubated for 4 h in OptiMEM in the absence (control) or presence of correcting CPP:ON complexes (10:1). Luciferase expression was quantified 20 h later and was expressed as RLU per microgram protein. (**d**) RT-PCR analysis of splice correction. Total RNA was extracted from the same cellular lysates as for luciferase assay and amplified by RT-PCR. PCR products from aberrant and correctly spliced luciferase pre-mRNA were analysed on a 2% agarose gel.

2. Materials

2.1. Cell Culture

1. Dulbecco's Modified Eagle Medium (500 mL D-MEM) supplemented (v/v) with 10% fetal bovine serum (FBS), 5 mL MEM Non-Essential Amino Acids (100×) (Invitrogen), 5 mL sodium pyruvate MEM (100 mM, Gibco) and 5 mL Penicillin-Streptomycin-Neomycin (PSN) antibiotic mixture (Invitrogen) for HeLa cells culture (supplemented D-MEM).
2. 175 cm^2 cell culture flasks
3. 24-well cell culture plates
4. MycoAlert® mycoplasma Detection Kit (Lonza)
5. Phosphate buffered saline (D-PBS) (Invitrogen)
6. Opti-MEM medium (Invitrogen) for serum-free incubations
7. Trypsin/EDTA 0.05% (Invitrogen) for cell dissociation
8. Forma Direct Heat CO_2 Incubator HEPA Class 100 (Thermo Electron Corporation) 37°C and 5% CO_2
9. Thoma chamber for cell counting
10. HeLa pLuc 705 are a kind gift from Dr. R. Kole (University of North Carolina, USA).

2.2. Synthesis of Peptides, CPP–PNA Conjugates and Stearylated CPP Derivatives

1. Fmoc-protected amino acids (Iris Biotech)
2. 2-(1H-Benzotriazole-1-yl)-1,1,3,3-tetramethyluronium tetrafluoroborate (TBTU) (Iris Biotech)
3. Fmoc-Rink-Amide-4-methylbenzhydrylamine resin (Iris Biotech)
4. Dimethylsulfoxide
5. Piperidine
6. Dimethylformamide (DMF)
7. Dichloromethane (DCM)
8. Acetic anhydride
9. *N*-methylpyrrolidone
10. Diisopropylethylamine (DIPEA)
11. Stearic acid
12. Diethylether
13. Polypropylene reactors, 5 mL, with polyethylene frits and plunger (Multisynthech)
14. 1-Hydroxybenzotriazole monohydrate (HOBt)
15. Trifluoroacetic acid (TFA)

16. Triisopropylsilane (TIS)
17. Ethanedithiol (EDT)
18. Boc-Cys-(S-3-nitro-2-pyridinesulfenyl)-OH (Bachem)
19. C18 chromatography column (250 mm × 25 mm i.d., 300 Å, 10 μm particle size)
20. Chromatography solvent A: 0.1% TFA/water (v/v)
21. Chromatography solvent B: 0.1% TFA/acetonitrile (v/v)

2.3. Complex Formation Between Stearylated CPPs and ONs

1. Sequence of the 2′O-Me phosphorothioate ON (RiboTask or Eurogentec) used to target the aberrant 705 splice site is given in **Figs. 5.3a** and **5.4a**. Store 20-μM aliquots (dissolved in Milli-Q water) at −20°C.
2. Sequences of stearylated peptides are given in **Figs. 5.3a** and **5.4a**. 20 μL of 1 mM peptide solutions in Milli-Q water are stored at −20°C.
3. Milli-Q water
4. Opti-MEM cell culture medium

2.4. FACS Analysis of CPP–PNA Conjugates and CPP: ON Complexes Following Cellular Uptake

1. Phosphate buffered saline (D-PBS) (Invitrogen)
2. Trypsin/EDTA 0.05% (Invitrogen) for cell dissociation
3. D-MEM (Invitrogen)
4. 24-well cell culture plates
5. Opti-MEM medium (Invitrogen)
6. FITC-labelled CPP–PNA conjugate (FITC ex = 488 nm; em = 520 nm) or Cy5-labelled 2′OMe-PS ON (Cy5 ex = 650 nm; em = 670 nm) (Eurogentec)
7. Phosphate buffered saline (D-PBS) (Invitrogen) with 5% (v/v) FBS
8. D-PBS with 0.5% FBS containing 0.05 μg/mL propidium iodide (PI) (Molecular Probes) for cell permeability quantification
9. FACSCanto™ flow cytometer (BD Biosciences) using FACS Diva® software

2.5. Luciferase Assay of Splicing Redirection

1. Trypsin/EDTA 0.05% (Invitrogen) for cell dissociation
2. Phosphate buffered saline (D-PBS) (Invitrogen)
3. 24-well cell culture plates
4. Opti-MEM medium (Invitrogen)
5. D-MEM supplemented with 10% (v/v) fetal bovine serum (FBS)

6. White 96-well plates (NUNC) for luciferase activity quantification
7. Transparent 96-well plates for protein quantification
8. 5× Reporter Lysis Buffer (Promega) for cell lysis
9. BCA™ Protein Assay Kit (Pierce) and Dynatech MR 5000 plate reader (Dynatech Labs) for cellular protein quantification
10. Berthold Centro LB 960 luminometer (Berthold Technologies) and Luciferase Assay System with Reporter Lysis Buffer (Promega) for luciferase activity quantitation

2.6. RT-PCR Evaluation of Splicing Redirection

1. Luciferase primers (Eurogentec): Forward 5′ TTG ATATGT GGA TTTCGA GTC GTC 3′ and reverse 5′ TGT CAA TCA GAG TGC TTT TGG CG 3′
2. TRI REAGENT™ (Sigma Aldrich), chloroform, isopropanol and ethanol for RNA extraction
3. Chloroform
4. Isopropanol
5. 75% (v/v) aqueous ethanol
6. Nuclease-free water
7. SuperScript III one-step RT-PCR system with Platinum® Taq polymerase (Invitrogen) and PTC200 Peltier Thermal cycler (MJ Research) for PCR amplification
8. BioPhotometer (Eppendorf) for RNA quantitation
9. Agarose (1% for RNA and 2% for DNA in TBE w/v (Tris Borate EDTA buffer)) and ethidium bromide for gel electrophoresis
10. Lumi imager F1 (Roche) for image acquisition using Lumianalyst software for image analysis

3. Methods

3.1. Cell Culture and Cell Dissociation

1. Culture HeLa pLuc 705 cells as exponentially growing subconfluent monolayer in supplemented D-MEM.
2. Wash cells twice with D-PBS and passage with Trypsin/EDTA every other day on 175-cm^2 flasks for routine maintenance for a maximum of 10 passages.
3. Test cells for the absence of mycoplasma contamination every month as described in the Lonza kit on 100 μL cell culture supernatant.

3.2. Synthesis of PNA, Peptides and CPP–PNA Conjugates

3.2.1. Conjugation of PNA with CPP

For laboratories with in-house capability for peptide synthesis, the preparation of PNAs and CPPs using t-Boc or Fmoc chemistries have been described (21, 22). Otherwise, they may be obtained from custom synthesis companies. To conjugate PNA and CPP different strategies may be employed, e.g. coupling via a disulfide bridge. For this purpose cysteine residues are incorporated into PNA and peptide sequences. To achieve a rapid and specific disulfide formation, one of the cysteine residues, in either PNA or CPP, is activated by a 3-nitro-2-pyridinesulfenyl group (Npys) (23). Commercially available Boc-Cys(Npys)-OH may be incorporated at any place in the sequence using t-Boc chemistry, or as the last amino acid residue using Fmoc chemistry due to the instability of Cys(Npys)-derivative towards treatment with piperidine.

1. Dissolve 1–2 mg of peptide in 100 μL 0.01 M acetate buffer, pH 5.5.
2. Dissolve an equal molar amount of PNA in 200 μL DMSO.
3. Add 200 μL of DMF to both solutions and mix thoroughly.
4. Combine both solutions and stir overnight in the dark.
5. Purify the construct by HPLC on a C18 column (250 mm × 25 mm i.d., 300 Å, 10 μm particle size) using a gradient from 20% of solvent B in solvent A to 100% solvent B in 45 min.
6. Monitor absorbance at 214 and 254 nm and collect UV-absorbing material.
7. Freeze-dry product and store at −20°C.
8. Analysis of products by mass spectrometry is recommended.

3.2.2. Stearylation of CPPs

1. Remove Fmoc-groups from the *N*-terminus of CPP (1 equivalent, 0.034 mmol) while it is still bound to the resin support used for synthesis. Treat with 20% piperidine in DMF for 30 min.
2. Wash resin three times with 5 mL DMF and three times with 5 mL DCM.
3. Dissolve 50 mg stearic acid (5 equivalents) in a mixture consisting of 340 μL 0.5 M TBTU (5 equiv.), 170 μL 1 M HOBt (5 equivalents), 125 μL DMSO and 1.3 mL DCM.
4. Add 58 μL of DIPEA (10 equiv.).
5. Transfer the solution to the tube with peptide on resin and shake overnight.
6. Wash three times with 5 mL DMF and three times with 5 mL DCM and check completeness of coupling with a standard assay for the detection of primary amines.
7. Dry the resin in vacuum.

8. Cleave stearylated CPP with 4 mL of cleavage cocktail (95% TFA, 2.5% TIS, 2.5% H_2O).
9. If the peptide contains a cysteine residue (which must retain a free thiol after cleavage) use a cocktail that contains EDT (94% TFA, 2.5% H_2O, 2.5% EDT, 1% TIS).
10. Precipitate the peptide with 40 mL diethylether and centrifuge.
11. Wash the precipitate twice with 30 mL diethylether, centrifuge and evaporate to dryness.
12. Purify the stearylated peptide by HPLC on a C18 column (250 mm × 25 mm i.d., 300 Å, 10 μm particle size) using a gradient from 20% of solvent B in solvent A to 100% solvent B in 45 min.
13. Monitor absorbance at 214 and 280 nm and collect UV-absorbing material.
14. Freeze-dry product and store at −20°C.
15. Analysis of the products by mass spectrometry is recommended.

3.3. Complexation of CPP Derivatives to Charged ON Analogues

1. Thaw 1 mM frozen stock solutions of stearylated CPPs (St-CPP) (20 μL) and add 180 μL of Milli-Q water to obtain a final concentration of 100 μM.
2. Thaw frozen 20 μM ON stock solution (antisense or scrambled version). (*see* **Notes 1** and **2**).
3. Prepare CPP/ON complexes at different molar ratios (MRs) and different ON concentrations (*see* **Notes 3** and **4**). The protocol below describes formation of complexes at MR 10:1 (CPP:ON) in triplicate using different ON concentrations.
4. Pipette 30 μL of ON solution into a 1.5-mL tube.
5. Add 210 μL of Milli-Q water.
6. Add 60 μL of St-CPP to obtain a final volume of 300 μL.
7. Allow complexes to form by incubating the solution for 1 h at room temperature. Meanwhile, add 150 μL of Milli-Q water into three new 1.5-mL tubes.
8. After incubation, make serial dilutions to obtain 150 μL solutions with final ON concentrations in the wells ranging from 25 to 200 nM of ON.

3.4. FACS Analysis of CPP–PNA Conjugates and St-CPP:ON Complexes Following Cellular Uptake

1. Wash exponentially growing HeLa pLuc705 cells with D-PBS, incubate with Trypsin/EDTA for 5 min, centrifuge at 900×g at 4°C for 5 min, rinse twice with D-PBS, centrifuge again, re-suspend in DMEM, plate on 24-well plates (3×10^5 cells/well for CPP–PNA or 2×10^5 for CPP:ON complex) and culture overnight.

2. Discard culture medium and rinse cells twice with D-PBS.
3. (a) Discard D-PBS and incubate cells with fluorescently labelled PNA conjugates diluted in Opti-MEM or D-MEM (*see* **Note 5**).
   (b) For CPP:ON complex add 450 μL OptiMEM to each well and then 50 μL pre-formed complex.
4. After incubation for appropriate period, rinse cells twice with D-PBS and incubate with Trypsin/EDTA for 5 min at 37°C (*see* **Note 6**).
5. Re-suspend cells in D-PBS with 5% FBS, centrifuge at 900×*g* for 5 min at 4°C and re-suspend in PBS with 0.5% FBS containing 0.05 μg/mL PI.
6. Analyse fluorescence by FACS for CPP-ON or St-CPP:ON cellular uptake and PI permeabilization using FACS Diva software. Exclude PI-stained cells from further analysis by appropriate gating. Analyse at least 10,000–20,000 events per sample (*see* **Figs. 5.3b** and **5.4b** for examples).

3.5. Luciferase Assay of Splicing Redirection

3.5.1. For CPP–PNA Conjugates

1. Detach exponentially growing HeLa pLuc705 cells with Trypsin/EDTA, plate on 24-well plates (1.75×10^5 cells/well) and culture overnight.
2. Rinse twice with D-PBS.
3. Incubate with splice redirecting conjugates (or their scrambled version) at appropriate concentrations usually between 0.5 and 4 h in Opti-MEM medium.
4. Rinse cells twice with D-PBS and continue incubation for 20 h in supplemented D-MEM.
5. Rinse cells twice with PBS and lyse with 300 μL 1× Reporter Lysis Buffer for 30 min at room temperature.
6. Quantify luciferase activity in a luminometer using Luciferase Assay System substrate. Use 10-μL sample lysate and add 100 μL luciferase agent. Perform all experiments in triplicate.
7. Measure cellular protein concentrations with BCA^{TM}Protein Assay Kit and read using an ELISA plate reader at 560 nm. Use 20 μL sample lysate, add 200 μL BCA-agent and incubate at room temperature 30 min. Perform all experiments in triplicate.
8. Express luciferase activities as relative luminescence units (RLU) per μg protein. Average each data point over three replicates (*see* **Fig. 5.2b** for an example).
9. Freeze down remainder of cellular lysate for further RT-PCR analysis (*see* **Section 3.6**).

3.5.2. For St-CPP:ON Complexes

1. Detach exponentially growing HeLa pLuc705 cells with Trypsin/EDTA, plate on 24-well plates (5×10^4 cells/well) and culture overnight.
2. Rinse twice with D-PBS.
3. Add 450 μL Opti-MEM in each well.
4. Add 50 μL (in triplicate) pre-formed complex and incubate 4 h at 37°C
5. Stop reaction by adding 1 mL supplemented D-MEM without rinsing step and continue incubation for 20 h.
6. Continue assay as described in **Section 3.5.1** Step 5 (*see* **Figs. 5.3c** and **5.4c** for examples).

3.6. RT-PCR Evaluation of Splicing Redirection

1. Extract total RNA using 1 mL TRI REAGENT™ per well after measurement of luciferase. Add 300 μL chloroform, mix vigorously and incubate for 10 min at room temperature.
2. Centrifuge at 12,000×*g* for 15 min at 4°C and add an equal volume of isopropanol to the aqueous phase. Mix well and incubate for 10 min at room temperature.
3. Centrifuge at 12,000×*g* for 10 min at 4°C and re-suspend pellet in 1 mL cold 75% (v/v) ethanol. Mix and centrifuge at 12,000×*g* for 5 min at 4°C. Discard supernatant. Evaporate ethanol for 1 min at 60°C.
4. Add 20 μL of nuclease-free water.
5. Quantify RNA samples using a BioPhotometer.
6. Analyse RNA samples on 1% (w/v) agarose gel.
7. Amplify 1 μg total RNA using SuperScript III one-step RT-PCR system with Platinum®Taq polymerase Luciferase-specific primers (*see* **Note 7**).
8. Analyse PCR products by electrophoresis using 2% (w/v) agarose gel. Use digestion products of plasmid pLuc705 restriction enzymes as molecular weight markers (*see* Figs. **5.3d** and **5.4d** for examples).
9. EC_{50} values can be deduced from experiments carried out over a range of concentrations upon gel analysis using the imaging Lumianalyst software.

4. Notes

1. ON should be always desalted on Sephadex G-25 column before use in complex formation with CPPs.

2. ONs can be used as non-labelled versions or as dye-labelled ones such as Cy5 without loosing any activity.
3. Protocols for CPP–PNA conjugates or CPP:ON complexes can be adapted to other CPPs and also to other ONs.
4. Try molar ratios ranging from 1:1 to 10:1 of CPP:ON, using an ON concentration of 200 nM. It is further advised to include a dose-response curve with each transfection reagent.
5. CPP-ON conjugates should preferably be used at low concentrations (below 1.0 μM) to avoid cell permeabilization.
6. Treatment with trypsin before FACS analysis is required to eliminate membrane-bound CPP-ON conjugates and CPP:ON complexes
7. Program used for reverse transcription and amplification:
   (a) Reverse transcription (1 cycle)
      - cDNA production: 30 min at 55°C
      - Denaturation: 2 min at 94°C
   (b) Amplification (30 cycles)
      - Denaturation: 20 s at 94°C
      - Hybridization: 30 s at 60°C
      - Elongation: 30 s at 68°C
   (c) Elongation: 1 cycle for 5 min at 68°C
   (d) Store PCR products at −20°C

Acknowledgements

We thank Ryszard Kole (University North Carolina) for providing the HeLa pLuc705 cell line. Fatouma Said Hassane holds a post-doctoral fellowship from the Association Française contre les Myopathies and Prisca Boisguerin holds a senior post-doctoral grant from Charité-Universtitaetsmedizin Berlin. Studies funded by AFM grant 14329 and ANR grant MyocarDaxx to Bernard Lebleu, by Swedish Research Council (VR-NT), Center for Biomembrane Research, Stockholm and Knut and Alice Wallenberg's Foundation to Ulo Langel.

References

1. Juliano, R., Bauman, J., Kang, H., and Ming, X. (2009) Biological barriers to therapy with antisense and siRNA oligonucleotides. *Mol Pharm.* 6, 686–695.
2. Egholm, M., Buchardt, O., Christensen, L., Behrens, C., Freier, S. M., Driver, D. A., Berg, R. H., Kim, S. K., Nordén, B., and Nielsen, P. E. (1993) PNA hybridizes

to complementary oligonucleotides obeying the Watson-Crick hydrogen-bonding rules. *Nature* **365**, 566–568.

3. Summerton, J. (1999) Morpholino antisense oligomers: the case for an RNase H-independent structural type. *Biochim Biophys Acta* **1489**, 141–158.
4. Ng, E. W., Shima, D. T., Calias, P., Cunningham, E. T., Jr., Guyer, D. R., and Adamis, A. P. (2006) Pegaptanib, a targeted anti-VEGF aptamer for ocular vascular disease. *Nat Rev Drug Discov.* **5**, 123–132.
5. Barchet, W., Wimmenauer, V., Schlee, M., and Hartmann, G. (2008) Accessing the therapeutic potential of immunostimulatory nucleic acids. *Curr Opin Immunol.* **20**, 389–395.
6. Aartsma-Rus, A., Fokkema, I., Verschuuren, J., Ginjaar, I., van Deutekom, J., van Ommen, G. J., and den Dunnen, J. T. (2009) Theoretic applicability of antisense-mediated exon skipping for duchenne muscular dystrophy mutations. *Hum Mutat.* **30**, 293–299.
7. Juliano, R., Alam, M. R., Dixit, V., and Kang, H. (2008) Mechanisms and strategies for effective delivery of antisense and siRNA oligonucleotides. *Nucleic Acids Res.* **36**, 4158–4171.
8. Derossi, D., Joliot, A. H., Chassaing, G., and Prochiantz, A. (1994) The third helix of the antennapedia homeodomain translocates through biological membranes. *J Biol Chem.* **269**, 10444–10450.
9. Vivés, E., Brodin, P., and Lebleu, B. (1997) A truncated HIV-1 Tat protein basic domain rapidly translocates through the plasma membrane and accumulates in the cell nucleus. *J Biol Chem.* **272**, 16010–16017.
10. Mäe, M., El-Andaloussi, S., Lehto, T., and Langel, Ü. (2009) Chemically modified cell-penetrating peptides for the delivery of nucleic acids. *Expert Opin Drug Deliv.* **6**, 1195–1205.
11. Nakase, I., Takeuchi, T., Tanaka, G., and Futaki, S. (2008) Methodological and cellular aspects that govern the internalization mechanisms of arginine-rich cell penetrating peptides. *Adv Drug Deliv Rev.* **60**, 598–607.
12. Fonseca, S. B., Pereira, M. P., and Kelley, S. O. (2009) Recent advances in the use of cell-penetrating peptides for medical and biological applications. *Adv Drug Deliv Rev.* **61**, 953–964.
13. Said Hassane, F., Saleh, A., Abes, R., Gait, M. J., and Lebleu, B. (2010) Cell penetrating peptides and their applications: an overview. *Cell Mol Life Sci.* **67**, 715–728.
14. Abes, S., Williams, D., Prevot, P., Thierry, A., Gait, M. J., and Lebleu, B. (2006) Endosome trapping limits the efficiency of splicing correction by PNA-oligolysine conjugates. *J Control Release* **110**, 595–604.
15. Lebleu, B., Moulton, H. M., Abes, R., Ivanova, G. D., Abes, S., Stein, D. A., Iversen, P. L., Arzumanov, A. A., and Gait, M. J. (2008) Cell penetrating peptide conjugates of steric block oligonucleotides. *Adv Drug Deliv Rev.* **60**, 517–529.
16. Kang, S. H., Cho, M. J., and Kole, R. (1998) Up-regulation of luciferase gene expression with antisense oligonucleotides: implications and applications in functional assay development. *Biochemistry* **37**, 6235–6239.
17. Abes, R., Moulton, H. M., Clair, P., Yang, S. T., Abes, S., Melikov, K., Prevot, P., Youngblood, D. S., Iversen, P. L., Chernomordik, L. V., and Lebleu, B. (2008) Delivery of steric block morpholino oligomers by (R-X-R)4 peptides: structure-activity studies. *Nucleic Acids Res.* **36**, 6343–6354.
18. Jearawiriyapaisarn, N., Moulton, H. M., Buckley, B., Roberts, J., Sazani, P., Fucharoen, S., Iversen, P. L., and Kole, R. (2008) Sustained dystrophin expression induced by peptide-conjugated morpholino oligomers in the muscles of mdx mice. *Mol Ther.* **16**, 1624–1629.
19. Yin, H., Moulton, H. M., Seow, Y., Boyd, C., Boutilier, J., Iverson, P., and Wood, M. J. (2008) Cell-penetrating peptide-conjugated antisense oligonucleotides restore systemic muscle and cardiac dystrophin expression and function. *Hum Mol Genet.* **17**, 3909–3918.
20. Lehto, T., Abes, R., Oskolkov, N., Suhorutsenko, J., Copolovici, D. M., Mäger, I., Viola, J. R., Simonson, O., Guterstam, P., Eriste, E., Smith, C. I., Lebleu, B., El-Andaloussi, S., and Langel, Ü. Delivery of nucleic acids with a stearylated (RxR)4 peptide using a non-covalent co-incubation strategy. *J Control Release* **141**, 42–51.
21. Kilk, K., and Langel, Ü. (2005) Cellular delivery of peptide nucleic acid by cell-penetrating peptides. *Methods Mol Biol.* **298**, 131–141.
22. Pooga, M., Soomets, U., Bartfai, T., and Langel, Ü. (2002) Synthesis of cell-penetrating peptide-PNA constructs. *Methods Mol Biol.* **208**, 225–236.
23. Bernatowicz, M. S., Matsueda, R., and Matsueda, G. R. (1986) Preparation of Boc-[S-(3-nitro-2-pyridinesulfenyl)]-cysteine and its use for unsymmetrical disulfide bond formation. *Int J Pept Protein Res.* **28**, 107–112.

Chapter 6

A Nanoparticle for Tumor Targeted Delivery of Oligomers

Xinrong Liu, Yi Wang, and Donald J. Hnatowich

Abstract

The tissue-specific delivery nanoparticle consists of an antisense oligomer, a cell-penetrating peptide, and an antitumor antibody, each biotinylated and each linked via streptavidin. Within the nanoparticle, the antibody provides specific targeted delivery and binding to the target cells, the peptide improves cell membrane transport, and the antisense oligomer, through its mRNA-binding ability, provides specific retention of the radioactivity in the target cell nucleus. The use of streptavidin as linker eliminates the need for covalent conjugation without appearing to interfere with the in vitro and in vivo properties of each component. The delivery nanoparticle is under development to improve tumor targeting with unlabeled siRNAs as well as radiolabeled antisense oligomers in a variety of tumor types. The anti-HER2 Trastuzumab (Herceptin) antibody, the tat peptide, and a radiolabeled antisense oligomer against the RIα mRNA have been used in this report as an example.

Key words: Tumor delivery, streptavidin, nanoparticle, antisense.

1. Introduction

When administered intravenously, free antisense oligomers accumulate at much higher levels in normal tissues than in tumor tissues (1). Thus, antisense application in clinical settings is still limited mainly due to the lack of efficient delivery systems. Antibody-mediated targeted drug delivery systems have attracted much attention due to their superior stability during systemic circulation and high selectivity toward a target protein on the cell surface (2). The HER2 transmembrane receptor is overexpressed in 25–30% of human breast cancer and is associated with a more aggressive tumor phenotype, poor prognosis, and faster relapse times at all stages of cancer development (3, 4). As a target antigen, HER2 is a readily accessible cell surface receptor

J. Goodchild (ed.), *Therapeutic Oligonucleotides*, Methods in Molecular Biology 764,
DOI 10.1007/978-1-61779-188-8_6, © Springer Science+Business Media, LLC 2011

and provides a basis for selective immunotargeting of tumor cells. Trastuzumab (Herceptin), a humanized monoclonal antibody against the extracellular domain of HER2, is approved for the adjuvant treatment of patients with HER2-positive metastatic breast cancer.

Our laboratory is developing streptavidin-delivery nanoparticles. If oligomers are biotinylated, they may be combined with cell-penetrating peptide carriers and with a variety of antitumor antibodies via streptavidin without the need for covalent conjugation (5–7). In this work, a three-component nanoparticle in which the Traszutumab anti-HER2 antibody has been added to the nanoparticle with the labeled anti-type I regulatory subunit α (RIα) phosphorodiamidate morpholino (MORF) oligomer and the tat peptide is described as an example of a delivery nanoparticle for improved tumor targeting in vitro and in vivo.

2. Materials (*see* Note 1)

2.1. Oligomer Conjugation and Radiolabeling

1. Size-exclusion HPLC analyses on a Superose 12 (Amersham Pharmacia Biotech) installed on a Waters 515 solvent delivery system equipped with an in-line radioactivity detector and a Waters UV2487 dual wavelength absorbance detector.
2. Running solution for HPLC analyses: 0.1 M Tris–HCl, pH 8.0, 20% acetonitrile (v/v).
3. HEPES buffer: 0.3 M HEPES. Adjust the pH to 7.5–8.0 by addition of 5 M sodium hydroxide or 5 M hydrochloric acid. Filter the solution through a 0.22-μm filter if sterilization is required.
4. Phosphorodiamidate morpholino (MORF) oligomer is obtained with a biotin group on the 3′ equivalent end via a 6-aminohexanoic acid linker and a primary amine on the opposite end (GeneTools) and dissolved in 0.3 M HEPES buffer, pH 8.0.
5. NHS-MAG3: Describing the synthesis of NHS-MAG3 is beyond the scope of this report. However, this bifunctional chelator can be readily synthesized in-house according to the procedure described by Winnard et al. (8) or by reacting *S*-acetylmercaptoacetyltriglycine with *N*-hydroxysuccinimide in the presence of N,N′-dicyclohexylcarbodiimide (DCC) as dehydration reagent in anhydrous *N*-methyl-2-pyrrolidinone (NMP). The product may be purified by precipitation in ethyl ether (*see* **Note 2**).

6. Cy3-NHS from GE Healthcare (Piscataway, NJ) is dissolved in anhydrous *N*-methyl-2-pyrrolidinone (NMP) at a concentration of 25 mg/mL.
7. Tartrate buffer I: Dissolve disodium tartrate dihydrate in the labeling buffer to a concentration of 100 mg/mL. Filter as above if necessary.
8. Tartrate buffer II: Dissolve disodium tartrate dihydrate in the labeling buffer to a concentration of 50 mg/mL and filter as above if necessary.
9. Dissolve ammonium acetate in deionized water to a concentration of 0.25 M and 2.0 M. Filter as above if necessary.
10. Ascorbic acid–HCl solution: Dissolve ascorbic acid in 10 mM HCl to a concentration of 1.0 mg/mL.
11. Freshly prepared 20 mg/mL $SnCl_2 \cdot 2H_2O$ in tartrate buffer I.
12. Freshly prepared 4 mg/mL $SnCl_2 \cdot 2H_2O$ in ascorbic acid–HCl solution.
13. 1 × 20 cm P4 open column (Bio-Rad).
14. Technetium-99m: ^{99m}Tc is obtained in its pertechnetate chemical form in saline from a commercially available ^{99}Mo-^{99m}Tc generator (Bristol-Myers Squibb Medical Imaging Inc.) (*see* **Note 3**).
15. Absolute ethanol.
16. Savant Speed Vac evaporating centrifuge.

2.2. Biotinylation of Trastuzumab and Testing

1. Phosphate buffered saline (PBS): Prepare 10× stock with 1.37 M NaCl, 27 mM KCl, 100 mM Na_2HPO_4, and 18 mM KH_2PO_4 (adjust to pH 7.4 with HCl if necessary) and autoclave before storage at room temperature. Prepare working solution by dilution of one part with nine parts water.
2. 0.3 M HEPES, pH 7.5 (*see* **Section 2.1**, Step 3).
3. Trastuzumab (Herceptin) is obtained from Genentech Inc. as the clinical drug.
4. Biotinylation reagent: Sulfo-NHS-LC-LC-Biotin (Pierce).
5. 1 × 20 cm G75 open column (Sigma).
6. Bicinchoninic acid reagent: BCA Protein Assay Reagent Kit (Pierce).
7. Slide-A-Lyzer Dialysis Cassettes, 10 K MWCO, 0.5–3 mL (Pierce).
8. Blocking and antibody dilution buffer: 3% (w/v) bovine serum albumin (BSA) in PBS.

9. Second antibody: Anti-human IgG conjugated with fluorescein isothiocyanate (FITC) (Sigma) and anti-biotin IgG antibody conjugated with FITC (Sigma).
10. SUM190 cells culture medium: Ham's F-12 medium with 5 μg/mL insulin, 1 μg/mL hydrocortisone, 5 mM ethanolamine, 10 mM HEPES, 5 μg/mL transferrin, 10 nM triiodo thyronine, 50 nM sodium selenite, and 0.5 g/L bovine serum albumin.
11. SUM149 cells culture medium: Ham's F-12 supplied with 5 μg/mL insulin, 1 μg/mL hydrocortisone, 10 mM HEPES, and 5% fetal bovine serum (FBS).
12. Becton Dickinson flow cytometer with CellQuest software.

2.3. Nanoparticle Preparation and Testing

1. PBS, G75 columns, and the BCA Protein Assay Reagent Kit are described in **Section 2.2**.
2. The tat (biotin-G-R-K-K-R-R-Q-R-R-R) is purchased HPLC purified as the native L isomer with the biotin attached to the amine end via a 6-aminohexanoic acid linker (21st Century Biochemicals).
3. 18-mer complementary phosphodiester DNA (Integrated DNA Technologies).
4. Streptavidin: 4 mg/mL in 0.15 M saline.
5. 60% (v/v) normal mouse serum (Jackson ImmunoResearch Lab) in PBS.
6. 0.7% (w/v) agarose gel containing 0.02% (w/v) ethidium bromide.
7. Kodak Imaging Station 440 CF.

2.4. Cellular Accumulation

1. SUM190 cells culture medium: Ham's F-12 medium (Gibco/BRL) with 5 μg/mL insulin, 1 μg/mL hydrocortisone, 5 mM ethanolamine, 10 mM HEPES, 5 μg/mL transferrin, 10 nM triiodo thyronine, 50 nM sodium selenite, and 0.5 g/L bovine serum albumin
2. SUM149 cells culture medium: Ham's F-12 medium (Gibco/BRL) with 5 μg/mL insulin, 1 μg/mL hydrocortisone, 10 mM HEPES, and 5% fetal bovine serum (FBS).
3. SK-BR-3 cells culture medium: McCoy's 5a medium (Invitrogen) with 10% FBS.
4. PBS is described in **Section 2.2**.
5. Lysis buffer: 0.2 M NaOH containing 1% SDS.
6. Counting tube (Sarstedt).
7. Automatic NaI (TI) well counter.

2.5. Confocal Immunofluorescence

1. Eight-well chambered coverglass.
2. McCoy's 5a medium without serum.
3. Fixing buffer: 3% (v/v) paraformaldehyde, 0.02% (v/v) glutaraldehyde in PBS.
4. Primary antibodies: Texas Red-labeled anti-cyanine antibody (Abcam), FITC-labeled transferrin (Molecular Probes), and FITC-labeled anti-human IgG (Sigma).
5. Nuclear stain: 100 nM 4-6-diamidino-2-phenylindole (DAPI) in PBS (Molecular Probes).
6. Mounting medium: Antifade (Molecular Probes).
7. Nuclei EZ Prep kit (Sigma).
8. Methanol.
9. Confocal microscope.

2.6. Subcellular Localization

Materials are described in **Sections 2.4** and **2.5**.

2.7. Biodistribution in Tumored Animals

1. Female nude mice: NIH Swiss, 30–40 g.
2. Matrigel: 5 mg/mL Matrigel (BD Biosciences) in SUM190 cell culture medium (described in **Section 2.4**).
3. PBS is described in **Section 2.2**.
4. NanoSPECT/CT (Bioscan) is a dual-modality small animal imager using multiplexed multi-pinhole SPECT technology.
5. Isoflurane.

2.8. Subcellular Localization in Tumored Animals

1. PBS is described in **Section 2.2**.
2. Sheep horseradish peroxidise-conjugated anticyanine antibody (Abcam).
3. Paraffin.
4. Xylene.
5. Ethanol.
6. 10% Phosphate buffered formalin, pH 7.0.
7. Retrievagen B solution (BD Biosciences).
8. Peroxidase and alkaline phosphatase dual endogenous enzyme blocking reagent (Dako).
9. All immunohistochemistry staining buffers (BD Biosciences) are provided by the UMMS DERC Morphology Core Facility.

3. Methods

3.1. Oligomer Conjugation and Radiolabeling

1. A solution of 1 mg of oligomer in 200 μL of 0.3 M HEPES buffer, pH 8.0, is added to a vial containing 0.7–1.0 mg *S*-acetyl NHS-MAG3 (9). The vial is vortexed immediately to a clear solution and incubated for 1 h at room temperature.
2. To 50 μL of 2 M ammonium acetate is added 120 μL of freshly prepared 20 mg/mL $SnCl_2 \cdot 2H_2O$ in tartrate buffer I, pH 9.2 (*see* **Note 4**), with agitation. After heating at 100°C for 25 min, the mixture is allowed to cool, and absolute ethanol is added to a concentration of 20% (v/v).
3. The mixture is purified on a 1 × 20 cm P4 column using 0.25 M ammonium acetate as eluant. The peak fractions are pooled and the oligomer concentration quantitated by UV absorbance at 265 nm. The purified samples are stored at 4°C (*see* **Note 5**). All solutions are passed through a 0.22-μm filter to ensure sterility.
4. Radiolabeling of each oligomer is achieved by introducing about 18.5–37 MBq (20 μL) of ^{99m}Tc-pertechnetate generator eluant into a combined solution consisting of about 15 μL MAG3-conjugated oligomer (10 μg) in 0.25 M ammonium acetate, 45 μL of 0.25 M ammonium acetate, 15 μL of 50 μg/μL tartrate buffer II, pH 8.7, and 5 μL of fresh 4 μg/μL $SnCl_2 \cdot 2H_2O$ in ascorbic acid–HCl solution. The final pH is about 8.5. After vortexing and then heating for 20 min in boiling water, the labeling is confirmed by size-exclusion HPLC with in-line UV and radioactivity detection using 20% acetonitrile in 0.1 M Tris–HCl, pH 8.0, as eluant at a flow rate 0.6 mL/min and radioactivity recovery is measured. Radiochemical purity of the labeled MORF oligomers is routinely shown by size-exclusion HPLC to be 90% or better.
5. The MORF oligomers are conjugated with the Cy3 fluorophore for confocal immunofluorescence. The Cy3-NHS is dissolved in NMP at a concentration of 25 mg/mL. About 400 μg of each MORF oligomer dissolved in 80 μL of 0.30 M HEPES (pH 8.0) and 25 μL of Cy3-NHS in anhydrous NMP (molar ratio Cy3-NHS to oligomer 17.8:1) is added to this solution. The mixture is vortexed thoroughly and incubated in the dark at room temperature for 1 h before purification on a 1 × 20 cm P4 column using 0.25 M ammonium acetate as eluant. Two pink-colored bands are observed due to free and conjugated Cy3. The latter faster-moving band is collected and concentrated on a Savant Speed Vac and the concentration is determined by measuring absorbance at 265 nm.

3.2. Biotinylation of Herceptin Antibody and Testing

1. The clinical Trastuzumab preparation is first dialyzed against 0.1 M phosphate-buffered saline (PBS) extensively using Slide-A-Lyzer Dialysis Cassettes (10 K MWCO), and then against 0.3 M HEPES, pH 7.5, for 3 h to remove all traces of histidine (6, 10). The antibody concentration is quantitated by BCA Protein Assay Reagent Kit.
2. The dialyzed Trastuzumab (25 mg) in pH 7.5, 0.3 M HEPES is added to a nonstick microfuge tube containing 0.52 mg of solid sulfo-NHS-LC-LC-biotin. The molar ratio of Trastuzumab to sulfo-NHS-LC-LC-biotin is therefore approximately 1:4.6 (*see* **Note 6**).
3. The mixture is gently vortexed to a clear solution and then incubated at room temperature for 1 h followed by purification on a 1×20 cm G75 open column using 0.15 M PBS, pH 7.4, as eluant. The fractions with peak absorbance at 280 nm are combined and the concentration is determined as above. The conjugated antibody is stored at 4°C in PBS containing 0.02% NaN_3 (*see* **Note 7**).
4. The biotinylated antibody is evaluated for epitope-binding ability in SUM190 and SUM149 cells. Cells are suspended in ice-cold 3% BSA blocking buffer at about 10^7 cells per mL and incubated with 2 μg/mL of biotinylated Trastuzumab or native Trastuzumab for 1 h at room temperature. The cells are then washed three times with PBS and incubated with FITC-labeled anti-human IgG antibody or anti-biotin IgG antibody in ice-cold antibody dilution buffer for 30 min at room temperature in the dark. After washing again with PBS, the cells are resuspended in ice-cold PBS and analyzed on a Becton Dickinson flow cytometer (BD Biosciences) using CellQuest software (BD Biosciences).

3.3. Nanoparticle Preparation and Testing

1. Since MORF oligomers are only about 50% biotinylated by the manufacturer, the molar ratios at addition must be adjusted accordingly (5–7, 10). As an example, the ^{99m}Tc-MORF is added very slowly and with continuous agitation into a 4 mg/mL solution of streptavidin in 0.15 M saline to a final MORF to streptavidin molar ratio of 2:1.
2. The solution is incubated at room temperature for 1 h followed by purification over a 1×20 cm G75 open column using PBS as eluant to remove unbound ^{99m}Tc-MORF without biotin. The early fractions with peak radioactivity are combined and the concentration of streptavidin determined colorimetrically using BCA Protein Assay Reagent Kit. Absorbance at 280 nm cannot be used for this determination since MORF also absorbs at this frequency.

3. The ^{99m}Tc-MORF/streptavidin nanoparticle is analyzed by size-exclusion HPLC as above. The appearance of a single peak in the UV trace at 280 nm corresponding to a single peak in the radioactivity trace is evidence of the complete complexation and absence of higher order products.
4. In the subsequent addition of biotinylated tat to the ^{99m}Tc-MORF/streptavidin nanoparticle, the absence of higher order MORF/tat nanoparticles is established in the same manner by HPLC analysis. With continuous stirring, the biotinylated tat is added slowly to the ^{99m}Tc-MORF/streptavidin nanoparticles to a final streptavidin to tat molar ratio of 1:1 and the preparation is incubated at room temperature for 1 h. An example of the results produced is shown in **Fig. 6.1**.
5. Finally, the biotinylated Trastuzumab is added to the ^{99m}Tc-MORF/streptavidin/tat nanoparticle in the same manner to a final streptavidin to antibody molar ratio of 1:1 but with gentle agitation to avoid denaturing the antibody and the preparation is incubated for 1 h at room temperature. The complete complexation and stability of biotinylated Trastuzumab to the MORF/tat/streptavidin nanoparticle are confirmed by agarose gel electrophoresis.

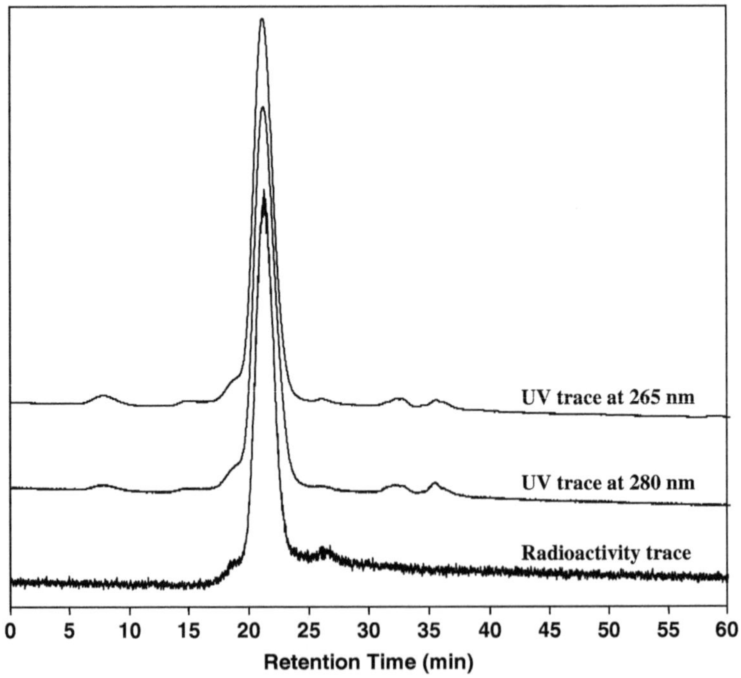

Fig. 6.1. Size-exclusion HPLC chromatographs by UV (265 and 280 nm) and radioactivity detection of the ^{99m}Tc-labeled antisense MORF/streptavidin/tat nanoparticle.

6. Nonradiolabeled MORF, the MORF/streptavidin one-component nanoparticle, the MORF/tat and the MORF/Trastuzumab two-component nanoparticles and the MORF/tat/Trastuzumab three-component nanoparticle are each analyzed and incubated in 37°C PBS or 60% normal mouse serum, in all cases at a final MORF concentration of 10 nM, and samples removed at 1 and 24 h. The 18-mer complementary phosphodiester DNA in PBS buffer is added to each sample to a final MORF to phosphodiester DNA molar ratio of 1:1 to form the charged duplex before analysis (*see* **Note 8**). Samples are developed by gel electrophoresis on a 0.7% agarose gel containing 0.02% ethidium bromide and photographed using Kodak Imaging Station 440 CF (New Haven, CT).

3.4. Cellular Accumulation Studies

1. The cells are plated in 24-well plates until about 60–70% confluence and then incubated for 0, 1, 3, 5, or 7 h with the ^{99m}Tc-labeled antisense (AS) or sense (S) MORFs as the one-, two-, or three-component nanoparticle in SUM190 culture medium with FBS-free or SUM149 culture medium with 0.5% FBS (5, 6).
2. At harvest, the radioactive medium is aspirated out and each well is rinsed with 1 mL PBS. The cells are then lysed with 1 mL of lysis buffer for 5 min at room temperature.
3. The celluar lysate from each well is collected into a counting tube along with 1 mL of rinse. The tubes are counted in an automatic NaI (TI) well counter along with a counting standard. Examples of the results produced are shown in **Fig. 6.2**.

3.5. Confocal Immunofluorescence

1. SK-BR-3 cells are seeded onto an 8-well chambered coverglass until about 60–70% confluence (6).
2. Cy3-labeled AS MORF/tat/Trastuzumab nanoparticles are added in serum-free medium at 50 nM and incubated for 2 h at 37°C.
3. One group of cells is incubated with FITC-labeled transferrin at 20 μg/mL for 1 h at 37°C. Thereafter the cells are washed with PBS twice and fixed in 100 μL of fixing buffer (*see* **Note 9**) in PBS for 15 min at room temperature before being placed at −20°C in methanol for 10 s. The cells are then washed three times with PBS and are incubated with Texas Red-labeled anti-Cyanine antibody (dilution of 1:200) (*see* **Note 10**) for 1 h at room temperature in the dark.
4. Another group of cells is washed with PBS twice and fixed as above. After washing three times with PBS, cells are incubated with Texas Red-labeled anti-Cyanine antibody and

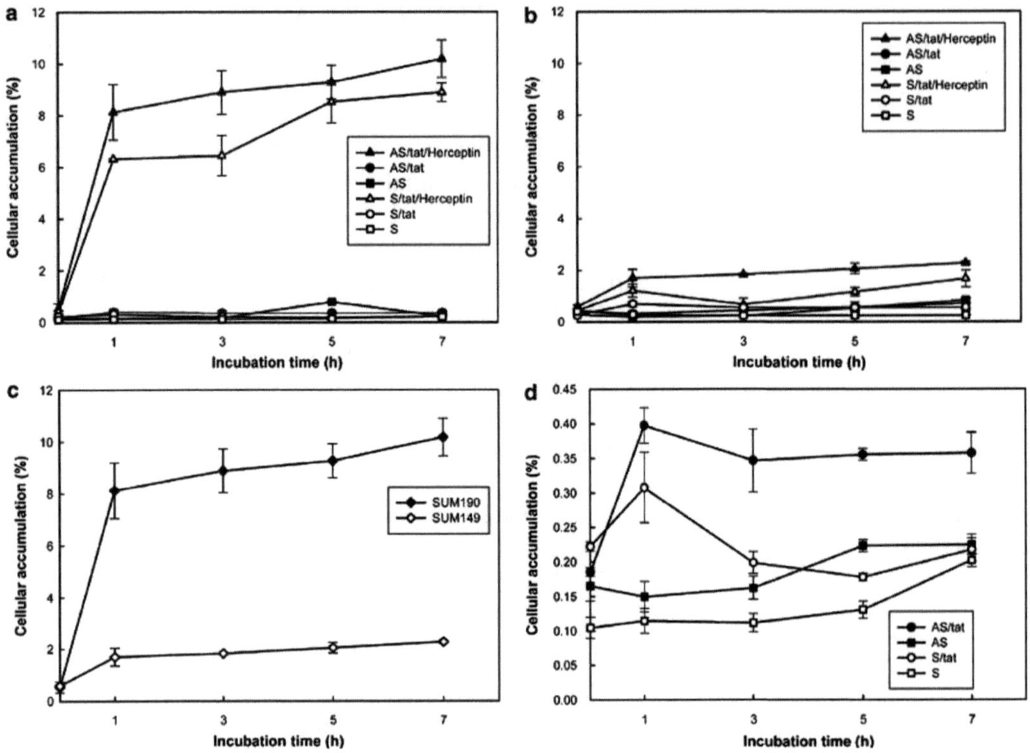

Fig. 6.2. Percent cellular accumulations over time of radiolabeled anti-RIa antisense (AS) and sense (S) MORF incubated as the MORF one-component nanoparticles alone, the two component with tat and the three component with tat and Herceptin (Trastuzumab) in SUM190 (**a**) and SUM149 (**b**) cells. The cellular accumulations of the AS MORF/tat/Herceptin nanoparticles in the SUM190 vs SUM149 cells are reproduced in (**c**) and the cellular accumulations of the one-component AS MORF and the two-component AS MORF/tat nanoparticles compared to the corresponding S MORF nanoparticles in SUM190 cells are reproduced in (**d**). Note the changes in scale. Results show that the Trastuzumab within the nanoparticle is able to bind to its determinant, the MORF within the nanoparticles may have escaped entrapment with preserved mRNA-binding ability, and the tat within the nanoparticle still exhibited its carrier function. (Reproduced from (6))

FITC-labeled anti-human IgG antibody at 2 μg/mL for 1 h at room temperature in the dark.

5. The antibody in each well is removed by rinsing the cells before the cells are incubated with DAPI solution at 100 nM in PBS for 5 min at room temperature for nuclear staining.
6. Cells are washed three times with PBS and then aspirated dry from one corner of the coverglass.
7. Mounting medium is added into each chamber.
8. The slides are viewed under confocal microscope. Excitation at 596 nm induces the Texas Red fluorescence (red emission) for MORF, excitation at 488 nm induces FITC fluorescence (green emission) for transferrin or Trastuzumab, and excitation at 358 nm induces DAPI fluorescence (blue emission). Examples of the signals for nanoparticles are shown in **Fig. 6.3**.

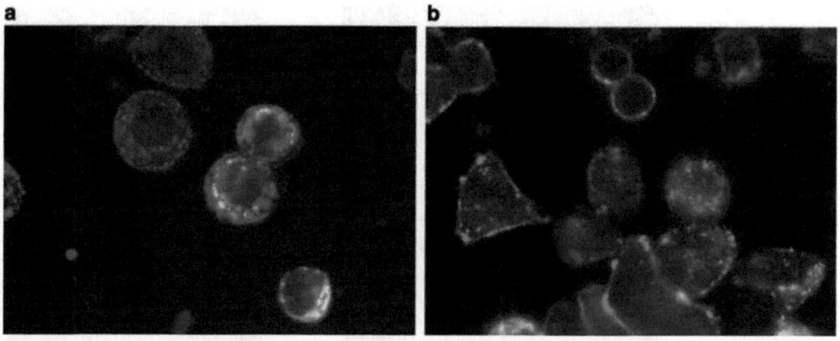

Fig. 6.3. Internalization and intracellular localization of Cy3-antisense MORF/tat/Trastuzumab nanoparticles in SK-BR-3 cells. The cells of (**a**) are incubated with Texas Red-labeled anti-Cyanine antibody (red stain) indicating the location of the MORF and FITC-labeled transferrin as a internalization marker (green stain). The cells of (**b**) are again incubated with Texas Red-labeled anti-Cyanine antibody (red stain) indicating the location of the MORF but now with FITC-labeled anti-human IgG antibody indicating the location of the Trastuzumab (green stain). All cells are also incubated with DAPI (blue stain) to show the location of the nucleus. (Magnification 600×). Imagings show that the antisense nanoparticle is internalized within 2 h of incubation with the Cy3-antisense MORF appearing in the nucleus. (Reproduced from (6))

3.6. Subcellular Localization

1. The SK-BR-3 cells are plated (4×10^6 cells per well) in 6-well plates as above (7).
2. Cells are incubated with ^{99m}Tc-labeled AS or S MORF as the three-component nanoparticles and as the free MORFs, each at 50 nM, in SK-BR-3 culture medium with 1% FBS for 3 h at 37°C.
3. The percentage of added radioactivity that became incorporated in the cells is first measured. Thereafter, the nuclear and other fractions are separated by Nuclei EZ Prep kit according to the manufacturer's instruction, and each is counted separately for radioactivity in an automatic NaI (T1) well counter along with a counting standard.

3.7. Biodistribution in Tumored Animals

1. Eight female nude mice at 7 weeks old are each injected subcutaneously in the left thigh with a 50-μL suspension containing 10^6 SUM190 cells mixed with 5 mg of Matrigel per milliliter and are used for imaging and biodistribution studies 4 weeks later when the tumors reached about 1 cm in diameter (7).
2. Four mice each receive 14.8 MBq of ^{99m}Tc-antiRIα MORF/tat/Trastuzumab nanoparticle containing 23 μg of Tastuzumab in 200 μL of PBS via a tail vein. The four remaining mice each receive the identical weight of Trastuzumab radiolabeled with 14.8 MBq of ^{99m}Tc in 200 μL of PBS but as the free antibody. One animal is selected from both groups for imaging on a NanoSPECT/CT camera at 2, 8, and 21 h after administration. Examples of tumored animal images are shown in **Fig. 6.4**.

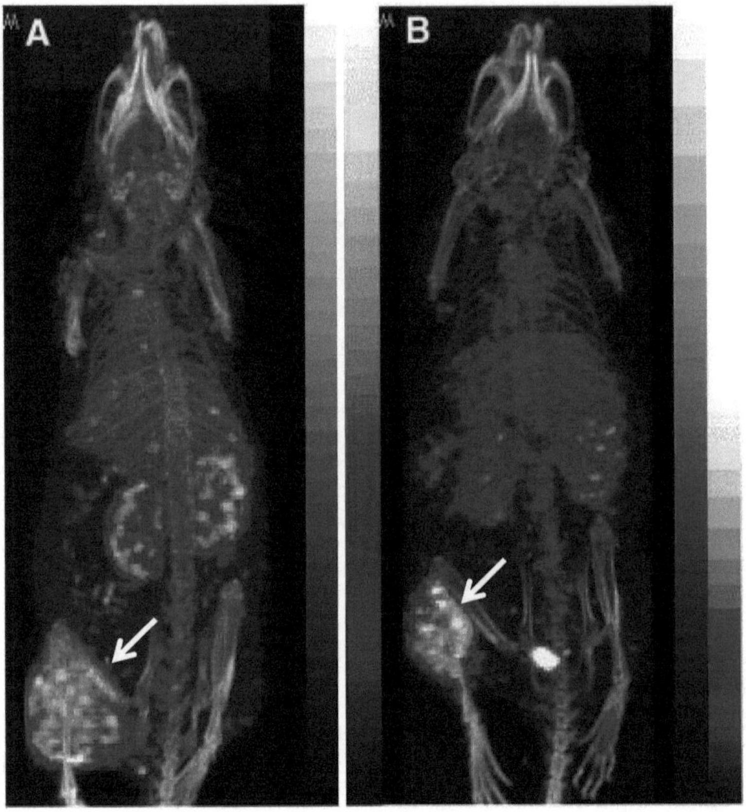

Fig. 6.4. Posterior projections of SPECT/CT acquisitions obtained on a NanoSPECT/CT camera of two SUM190 tumored mice 8 h post administration. The biodistribution of the ^{99m}Tc-antiRIαMORF/tat/Trastuzumab nanoparticle (**a**) shows higher liver, spleen, and kidney levels but comparable accumulations in tumor (*white arrow*) to that of the ^{99m}Tc-Trastuzumab antibody (**b**). (Reproduced from (7))

3. The mice are sacrificed by cervical dislocation after anesthesia with inhalation of isoflurane at 32 h.
4. The radioactivity accumulated in tissues or organs of interest is measured by removing the tissues or organs, followed by counting on an automatic NaI (T1) well counter along with a counting standard.

3.8. Subcellular Localization in Tumored Animals

1. The tumor mouse model is set up exactly as above (7).
2. In groups of 3, female nude mice receive either the AS or the S Cy3-MORF as the MORF/tat/Trastuzumab nanoparticle containing 1 μg of MORF in 100 μL of PBS via a tail vein. An additional three mice receive only 100 μL of PBS.
3. The mice are sacrificed as above at 24 h and tissues and organs of interest are removed.
4. Tissues are fixed in 10% buffered formalin for 24 h and then embedded in paraffin.

5. Paraffin sections are cut to 5 μm and deparaffinized by two washes of xylene for 10 min each and then in 100% ethanol twice for 2 min each.
6. The slides are heated in a microwave at 90°C for 10 min in Retrievagen B solution, pH 9.5 (*see* **Note 11**), and then cooled to room temperature.
7. After rinsing in PBS, endogenous peroxidase activity is blocked by incubation with dual endogenous enzyme blocking reagent for 10 min.
8. The slides are then washed with 2–3 changes of distilled water and incubated with sheep horseradish peroxidise-conjugated anticyanine antibody from Abcam (dilution of 1:50 of 1 mg/mL stock) overnight at room temperature.
9. All samples are counterstained with hematoxylin (blue stain) to stain the nucleus and mounted in water-soluble medium.
10. Slides are viewed under microscope equipped with CCD camera. Examples of the signals for nanoparticles are shown in **Fig. 6.5**.

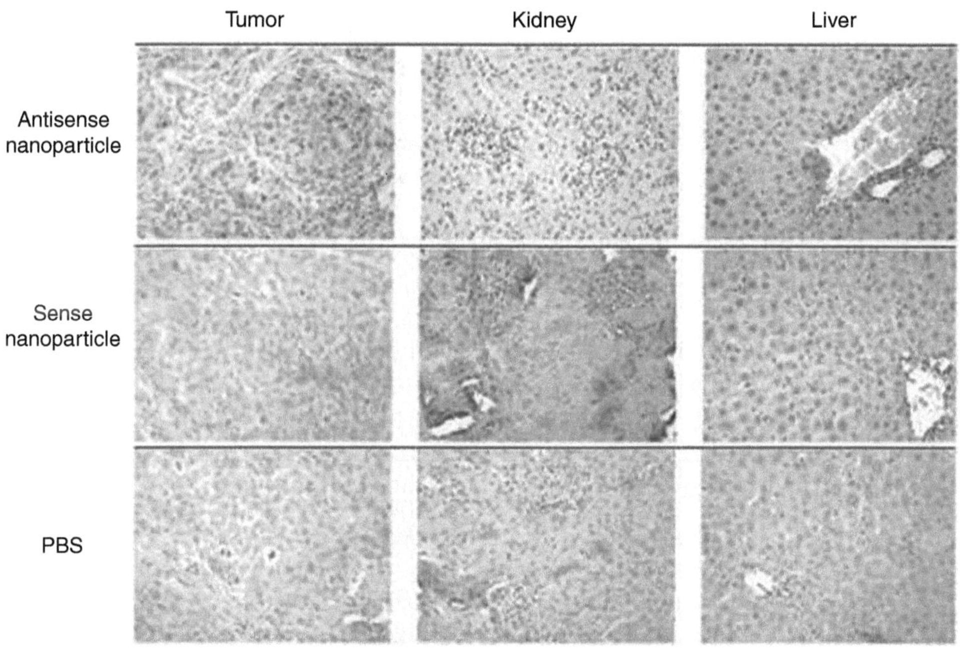

Fig. 6.5. Immunohistochemical staining of formalin fixed, paraffin embedded tumor (*left panels*), kidney (*middle panels*) and liver (*right panels*) from SUM190 tumored mice administered Cy3 labeled antisense (*upper panels*) or sense (*middle panels*) MORF/tat/Trastuzumab or PBS (*bottom panels*) after 24 h. (Magnification 400×). Microscopic examination shows nuclear staining in approximately 20% of the tumor cells in animals injected with the antisense nanoparticle and 10% of the tumor cells in animals receiving the sense nanoparticle, whereas no nuclear staining is seen in the tumor cells of mice given the PBS injection. No nuclear staining is seen in the parenchymal cells of kidney and liver from the mice receiving the antisense nanoparticle, sense nanoparticle, or PBS. (Reproduced from (7))

4. Notes

1. All reagents should be analytical grade and, if so, may be used without additional purification.
2. The NHS-MAG3 should be stored as a solid in tightly sealed vials at below −20°C. The NHS-MAG3 should be brought to room temperature before exposing to atmospheric moisture. If the biomolecule has a tendency to stick to surfaces, use a nonstick centrifuge tube.
3. As with all radionuclides emitting penetrating radiation, standard shielding and radionuclide handling procedures must be employed with ^{99m}Tc. Typically, labeling is performed in a leaded enclosure in a lead-shielded container with wall thicknesses typically of 0.5–1 cm lead. Direct exposure to the radioactivity should be kept to a minimum. Individuals working with the material should monitor their radiation exposure with appropriate devices.
4. The $SnCl_2$ solution should be prepared fresh and used only on the day of preparation.
5. The MAG3 conjugated MORF may be stored in the 0.25 M ammonium acetate solution at −20°C for at least 2 years without loss of radiolabeling efficiency.
6. The longer LC–LC linker is used in the antibody conjugation in place of the shorter LC linker used in the MORF conjugation to help avoid steric hindrance when bound to streptavidin.
7. The biotinylated Trastuzumab may be stored in PBS containing 0.02% NaN_3 at 4°C at least 12 months to preserve epitope-binding activity. Nevertheless, testing the epitope-binding ability before each experiment is strongly recommended.
8. Since MORF oligomers are uncharged, the complementary negatively charged phosphodiester (PO) DNA is added to each sample before gel electrophoresis.
9. Different fixation procedures may be tried to optimize the immunostaining. The choice of fixatives will also depend on the subcellular localization of the antigen (soluble, membrane bound, cytoskeleton associated). Paraformaldehyde/glutaraldehyde fixation is used in our protocol for double labeling of membrane bound and celluar antigens.
10. Because of the low intensity of the Cy3 fluorophore in these cell studies, it is necessary to use the Texas Red-labeled anti-Cyanine antibody as probe to establish the cellular distribution of the MORF.

11. The pH value of Retrievagan solution is a key factor to optimize the immunostaining. We have found that the optimum pH value is from 8.0 to 9.5 and pH < 7.0 is avoided.

Acknowledgments

This work was supported in part by the Office of Science (BER), U.S. Department of Energy (grant DE-FG02-03ER63602); the National Institutes of Health (grant R01CA94994 and R21CA129591); and congressionally directed Medical Research Programs (grant W81XWH-06-1-0649).

References

1. Nakamura, K., Wang, Y., Liu, X., Kitamura, N., Kubo, A., and Hnatowich, D. J. (2007) Cell culture and xenografted animal studies of radiolabeled antisense DNA/carrier nanoparticles using streptavidin as linker. *J. Nucl. Med.* **48**, 1845–1852.
2. Song, E., Zhu, P., Lee, S. K., Chowdhury, D., Kussman, S., Dykxhoorn, D. M., et al. (2005) Antibody mediated in vivo delivery of small interfering RNAs via cell-surface receptors. *Nat. Biotechnol.* **23**, 709–717.
3. Ross, J. S., Fletcher, J. A., Bloom, K. J., Linette, G. P., Stec, J., Symmans, W. F., et al. (2004) Targeted therapy in breast cancer: the HER-2/neu gene and protein. *Mol. Cell Proteomics* **3**, 379–398.
4. Slamon, D. J., Clark, G. M., Wong, S. G., Levin, W. J., Ullrich, A., and McGuire, W. L. (1987) Human breast cancer: correlation of relapse and survival with amplification of the HER-2/neu oncogene. *Science* **235**, 177–182.
5. Wang, Y., Nakamura, K., Liu, X., Kitamura, N., Kubo, A., and Hnatowich, D. J. (2007) Simplified preparation via streptavidin of antisense oligomers/carriers nanoparticles showing improved cellular delivery in culture. *Bioconjug. Chem.* **18**, 1338–1343.
6. Liu, X., Wang, Y., Nakamura, K., Kubo, A., and Hnatowich, D. J. (2008) Cell studies of a three-component antisense MORF/tat/Herceptin nanoparticle designed for improved tumor delivery. *Cancer Gene Ther.* **15**, 126–132.
7. Liu, X., Wang, Y., Nakamura, K., Kawauchi, S., Akalin, A., Cheng, D., et al. (2009) Auger radiation-induced, antisense-mediated cytotoxicity of tumor cells using a 3-component streptavidin-delivery nanoparticle with ^{111}In. *J. Nucl. Med.* **50**, 582–590.
8. Winnard, P., Chang, F., Rusckowski, M., Mardirossian, G., and Hnatowich, D. J. (1997) Preparation and use of NHS-MAG3 for technetium-99m labeling of DNA. *Nucl. Med. Biol.* **24**, 425–432.
9. Wang, Y., Liu, G., and Hnatowich, D. J. (2006) Methods for MAG3 conjugation and 99mTc radiolabeling of biomolecules. *Nat. Protoc.* **1**, 1477–1480.
10. Wang, Y., Liu, X., Chen, L., Cheng, D., Rusckowski, M., and Hnatowich, D. J. (2009) Tumor delivery of antisense oligomer using trastuzumab within a streptavidin nanoparticle. *Eur. J. Nucl. Med. Mol. Imaging* **36**, 1977–1986.

Chapter 7

Light-Directed Delivery of Nucleic Acids

Sigurd Bøe, Lina Prasmickaite, Birgit Engesæter, and Eivind Hovig

Abstract

A major barrier within the field of non-viral gene therapy toward therapeutic strategies, e.g., tumor therapy, has been lack of appropriate specific delivery strategies to the intended target tissues or cells. In this chapter, we describe a protocol for light-directed delivery of nucleic acids through the use of photochemical internalization (PCI) technology. PCI is based on a photosensitizing compound that localizes to endocytic membranes. Upon illumination, the photosensitizing compound induces damage to the endocytic membranes, resulting in release of endocytosed material, i.e., nucleic acids into cytosol. The main benefit of the strategy described is the possibility for site specific delivery of nucleic acids to a place of interest.

Key words: Light-directed delivery, PCI, nucleic acids, photosensitizer, endosomal pathway, singlet oxygen.

1. Introduction

Natural (deoxyribonucleic or ribonucleic acid) or artificial (e.g., peptide nucleic acid, locked nucleic acid, or morpholino) nucleic acids can be delivered into cells in order to obtain a biological effect by altering gene expression. By delivery of nucleic acids into the cytosol or nucleus of a cell, gene expression may be modulated by increasing or decreasing levels of proteins. In general, cellular uptake of nucleic acids occurs by a variety of different mechanisms collectively termed endocytosis (1). The internalized nucleic acids are trapped in the endosomal vesicles, which eventually fuse with lysosomes where the endosomal content usually is degraded [reviewed in (2)]. To improve cytosolic delivery of nucleic acids, carrier systems with fusiogenic properties or endosomolytic activity have been developed (3, 4). Although carrier

J. Goodchild (ed.), *Therapeutic Oligonucleotides*, Methods in Molecular Biology 764,
DOI 10.1007/978-1-61779-188-8_7, © Springer Science+Business Media, LLC 2011

systems have improved, lack of efficient targeting, low cellular and intracellular delivery, and non-specific effects have hampered therapeutic success [reviewed in (5–7)]. Among recent new technologies for intracellular delivery, photochemical internalization (PCI) represents a useful strategy for light-directed delivery of nucleic acids located in the endocytic pathway (8–18) (**Fig. 7.1**).

The PCI strategy is based on illumination of target cells preloaded with a photosensitizing compound (photosensitizer) that localizes preferably in endosomal and lysosomal membranes. Illumination activates the photosensitizer, which further leads to the formation of mainly singlet oxygen that destroys the endosomal membranes, leading to a release of endocytosed material. Importantly, PCI represents a method for targeted delivery of nucleic acids, since the biological effect of nucleic acids is primarily achieved in only light-exposed areas. In non-exposed areas, nucleic acids tend to be degraded in the lysosomes.

1.1. PCI Technology

1.1.1. PCI in General

PCI, a method for light-directed delivery of various macromolecules, was developed by Berg and colleagues in 1999 (19) and is based on the principle of photodynamic therapy (PDT). PDT can be defined as the use of a photosensitizer in combination with light and oxygen to induce photochemical reactions lethal to cells in the body, with the purpose to treat diseases. In contrast, PCI uses a photosensitizer, light, and oxygen to liberate macromolecules trapped inside endosomal and lysosomal vesicles potentiating the biological effect of these macromolecules. Various photosensitizers, depending on their physicochemical properties, enter and localize inside the cell differently. Photosensitizers used in PCI localizes preferentially in the endosomal and lysosomal membranes (20, 21) while photosensitizers used in PDT usually target a number of organelles, including the cell

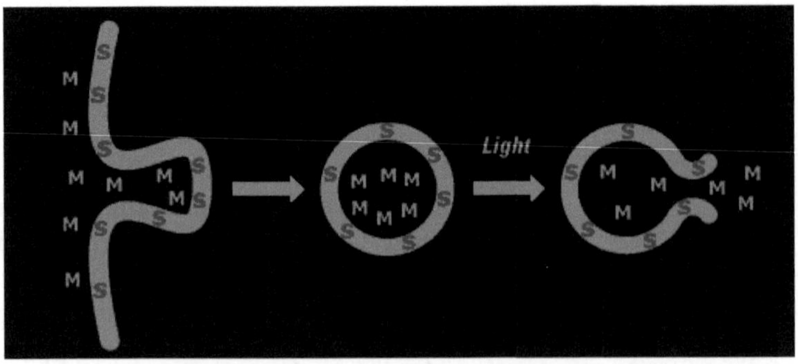

Fig. 7.1. The principle of PCI. Both the photosensitizer (S) and the macromolecule (M) are endocytosed and localize to the same endosomal vesicle. Light exposure induces photochemical reactions, leading to disruption of vesicular membranes and facilitated endosomal release of the endosomal content into cytosol. Adapted with permission from PCI Biotech AS.

membrane (22, 23), nucleus (24, 25), and mitochondria (26, 27). Noteworthy is the conjugation of photosensitizers with tumor-specific ligands for enhanced uptake in malignant tissue [reviewed in (28)]. Photosensitizers that enter the cell via endocytosis tend to accumulate in endocytic vesicles and, therefore, are suitable for PCI. These include photosensitizers that aggregate and hydrophilic photosensitizers such as di- and tetrasulfonated tetraphenyl porphines ($TPPS_n$) (20, 29) and aluminum phthalocyanines ($AlPcS_n$) (30, 31). However, not all photosensitizers that localize in endocytic vesicles are equally efficient in PCI, and the most potent photosensitizers found so far are $TPPS_{2a}$ and $AlPcS_{2a}$ (32). $TPPS_{2a}$ and $AlPcS_{2a}$ have two sulfonate groups on adjacent phenyl or phthalate rings (**Fig. 7.2**), making the photosensitizer molecules amphiphilic; therefore, they localize primary in the membranes of endocytic vesicles rather than the lumen (20, 33). Upon light exposure, such photosensitizers destroy mainly vesicular membranes, whereas the content of the organelles (e.g., transfected nucleic acids) remains mostly unaffected due to the extremely short diffusion length (<20 nm) and a short lifetime (<40 ns) of singlet oxygen (34). Endosomaly and lysosomaly localized photosensitizers used in PCI will result in photochemical cell killing (PDT) if exposed to adequate illumination. The light source used in PDT and PCI is commonly a broad-spectrum lamp or a laser. Lasers are preferable, as they generate a specific wavelength of light and can be set to match the absorption peak of the specific photosensitizer used. Another benefit of using a laser is the possibility of using optical fibers for treatment of internal organs [reviewed in (35, 36)]. In vitro, PCI has been shown to potentiate biological effects of a variety of delivered molecules, for instance peptides (19), plasmids (37, 38), viruses (39, 40), standard DNA oligodeoxynucleotides (41), standard messenger RNA (mRNA) molecules (11), peptide nucleic acid (PNA) molecules (9, 13–16), siRNA molecules combined with lipids (10, 12, 17), and polymers (8). In vivo, PCI has been shown to enhance delivery of plasmid DNA (37, 42), protein toxin gelonin (43, 44), and chemotherapeutic bleomycin

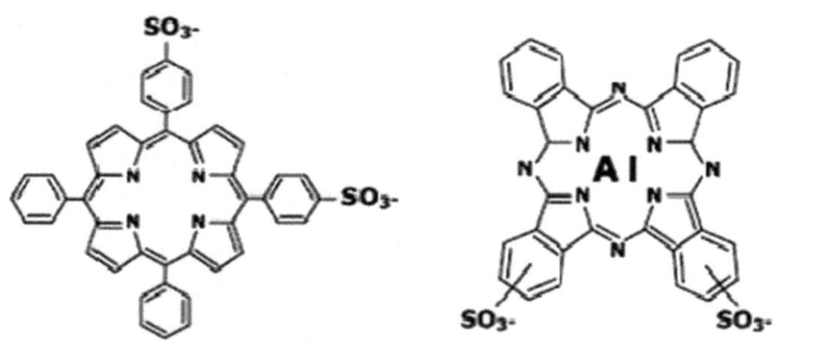

Fig. 7.2. Chemical structure of $TPPS_{2a}$ and $AlPcS_{2a}$.

(45). Interestingly, PCI-mediated delivery of siRNA molecules in vivo has recently been documented in an EGFP model system after intratumoral injection in mice (18). Previously it has been reported that maximum PCI-mediated delivery is achieved by the use of a light dose that reduces cell viability by approximately 50% (41). However, recent studies have shown that the photocytotoxic effect of PCI can be minimized by using appropriate conditions (8, 10). Although the toxicity is in most cases a disadvantage, it may in some cases be beneficial, i.e., in cancer therapy.

1.1.2. Photochemical Reactions

PCI relies on the use of a photosensitizer, illumination, and oxygen. A photosensitizer is able to absorb photons when exposed to light with frequency overlapping the absorption spectra of the photosensitizer. This light exposure excites the photosensitizer to a singlet state, which is rapidly converted to a triplet state [reviewed in (46)] and further induces photochemical reactions. There are two types of photochemical reactions, type I and type II, where type II is the most common in PCI (47). In the type I process, the substrate or solvent reacts with the excited photosensitizer (in either singlet or triplet state) to give radicals or radical ions, respectively, due to hydrogen or electron transfer. These radicals or radical ions can again react with molecular oxygen to form reactive oxygen species (ROS), e.g., peroxides. In the type II reactions, the excited photosensitizer reacts with molecular oxygen to form singlet oxygen. The types I and II are always in competition; factors which govern the competition include oxygen concentration, the concentration and the reactivity of the substrate, the excited state of the photosensitizer, and the singlet oxygen lifetime. Singlet oxygen has an extremely short diffusion length (<20 nm) and a short lifetime (<40 ns) (34). Nevertheless, singlet oxygen is highly reactive and can introduce damage to nearby located molecules such as unsaturated fatty acids, some amino acids, and the nucleic acid component guanine.

1.1.3. PCI In Vivo – Advantages

1. PCI can be combined with molecules of various sizes. The only restriction regarding the size of the given macromolecule/complex is the ability of endosomal entrapment, making PCI very flexible for delivery of many different macromolecules/complexes [reviewed in (48)].
2. Due to the local and focused light-dependent activation, PCI is a method with high site-specificity, limiting the biological effect primarily to illuminated areas, which should reduce systemic side effects of the delivered molecule.
3. PCI is a method of high efficiency with the possibility of using lower photochemical doses, which is favorable for reducing possible non-specific effects, i.e., photocytotoxicity (8, 10, 11).

4. PCI may be combined with other treatment regimens, such as surgery (49).
5. Recent developments in fiber optics and laser technology provides the opportunity to illuminate tissue inside the human body using fiberscopes [reviewed in (36, 50)].
6. Compared to ionizing radiation and some chemotherapeutics, the PCI technology has low or no oncogenic effects. Therefore, as for PDT of cancer, PCI treatment may be repeated several times.
7. As opposed to radiation and cytostatics, PCI can be applied efficiently to non-dividing cells, which could be essential for killing resting malignant cells.
8. PCI allows us to penetrate to deeper tissue levels than conventional PDT, since lower light doses are sufficient to induce PCI and to initiate a biological effect.

1.1.4. PCI In Vivo – Limitations and Challenges

1. For PCI to become a successful therapeutic approach, nucleic acids and the photosensitizer all need to be co-localized in endosomal or lysosomal compartments of the target tissue.
2. A restriction of PCI in vivo is penetration of light in tissue. The light penetration is tissue-dependent and decays approximately exponentially for every 2–3 mm of depth. Maximum depth of necrosis observed for PDT is about 1 cm (51). However, this may also be viewed as an advantage, since illumination restricts the biological effect to light-exposed areas.
3. The fact that PCI treatment of cells can induce phototoxicity could be seen as both an advantage and a disadvantage. In a cancer therapy, where the obvious goal is to kill tumor cells, this is an advantage. However, in PCI-based cell therapy without the need for reduced cell viability, the PCI-mediated toxicity would be a disadvantage. In fact, PCI can prove effective in vitro without any reduction in cell viability when using appropriate materials and conditions (8, 10).

2. Materials

2.1. Cell Culture

1. Adherent cells to be transfected with nucleic acids by PCI are cultured according to recommendations for the specific cell lines. Non-adherent cells may also be transfected by PCI, but have so far rarely been used.
2. Cell growth medium supplemented with 10% fetal calf serum (FCS), 100 U/mL penicillin, 100 μl/mL streptomycin, and 2 mM glutamine (all Bio Whittaker, Wakersville, MD). Store at 4°C. (We routinely use RPMI-1640 medium; however,

other culture media can be used if recommended for the cells to be transfected.)

2.2. Stock Solution of the Photosensitizer $TPPS_{2a}$

1. The photosensitizer meso-tetraphenylporphine with two sulfonate groups on adjacent phenyl rings ($TPPS_{2a}$; Porphyrin Products, Logan, UT) (*see* **Note 1**).
2. 0.1 M NaOH.
3. Phosphate-buffered saline (PBS): 0.2 g/L KH_2PO_4, 8.0 g/L NaCl, 1.15 g/L Na_2HPO_4. Sterilize by filtration and store at 4°C.

2.3. Preparation of siRNA/Branched Polyethylenimine Complexes

1. Pre-annealed siRNA stock solution at 20 μM, stored at −20°C.
2. Transfection agent branched polyethylenimine (B-PEI), MW 25,000, (Sigma, St. Louis, MO) (*see* **Note 2**). To make stock solution (1 μg/mL), dissolve 1 μg B-PEI in 1 mL distilled water. Sterilize by filtration and store at 4°C.
3. Distilled water. Autoclave and store at 4°C.
4. Sterile polypropylene microcentrifuge tubes.

2.4. Photochemical Transfection

1. 6-well culture plates (cat. no. 140675 Nunc).
2. Sterile tubes.
3. Light source for excitation of $TPPS_{2a}$ at 435 nm (*see* **Note 3**).

3. Methods

The following protocol is for photochemical transfection of adherent cells in a 6-well culture plate (well diameter 3.5 cm). Amounts and volumes presented below are calculated for one such well, but can be adjusted accordingly to any other cell culture plate, cell number, and volume.

3.1. Preliminary Experiments to Find Suitable Light Dose

The light dose to be used for induction PCI-mediated delivery has to be found in advance. Thus, cell survival as a function of light dose has to be measured individually for every cell line, cell culture plate, and light source. As a starting point, the light dose should be adjusted to just below the initiation of cell killing (*see* **Note 3**). Cell survival can be measured by one of the common cell survival tests such as the 3-(4,5-dimethylthiazol-2-yl)-5-(3-carboxymethoxyphenyl)-2-(4-sulfophenyl)-2H-tetrazolium)

(MTS) test, protein synthesis, or another test established in an individual laboratory.

3.2. Preparation of Adherent Cells for Photochemical Transfection

The cells are seeded out 2 days before the experiment. In a 6-well culture plate, seed cells in 2 mL/well of growth medium containing 10% FCS. The number of cells seeds out per well depends on the cell size and cell growth rate. The cells should usually be at 20–40% confluency at the start of the incubation with siRNA/B-PEI complexes. (Higher and lower cell densities may also be used: Therefore, adjust the seeding density for every cell line and incubation time.)

3.3. Preparation of Photosensitizer $TPPS_{2a}$ Solutions

Work in subdued light (*see* **Note 4**).

3.3.1. Preparation of Stock Solution (1 mg/mL)

1. Dissolve 1 mg of $TPPS_{2a}$ in a small volume (approx. 0.1–0.2 mL) of 0.1 M NaOH.
2. Dilute with PBS to a final volume of 1 mL. The photosensitizers usually dissolve well in this way. If higher concentrations are needed or it is difficult to completely dissolve the photosensitizer, we recommend sonication bath treatment of the solution.
3. Sterilize by filtration, and store at –20°C in small aliquots for up to 6 months. Can be used several times after thawing and freezing (*see* **Note 4**).

3.3.2. Preparation of Working Solution (0.5 μg/mL)

Working solution should be freshly made before application to the cells. Add 0.63 μL of $TPPS_{2a}$ stock solution (1 mg/mL) to 0.8 mL of growth medium containing 10% FCS. This solution is ready to be applied to the cells.

3.4. Preparation of the siRNA/B-PEI Complexes

The complexes should be prepared just before addition to the cells.

As an example we here make complexes with an N/P ratio of 5 (*see* **Note 5**); therefore, the amounts presented below correspond to complexes with an N/P ratio of 5.

1. Prepare siRNA solution in a separate sterile microcentrifuge tube: 5 μL of siRNA (we use 20 μM of siRNA stock solution) diluted to 100 μL in sterile water. Gently mix the solution by pipetting several times. Do not vortex.
2. Prepare B-PEI solution in a separate sterile microcentrifuge tube: 1 μL of B-PEI stock solution (1 μg/mL) diluted to 100 μL in sterile water. Gently mix the solution by pipetting several times. Do not vortex.

3. Slowly transfer the B-PEI solution into the microcentrifuge tube containing the siRNA solution. The final volume of the mixture is 200 μL. Gently mix the mixture by pipetting several times. Do not vortex.
4. Leave the mixture on the bench at room temperature for 30 min to allow formation of siRNA/B-PEI complexes.
5. Transfer the whole siRNA/B-PEI mixture (200 μL) into the well containing 800 μL growth medium making a total volume of 1 mL (*see* **Note 6**).

3.5. Photochemical Transfection In Vitro

All the procedures starting from here should be carried out in subdued light (*see* **Note 4**). A simplified experimental scheme is presented in **Fig. 7.3**.

1. Seed out the cells 2 days before the experiment (*see* **Section 3.2**)
2. Remove the growth medium from cells and add 800 μL of growth medium containing 0.63 μg/mL $TPPS_{2a}$ (*see* **Section 3.3.2**).
3. Add 200 μL growth medium containing freshly prepared siRNA/B-PEI mixture (*see* **Section 3.4**).
4. Incubate overnight, i.e., for 16–18 h (*see* **Note 7**) at 37°C in a CO_2 incubator.
5. Remove the medium and wash the cell monolayer three times with growth medium to remove $TPPS_{2a}$ on the cell surface.
6. Incubate at 37°C in a CO_2 incubator for the desired time. We recommend 4 h (*see* **Note 8**).
7. Expose the cells to light. The optimal light exposure time depends on the cell line used; the light dose should therefore be determined individually (*see* **Note 3**).
8. Grow the cells for desired time and analyze the biological effect, i.e., change in target gene expression (*see* **Note 9**).

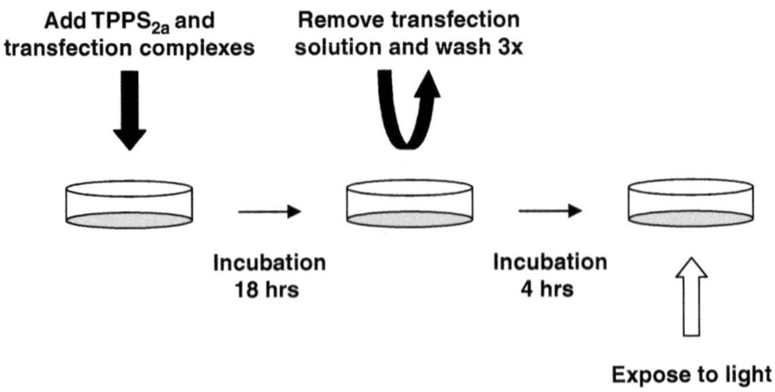

Fig. 7.3. Experimental scheme.

4. Notes

1. Here we have described the PCI procedure based on the use of $TPPS_{2a}$, the photosensitizer that we have most experience with in respect to photochemical transfection of nucleic acids. However, some other photosensitizers, localized in the membranes of endocytic vesicles, may also be used. Thus, we have shown that photochemical transfection works equally well with $AlPcS_{2a}$, whereas $TPPS_4$ and Photofrin are less efficient (32). If other photosensitizers than TPPS2a are to be used, it may, however, be necessary to use other light sources, to match the spectral characteristics of the specific photosensitizer.
2. Other non-viral cationic transfection vectors than B-PEI could also be tested for application in photochemical transfection. We have investigated a number of carriers (**Table 7.1**).
3. The light source used in our laboratory was a LumiSource lamp (PCI Biotech) consisting of four light tubes, which deliver blue light with a peak at 435 nm and a fluence rate of 5.2 mW/cm^2. However, any light source with suitable characteristics (*see* below) could be used. For every cell line to be transfected the illumination time should be adjusted, since the light delivered from different lamps will vary, and also the optimal light dose will vary between cell lines, cell plate format, etc. We routinely calibrate by measuring cell viability using one of the common cell survival tests, such as the MTS test, or measurement of protein synthesis. In general, a light dose close to initiation of cell killing would be a good starting point for photochemical transfection. When choosing the light source, the following points should be taken into consideration:
   a. The light source should deliver light of a wavelength suitable for excitation of $TPPS_{2a}$, which has an absorption peak at 435 nm (**Fig. 7.4**). It is also an advantage to avoid too much UV light, since UV light may induce cytotoxic effects. We have used the LumiSource lamp from PCI Biotech (Oslo, Norway), which has a filter for excluding UV light. Furthermore, the LumiSource lamp is equipped with a fan to prevent hyperthermia.
   b. The light source should generate a homogeneous light field in the area where the cells are illuminated.
   c. Although light has to penetrate the plastic dish before reaching the cell monolayer (**Fig. 7.4c**), the type of plastic dish does not seem to influence the results, but the well

Table 7.1
An overview over various carriers evaluated for PCI-induced delivery in the OHS cell line (Ph.D thesis of Bøe, S)

Carrier	Molecule	Cell viability (%)	Light-dose (mJ/cm^2)	Silencing/ expression without PCI	Silencing/ expression with PCI
Peptides	PNA	5–70	420	0–15%	80–90%
Lipofectamine 2000	siRNA	100	280	80–90%	80–90%
Lipofectin	siRNA	100	280	80–90%	80–90%
siPORT	siRNA	100	280	No gene silencing	No gene silencing
FuGene6	siRNA	100	280	No gene silencing	No gene silencing
DOTAP (unpublished)	siRNA	100	280	No gene silencing	No gene silencing
jetSI	siRNA	100	280	20–30%	80–90%
jetSI-ENDO	siRNA	100	280	20–30%	80–90%
INTERFERIN (unpublished)	siRNA	100	280	80–90%	80–90%
jetPEI (unpublished)	siRNA	100	280	60–70%	70–80%
MPEI 423	siRNA	100	210	No gene silencing	No gene silencing
BPEI 0.8kDa	siRNA	100	210	0–20%	50–80%
BPEI 1.2 kDa	siRNA	100	210	0–20%	80–90%
BPEI 1.3 kDa	siRNA	100	210	0–20%	80–90%
BPEI 1.8 kDa	siRNA	100	210	0–20%	80–90%
BPEI 2.0 kDa	siRNA	100	210	0–20%	80–90%
LPEI 2.5 kDa	siRNA	100	210	0–20%	80–90%
LPEI 25 kDa	siRNA	100	210	0–20%	80–90%
BPEI 25 kDa	siRNA	100	210	0–20%	80–90%
BPEI 25 kDa	EGFP mRNA	100	210–280	5–15% EGFP protein expression	80–90% EGFP protein expression

Abbreviations: MPEI, LPEI, BPEI – mixed (M), linear (L), and branched (B) polyethylenimine (PEI), kDa – kilo Dalton, mJ/cm^2 – millijoules per square centimeter.

format may influence on the recommended illumination time.

4. $TPPS_{2a}$ is a relatively photostable photosensitizer. However, we recommend protection of the photosensitizer solutions from light by aluminum foil to avoid possible light-inducible

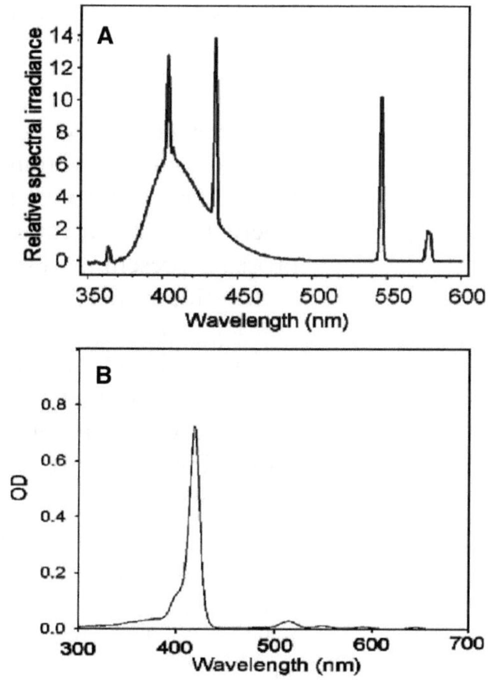

Fig. 7.4. (**a**) Lamp spectrum of light source. (**b**) $TPPS_{2a}$ absorption spectrum.

damage to the photosensitizer itself. The photosensitizers may also aggregate after long-time storage or repeated freezing and thawing. Aggregation will reduce the efficacy of the photosensitizer and should be avoided. Usually the color of the stock solution of $TPPS_{2a}$ is dark brown, and change of the color is usually a bad sign indicating aggregation or decomposition of the photosensitizer. In this case, the quality of the stock solution should be checked by measuring the phototoxic effect on cells using cell viability tests.

5. Here, we have described the use of the siRNA/B-PEI complexes with a nitrogen/phosphate (N/P) ratio of 5 (*see* **Section 3.4**). However, the N/P ratio should be optimized for photochemical transfection. Different approaches are used to define the exact composition of lipo- and polyplexes, such as the weight/weight (w/w) ratio, the charge ratio, and N/P ratio. We have used the N/P ratio in our work, as this is the most common way to define PEI polyplexes. The N/P ratio is expressed as nitrogen in PEI per phosphate group in the nucleic acid molecule. The N/P ratio of PEI/siRNA polyplexes = [Amount of PEI × Monomer size of siRNA]/[Amount of siRNA × Monomer size of PEI]. An example of N/P calculation:
   I. 1 μg PEI with monomer size = 43 g/mol
   II. 1.4 μg siRNA with monomer size of 316.7 g/mol

III. Using the formula above: N/P = [1 × 316.7]/[1.4 × 43]

IV. N/P = 5.26

6. All experiments were performed in FCS containing medium, but PCI-mediated transfection also works in serum-free media. This fact can be important when using cationic lipids where FCS-free medium is usually recommended. Anionic proteins present in FCS containing medium may interact with cationically charged complexes and reduce transfection efficacy. However, with respect to possible in vivo applications it is important to be able to target cells or tissue in a model that contains anionic fluids. In this setting, the presence of anionic compounds in vitro is beneficial for mimicking a strategy developed with the aim of therapeutic applications.

7. To ensure high amounts of macromolecules in endosomal vesicles we have incubated the cells with the nucleic acids like PNA–peptide conjugates, siRNA molecules and EGFP mRNA for 18 h before three washing steps. A transfection time of 18 h was chosen based on microscopy data, implying a higher PNA–peptide conjugate concentration present in the endosomal vesicles after 18 h compared to 6 h (Bøe, unpublished data). Similar results were observed after EGFP mRNA/PEI transfection, giving a higher amount of EGFP positive cells after 18 h of transfection compared to 6 h of transfection (**Fig. 7.5**, Bøe, unpublished data). Our results show that natural nucleic acids like siRNA and mRNA may be functional even after ≥22 h in the endosomal pathway (8, 10, 11). Together, these results indirectly demonstrate the capacity of the carrier to protect the nucleic acid from degradation in late endosomes and lysosomes.

8. To reduce damage to the plasma membrane, it is not recommended to expose the cells to light until 4 h after the removal of $TPPS_{2a}$. Therefore, we routinely incubate the cells with the siRNA/B-PEI complexes in $TPPS_{2a}$-free medium for 4 h (*see* **Section 3.5**, Step 5) to ensure that most of the $TPPS_{2a}$ bound to the plasma membrane is washed out or internalized into the cell before irradiation. This ensures that the light exposure does not induce extensive photochemical damage to the plasma membrane, which is lethal for cells, and that the main effect will be due to induced rupture of endocytic vesicles, which is where $TPPS_{2a}$ localizes. Shorter incubation than 16–18 h with the siRNA/PEI complexes and/or $TPPS_{2a}$ is possible, but the total incubation time in $TPPS_{2a}$-free medium before irradiation should not be much shorter than 4 h. It should be noted that the time needed to remove the bulk of photosensitizer from the plasma membrane varies between cell lines, but usually 4 h of

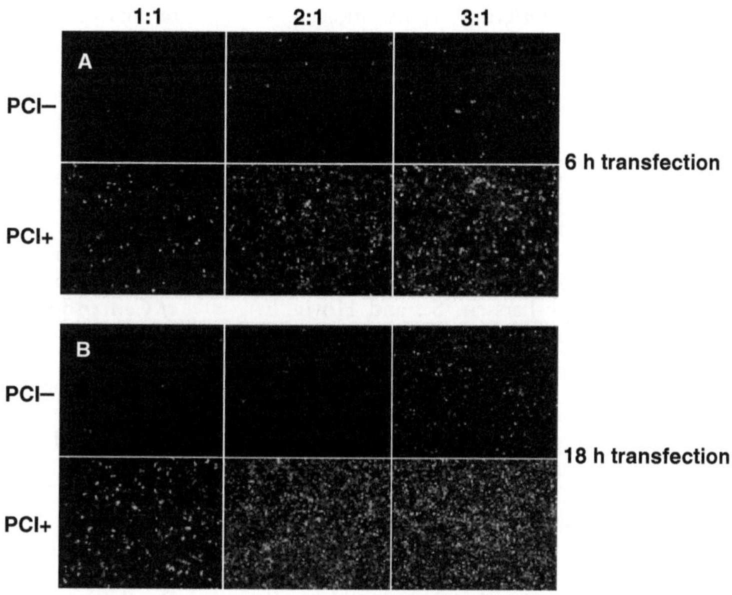

Fig. 7.5. The effect of 6 h (**a**) vs. 18 h (**b**) EGFP mRNA transfection time upon EGFP expression in the OHS cell line. Pictures were taken 24 h after light exposure without (*upper panel*) or with (*lower panel*) PCI. EGFP expression is shown at N/P ratios of 1:1, 2:1, and 3:1 (*from left to right*).

incubation in photosensitizer-free medium has been found to induce efficient photochemical internalization of nucleic acids.

9. It should be noted that photochemical treatment might temporarily inhibit transcription and translation (52, 53); thus it may delay modulation of gene expression. This should be kept in mind when deciding the time point after the treatment to measure the expression. Generally, we analyze change in gene expression 1–4 days after light treatment.

References

1. de Duve, C. (1963) Lysosomes. In: A. V. S. de Reuck and M. P. Cameron (eds) *Ciba Foundation Symposium*. Churchill, London, pp. 411–412.
2. Lloyd, J. B. (2000) Lysosome membrane permeability: implications for drug delivery. *Advanced Drug Delivery Reviews* **41**, 189–200.
3. Felgner, P. L., Gadek, T. R., Holm, M., et al. (1987) Lipofection: a highly efficient, lipid-mediated DNA-transfection procedure. *Proceeding of. National Academy of Science USA* **84**, 7413–7417.
4. Boussif, O., Lezoualc'h, F., Zanta, M. A., et al. (1995) A versatile vector for gene and oligonucleotide transfer into cells in culture and in vivo: polyethylenimine. *Proceeding of National Academy of Science USA* **92**, 7297–7301.
5. Read, M. L., Logan, A., and Seymour, L. W. (2005) Barriers to gene delivery using synthetic vectors. *Advances in Genetics* **53PA**, 19–46.
6. Sioud, M. (2007) RNA interference and innate immunity. *Advanced Drug Delivery Reviews* **59**, 153–163.
7. Pirollo, K. F., and Chang, E. H. (2008) Targeted delivery of small interfering RNA: approaching effective cancer therapies. *Cancer Research* **68**, 1247–1250.
8. Bøe, S., Longva, A. S., and Hovig, E. (2008) Evaluation of various polyethylenimine formulations for light-controlled gene silencing

using small interfering RNA molecules. *Oligonucleotides* **18**, 23–32.
9. Bøe, S., and Hovig, E. (2006) photochemically induced gene silencing using PNA-peptide conjugates. *Oligonucleotides* **16**, 145–157.
10. Bøe, S., Longva, A. S., and Hovig E. (2007) Photochemically induced gene silencing using small interfering RNA molecules in combination with lipid carriers. *Oligonucleotides* **17**, 166–173.
11. Bøe, S., Saeboe-Larssen, S., and Hovig E. (2010) Light-induced gene expression using messenger RNA molecules. *Oligonucleotides* **20**(1), 1–6.
12. Bøe, S., and Hovig E. (2008) Method for introducing siRNA into cells by photochemical internalisation. WO/2008/007073
13. Bøe, S., Hovig, E., and Fodstad, O. (2006) Method for introducing a PNA molecule conjugated to a positively charged peptide into the cytosol and/or the nucleus by photochemical internalisation (PCI). WO/2006/003463, PCT/GB2005/002679.
14. Folini, M., Berg, K., Millo E, et al. (2003) Photochemical internalization of a peptide nucleic acid targeting the catalytic subunit of human telomerase. *Cancer Research* **63**, 3490–3494.
15. Folini, M., Bandiera, R, Millo E, et al. (2007) Photochemically enhanced delivery of a cell-penetrating peptide nucleic acid conjugate targeting human telomerase reverse transcriptase: effects on telomere status and proliferative potential of human prostate cancer cells. *Cell Proliferation* **40**, 905–920.
16. Shiraishi, T., and Nielsen, P. E. (2006) Photochemically enhanced cellular delivery of cell penetrating peptide-PNA conjugates. *FEBS Letters* **580**, 1451–1456.
17. Oliveira, S., Fretz, M. M., Hogset, A., Storm, G., and Schiffelers, R. M. (2007) Photochemical internalization enhances silencing of epidermal growth factor receptor through improved endosomal escape of siRNA. *Biochimica et Biophysica Acta* **1768**, 1211–1217.
18. Oliveira, S., Hogset, A., Storm, G., and Schiffelers, R. M. (2008) Delivery of siRNA to the target cell cytoplasm: photochemical internalization facilitates endosomal escape and improves silencing efficiency, in vitro and in vivo. *Current Pharmaceutical Design* **14**, 3686–3697.
19. Berg, K., Selbo, P. K., Prasmickaite, L., et al. (1999) Photochemical internalization: a novel technology for delivery of macromolecules into cytosol. *Cancer Research* **59**, 1180–1183.
20. Berg, K., and Moan, J. (1994) Lysosomes as photochemical targets. *International Journal of Cancer* **59**, 814–822.
21. Moan, J., Berg, K., Anholt, H., and Madslien, K. (1994) Sulfonated aluminium phthalocyanines as sensitizers for photochemotherapy. Effects of small light doses on localization, dye fluorescence and photosensitivity in V79 cells. *International Journal of Cancer* **58**, 865–870.
22. Gibson, S. L., Murant, R. S., and Hilf, R. (1988) R. Photosensitizing effects of hematoporphyrin derivative and photofrin II on the plasma membrane enzymes 5′-nucleotidase, Na+K+-ATPase, and Mg2+-ATPase in R3230AC mammary adenocarcinomas. *Cancer Research* **48**, 3360–3366.
23. Moan, J, Johannessen, J. V., Christensen, T., Espevik, T., and McGhie, J. B. (1982) Porphyrin-sensitized photoinactivation of human cells in vitro. *The American Journal of Pathology* **109**, 184–192.
24. Boegheim, J. P., Dubbelman, T. M., Mullenders, L. H., and Van Steveninck, J. (1987) Photodynamic effects of haematoporphyrin derivative on DNA repair in murine L929 fibroblasts. *The Biochemical Journal* **244**, 711–715.
25. Gomer, C. J., Rucker, N., Banerjee, A., and Benedict, W. F. (1983) Comparison of mutagenicity and induction of sister chromatid exchange in Chinese hamster cells exposed to hematoporphyrin derivative photoradiation, ionizing radiation, or ultraviolet radiation. *Cancer Research* **43**, 2622–2627.
26. Berns, M. W., Dahlman, A., Johnson, F. M., et al. (1982) In vitro cellular effects of hematoporphyrin derivative. *Cancer Research* **42**, 2325–2329.
27. Hilf, R., Murant, R. S., Narayanan, U., and Gibson, S. L. (1986) Relationship of mitochondrial function and cellular adenosine triphosphate levels to hematoporphyrin derivative-induced photosensitization in R3230AC mammary tumors. *Cancer Research* **46**, 211–217.
28. van Dongen, G. A., Visser, G. W., and Vrouenraets, M. B. (2004) Photosensitizer-antibody conjugates for detection and therapy of cancer. *Advanced Drug Delivery Reviews* **56**, 31–52.
29. Berg, K., Western, A., Bommer, J. C., and Moan, J. (1990) Intracellular localization of sulfonated meso-tetraphenylporphines in a human carcinoma cell line. *Photochemistry and Photobiology* **52**, 481–487.

30. Moan, J., Berg, K., Kvam, E., et al. (1989) Intracellular localization of photosensitizers; discussion 11. *Ciba Foundation Symposium* **146**, 95–107.
31. Moan, J., Berg, K., Bommer, J. C., and Western, A. (1992) Action spectra of phthalocyanines with respect to photosensitization of cells. *Photochemistry and Photobiology* **56**, 171–175.
32. Prasmickaite, L., Hogset, A., and Berg, K. (2001) Evaluation of different photosensitizers for use in photochemical gene transfection. *Photochemistry and Photobiology* **73**, 388–395.
33. Maman, N., Dhami, S., Phillips, D., and Brault, D. (1999) Kinetic and equilibrium studies of incorporation of di-sulfonated aluminum phthalocyanine into unilamellar vesicles. *Biochimica et Biophysica Acta* **1420**, 168–178.
34. Moan, J., and Berg, K. (1991) The photodegradation of porphyrins in cells can be used to estimate the lifetime of singlet oxygen. *Photochemistry and Photobiology* **53**, 549–553.
35. Pass, H. I. (1993) Photodynamic therapy in oncology: mechanisms and clinical use. *Journal of the National Cancer Institute* **85**, 443–456.
36. Mang, T. S. (2004) Lasers and light sources for PDT: past, present and future. *Photodiagnosis and Photodynamic Therapy* **1**, 43–48.
37. Ndoye, A., Dolivet, G., Hogset, A., et al. (2006) Eradication of p53-mutated head and neck squamous cell carcinoma xenografts using nonviral p53 gene therapy and photochemical internalization. *Molecualr Therapy* **13**, 1156–1162.
38. Prasmickaite, L., Hogset, A., Tjelle, T. E., Olsen. V. M., and Berg, K. (2000) Role of endosomes in gene transfection mediated by photochemical internalisation (PCI). *The Journal of Gene Medicine* **2**, 477–488.
39. Engesaeter, B. O., Bonsted, A., Lillehammer, T., Engebraaten, O., Berg, K., and Maelandsmo, G. M. (2006) Photochemically mediated delivery of AdhCMV-TRAIL augments the TRAIL-induced apoptosis in colorectal cancer cell lines. *Cancer Biology & Therapy* **5**, 1511–1520.
40. Hogset, A., Ovstebo Engesaeter, B., Prasmickaite, L., Berg, K., Fodstad, O., and Maelandsmo, G. M. (2002) Light-induced adenovirus gene transfer, an efficient and specific gene delivery technology for cancer gene therapy. *Cancer Gene Therapy* **9**, 365–371.
41. Hogset, A., Prasmickaite, L., Tjelle, T. E., and Berg, K. (2000) Photochemical transfection: a new technology for light-induced, site-directed gene delivery. *Human Gene Therapy* **11**, 869–880.
42. Nishiyama, N., Iriyama, A., Jang, W. D., et al. (2005) Light-induced gene transfer from packaged DNA enveloped in a dendrimeric photosensitizer. *Nature Materials* **4**, 934–941.
43. Dietze, A., Peng, Q., Selbo, P. K., et al. (2005) Enhanced photodynamic destruction of a transplantable fibrosarcoma using photochemical internalisation of gelonin. *British Journal of Cancer* **92**, 2004–2009.
44. Selbo, P. K., Sivam, G., Fodstad, O., Sandvig, K., and Berg, K. (2001) In vivo documentation of photochemical internalization, a novel approach to site specific cancer therapy. *International Journal of Cancer* **92**, 761–766.
45. Berg, K., Dietze, A., Kaalhus, O., and Hogset, A. (2005) Site-specific drug delivery by photochemical internalization enhances the antitumor effect of bleomycin. *Clinical Cancer Research* **11**, 8476–8485.
46. Foote, C. (1987) Type I and Type II mechanisms of photodynamic action. In: J. R. Heitz and K. R. Downnum (eds) *Light-Activated Pesticides.* American Chemical Society, Washington, DC, pp. 22–38.
47. Spikes, J. D. (1984) Photobiology of porphyrins. *Progress in Clinical and Biological Research* **170**, 19–39.
48. Selbo, P. K., Hogset, A., Prasmickaite, L., and Berg, K. (2002) Photochemical internalisation: a novel drug delivery system. *Tumour Biology* **23**, 103–112.
49. Norum, O. J., Giercksky, K. E., and Berg, K. (2009) Photochemical internalization as an adjunct to marginal surgery in a human sarcoma model. *Photochemical and Photobiological Sciences* **8**, 758–762.
50. Brown, S. B., Brown, E. A., and Walker, I. (2004) The present and future role of photodynamic therapy in cancer treatment. *The Lancet Oncology* **5**, 497–508.
51. Berg, K., and Moan, J. (1998) Optimization of wavelenghts in photodynamic therapy. In: J. G. Moser (ed) *Photodynamic Tumor Therapy, 2nd and 3rd Generation Photosensitizers.* Harwood Academic Publishers, London, pp. 151–68.
52. Moan, J., McGhie, J., and Jacobsen, P. B. (1983) Photodynamic effects on cells in vitro exposed to hematoporphyrin derivative and light. *Photochemistry and Photobiology* **37**, 599–604.
53. Davies, C. L., Ranheim, T., Malik, Z., Rofstad, E. K., Moan, J., and Lindmo, T. (1988) Relationship between changes in antigen expression and protein synthesis in human melanoma cells after hyperthermia and photodynamic treatment. *British Journal of Cancer* **58**, 306–313.

Chapter 8

Antibody Targeted siRNA Delivery

Masoud M. Toloue and Lance P. Ford

Abstract

It is very clear that RNA interference (RNAi) is a potent and versatile tool for gene silencing. One of the hurdles to making siRNA/miRNA a human therapeutic includes effective in vivo delivery and being able to deliver drugs to target cells only. The commercial success of in vivo applications of RNAi hinges on the development of new delivery methods. Our strategy involves the use of antibody-based delivery agents to target and deliver siRNA into specific cell types. We have developed antibody-based agents for directed delivery into cultured cells and animal disease models. Using antibodies against various cell surface receptors, modified siRNAs are attached to antibody complexes using RNA carrier proteins. The complex can then be intravenously administered to in vivo models and taken up by specific cells via receptor-mediated endocytosis. The labile structure of the linking agents enables release of siRNA molecules post internalization. Using this targeting strategy, we have developed a method that allows any commercially available or recombinant antibody to be conjugated to siRNA for delivery purposes.

Key words: siRNA, miRNA delivery, antibody directed delivery, conjugation kit, therapeutic monoclonal antibodies, receptor-mediated endocytosis, in vivo models, therapeutic small RNA.

1. Introduction

Double stranded RNAs (siRNAs) consist of 19–23 base pairs (bp) and induce cellular post-transcriptional gene silencing (1–3). Unlike other gene targeting technologies, RNAi mediators (molecules capable of inducing RNA interference, including short interfering (siRNA) and microRNA (miRNA)) can target the knockdown of almost every gene in a cell, with high specificity and potency. There has been increasing interest in the use of synthetic siRNAs for in vivo gene expression studies and as a new class of human therapeutic. Despite the therapeutic potential of siRNA, its application in a clinical setting and in most in vivo models is limited due to the lack of an efficient delivery

J. Goodchild (ed.), *Therapeutic Oligonucleotides*, Methods in Molecular Biology 764,
DOI 10.1007/978-1-61779-188-8_8, © Springer Science+Business Media, LLC 2011

system (4). Several efforts including high-pressure intravenous injection, injection of naked siRNAs and siRNA–liposomal formulations have not met with great success. Efforts to deliver siRNAs with high-pressure tail-vein injection result in target specific knockdown of reporter genes in the liver but require rapid injection of solutions two and a half times the blood volume of the animal (5). Hydrodynamic intravenous injections can cause right-sided heart failure and are not practical for experimentation or systemic human use (6). Improvements to in vivo delivery have been made using lipid conjugations introduced intravenously by low pressure injection. These tend to be effective in the knockdown of genes in the liver (5, 7, 8). Typically, the majority of liposomal formulations are sequestered into the cells of the reticuloendothelial system and end up in the liver within hours of delivery (9). In addition, systemic distribution of siRNA often requires very high doses which lends to its distribution in unintended locations, often inducing side effects including immune receptor activation, interferon synthesis and potential sequestering of miRNA from RISC (10, 11). In order for siRNA to become a better in vivo tool/therapeutic it will require significantly reduced dosage, development of less-harsh delivery methods and targeted rather than distributed delivery. One approach which meets all of these requirements is to target siRNA to specific (disease-related) cell types using affinity reagents such as antibodies. Described in this chapter, siRNAs are electrostatically associated with a carrier that is chemically conjugated to an antibody. The carrier-conjugated antibody can be chosen or designed to bind to antigens expressed on the surface of targeted cells. After injection, the siRNA–antibody conjugate specifically locates and binds to target cells preventing the delivery of siRNA to non-targeted cell types (e.g., normal cells). Once bound, the complex becomes internalized (with the antigen) via clathrin-mediated endocytosis. siRNA then dissociates into the cytoplasm and engages RISC machinery. This not only satisfies the needed delivery improvements described above but also addresses all the primary causes of poor siRNA bioavailability: (1) it improves cell penetration, (2) effectively sequesters siRNA from the plasma matrix and prevents nuclease degradation, and (3) increases the size of the siRNA complex, preventing kidney excretion or removal by the reticuloendothelial system.

Song et al. (12) and Peer et al. (13) have already demonstrated that the siRNA–antibody approach is efficacious in gene knockdown in mice. In its first documentation (12), the gene encoding a nucleic acid binding protein was fused to a gene encoding a Fab fragment of an antibody directed against the HIV gp160 (env) protein. The Fab–single protamine fusion protein was then loaded with siRNA and delivered into cells expressing the env gene. In a following report (13), a recombinant antibody–protamine fusion was used to specifically deliver siRNA into cells expressing the integrin lymphocyte function associated antigen-1 (LFA-1) protein. While exciting progress toward in vivo delivery has been achieved using the recombinant immunoconjugate approach, there are several barriers that limit its development. For one, the size of the RNA-binding portion of the protein is small, limiting the molar amount of siRNA that can be bound. Song et al. (12) used an immunoconjugate containing a single protamine fragment per antibody fragment. The protamine domain (3 kDa) is very small compared to the antibody component of the protein, consequently allowing only less than 6 molecules of siRNA to bind per molecule of antibody. Recombinant modified antibody fusions are notoriously difficult to express in a soluble active form, often requiring inefficient denaturation and refolding procedures that result in low yield of quality antibody. Finally, the recombinant delivery approach lacks the ability to effectively release siRNA from the protein domain of the antibody conjugate as the two are genetically fused. This significantly limits the bioavailability of siRNA to the host cell RISC machinery. In this chapter we describe a versatile method for antibody–siRNA conjugation that is not limited by the bounds of genetic fusion. This technique involves the chemical conjugation of several siRNA binding domains to any commercially available antibody by intracellular labile immobilizing attachments. This enables the delivery vehicle to hold and efficiently deliver an efficacious siRNA dose to target cells only.

2. Materials

2.1. Antibody Endocytosis Assay

1. Mammalian cultured cells with surface receptor or antigen recognized by antibody paratope.
2. Fluorescently conjugated monoclonal antibody or antibody fragment.
3. Flow cytometer with matching excitation laser and filters.
4. 1 × HBSS: 0.137 M NaCl, 5.4 mM KCl, 0.25 mM Na_2HPO_4, 0.44 mM KH_2PO_4, 1.3 mM $CaCl_2$, 1.0 mM $MgSO_4$, 4.2 mM $NaHCO_3$
5. Chymotrypsin: 50 μg/mL in HBSS
6. Proteinase K: 50 μg/mL in HBSS
7. Paraformaldehyde: 2%

2.2. Antibody–siRNA Conjugation

1. Monoclonal antibody or antibody fragment ranging in size from 20 to 150 kDa from any commercial source or recombinant design.

2. siRNA, miRNA, or antisense oligonucleotide with or without base or bond modifications for serum stability.
3. Dimethylformamide (DMF) used to solubilize conjugation linkers.
4. T3 carrier (Bioo Scientific, Austin, TX), the carrier of the delivery vehicle that binds to the negatively charged siRNA or miRNA backbone.
5. Immobilizing attachment A and 1 × modification buffer (Bioo Scientific, Austin, TX) to bind to a monoclonal antibody. Immobilizing attachment A should be stored at 4°C in desiccant until reconstituted and prepared fresh for each experiment.
6. Immobilizing attachment B and 10 × conjugation buffer (Bioo Scientific, Austin, TX) for binding to the T3 carrier. Immobilizing attachment B should be stored at 4°C in desiccant until reconstituted and prepared fresh for each experiment.
7. Bovine IgG control: 1 mg/mL. Used as a positive control to compare binding efficiency.
8. PD-10 desalting columns (2.5 and 1 mL) (GE Healthcare, Piscataway, NJ) are used to stop the conjugation reaction and remove unbound immobilizing attachments.
9. Centriprep® column (Millipore, Bedford, MA) used to concentrate antibody–siRNA complex after conjugation.
10. RNase-free pipette tips.
11. RNase-free water.

2.3. SDS-Polyacrylamide Gel Electrophoresis (SDS-PAGE)

1. 4 × Separating buffer: 18.17 g Tris base, 4 mL 10% SDS. Dissolve in water, adjust pH to 8.8, and dilute to 100 mL with distilled water.
2. 4 × Stacking buffer: 6.06 g Tris base, 4 mL 10% SDS. Dissolve in water, adjust pH to 6.8, and dilute to 100 mL with distilled water.
3. 1 × Running buffer: 3.03 g Tris base, 14.4 g glycine, 1.0 g SDS. Dissolve in water, adjust pH to 8.3–8.4 with sodium hydroxide, and dilute to 1 L with distilled water.
4. SDS Sample buffer: 50 mM Tris-Cl (pH 6.8), 10% glycerol, 2% SDS, 0.1% bromophenol blue, 100 mM dithiothreitol (DTT). Mix all ingredients together except for DTT. SDS sample buffer lacking DTT should be stored at room temperature. The DTT should be added just before the buffer is used.
5. 30% 29:1 acrylamide/bis-acrylamide (Sigma Aldrich).

6. Ammonium persulfate (APS). Prepare a 10% solution in water and aliquot for single use.
7. *N, N, N, N′*-Tetramethylethylenediamine (TEMED).
8. Water-saturated butanol.
9. Molecular weight markers: Kaleidoscope (Bio-Rad, Hercules, CA).
10. Bradford reagent kit (Bioo Scientific, Austin, TX).
11. Coomassie brilliant blue R-250 staining solution: 500 mL water, 400 mL methanol, 100 mL acetic acid, 2 g Coomassie brilliant blue.
12. Coomassie destain solution (50% methanol in water).

2.4. Conjugated siRNA–Antibody Delivery in Cell Culture

1. 96-well tissue culture plates.
2. Centricon® centrifugal filter (Millipore, Bedford, MA). Size cut-off of filter will depend on your antibody size.
3. Dulbecco's modified Eagle's medium (DMEM) supplemented with 10% FBS fetal bovine serum.
4. Trypsin: 0.25% in 1 mM ethylenediaminetetraacetic acid (EDTA).
5. 10 × PBS (RNase-free, sterile): 1.37 M NaCl, 27 mM KCl, 100 mM Na_2HPO_4, and 18 mM KH_2PO_4 in RNase-free water. Adjust the pH to 7.4 with HCl and autoclave to sterilize.

2.5. In Vivo siRNA Delivery

1. Insulin syringe, 29 gauge, for conjugate injections.
2. Phenobarbitol or ketamine as mouse anesthetic.
3. Protein extraction buffer and Bradford reagent kit (Bioo Scientific, Austin, TX).
4. PBS (RNase-free, sterile): Prepare a 10 × stock with 1.37 M NaCl, 27 mM KCl, 100 mM Na_2HPO_4, and 18 mM KH_2PO_4 using RNase-free water. Adjust the pH to 7.4 with HCl and autoclave to sterilize.
5. BiooPure™ RNA isolation reagent (Bioo Scientific, Austin, TX).
6. Mice (Jackson Labs, Bar Harbor, ME).

3. Methods

Described in detail is a novel method in which siRNA is electrostatically associated with a carrier molecule chemically conjugated to an antibody. This new siRNA–antibody targeted delivery technique allows the user to complex several siRNA binding proteins (carrier) to the antibody enabling a large siRNA carrying capacity. This technique is not limited by the manipulation or production of difficult-to-express recombinant antibodies, but allows one to use commercially available or conventionally produced custom monoclonal antibodies for siRNA conjugation. This allows the user to match any antibody with siRNA, making it a versatile gene regulation tool, and when associated with humanized antibodies, a potential therapeutic drug vehicle. By using intracellular labile linking agents, the delivery method described in this chapter allows siRNA to be released from its delivery moiety once inside the cell.

3.1. Antibody Endocytosis Assay

1. Count and aliquot two groups of 10^6 mammalian cells.
2. Wash both groups of cells separately with HBSS and incubate one group of cells with a fluorescently conjugated antibody for 45 min at 4°C and the other group at 37°C.
3. Split cells from step two and treat second set (those incubated at 4°C and 37°C) for 45 min with 50 μg/mL chymotrypsin and 50 μg/mL proteinase K in HBSS to remove surface-bound fusion proteins.
4. Resuspend cells in 2% paraformaldehyde or analyze immediately by flow cytometry.
5. Use flow cytometer settings to measure relative fluorescence vs. cell number to test the internalization ability of the fluorescently labeled antibody.
6. Percentage of internalized antibody for cells incubated at 37°C should be calculated using the following equation: (mean fluorescence intensity treated with proteases – fluorescent background) × 100/(mean fluorescence intensity not treated with proteases – fluorescent background).
7. The percent of internalized antibody will vary according to antibody type, cell line, and receptor. Choose an antibody that readily internalizes for delivery preparation in **Section 3.2**.

3.2. Preparation of Samples for Antibody–siRNA Carrier Conjugation

1. The procedure described here allows for three 1 mg antibody conjugation reactions to be performed, or one 3 mg antibody conjugation, or one 2 mg antibody conjugation and one 1 mg antibody conjugation (*see* **Note 1**).
2. Make a 1 × conjugation buffer solution by adding 30 mL of 10 × conjugation buffer to 270 mL sterile, RNase-free water. Mix well and place in a sterile bottle.
3. Allow immobilizing attachment A and immobilizing attachment B to reach room temperature to avoid condensation. Microfuge to ensure material does not remain in cap.

4. Reconstitute the immobilizing attachments in DMF by adding 50 μL to immobilizing attachment A. Vortex for 15 s; spin down in microfuge for 5 s. Add 25 μL of DMF to immobilizing bttachment B. Vortex for 15 s; spin down in microfuge for 5 s.
5. Reconstitute the (user supplied) lyophilized antibody you wish to conjugate in 2.5 mL of 1 × modification buffer. Use a minimum of 1 mg of antibody. Amounts less than 1 mg or greater than 3 mg may not result in a successful conjugation (*see* **Note 2**).
6. Refer to **Tables 8.1**, **8.2**, and **8.3** below and match the approximate molecular weight of your antibody in the left hand column of the table to determine the corresponding amount of T3 carrier needed. Pipette appropriate amount of T3 carrier into a conical tube. Bring the volume to 2.5 mL using 1 × modification buffer. The bovine IgG control antibody and T3 carrier are ready-to-use from the manufacturer and do not need further dilution.
7. The amount of immobilizing attachments used is determined by the quantity and molecular weight of the antibody and T3 carrier protein. **Table 8.1** provides pre-calculated amounts of immobilizing attachments A and B. Depending on whether you start with 3, 2, or 1 mg of antibody, use **Table 8.1** to match the approximate molecular weight of your antibody in the left hand column to determine the corresponding amount of immobilizing attachments needed. If the molecular weight of your antibody falls between the values listed in the table, the amount of immobilizing attachment A to use should be between coinciding values listed.

Table 8.1
Conjugation of carrier protein with 3 mg of antibody

Antibody MW (kDa)	Amount of immobilizing attachment A (after reconstituted in DMF) for antibody (μL)	Amount of immobilizing attachment B (after reconstitution in DMF) for carrier protein (μL)	Amount of carrier protein (mg)
150	3.7	17.4	4.8
100	5.6	17.4	4.8
75	7.5	17.4	4.8
50	11.1	17.4	4.8
30	18.5	17.4	4.8
20	27.8	17.4	4.8

Table 8.2
Conjugation of carrier protein with 2 mg of antibody

Antibody MW (kDa)	Amount of immobilizing attachment A (after reconstituted in DMF) for antibody (μL)	Amount of immobilizing attachment B (after reconstitution in DMF) for carrier protein (μL)	Amount of carrier protein (mg)
150	2.5	11.6	3.2
100	3.7	11.6	3.2
75	5.0	11.6	3.2
50	7.4	11.6	3.2
30	12.4	11.6	3.2
20	18.5	11.6	3.2

Table 8.3
Conjugation of antibody with 1 mg of antibody

Antibody MW (kDa)	Amount of immobilizing attachment A (after reconstituted in DMF) for antibody (μL)	Amount of immobilizing attachment B (after reconstitution in DMF) for carrier protein (μL)	Amount of carrier protein (mg)
150	1.2	5.8	1.6
100	1.9	5.8	1.6
75	2.5	5.8	1.6
50	3.7	5.8	1.6
30	6.2	5.8	1.6
20	9.3	5.8	1.6

8. The procedure described from this point on is for the conjugation of 3.0 mg of a 75-kDa monoclonal antibody. If using less than 3 mg of antibody or if your antibody is a different molecular weight, refer to **Table 8.1** to determine the optimal amount of immobilizing attachment A, T3 carrier, and immobilizing attachment B.
9. Add 7.5 μL of immobilizing attachment A to the antibody.
10. Incubate for 4 h at room temperature (20–25°C).
11. Desalt antibody as described in Steps 17–19.
12. To perform a control antibody conjugation using 1 mg bovine IgG antibody control (MW 150,000), add 1.2 μL of immobilizing attachment A to the bovine IgG. Incubate for 4 h at room temperature. Desalt control antibody as described in Step 20.

13. Very slowly add and mix 17.4 μL of reconstituted immobilizing attachment B to 4.8 mg of T3 carrier (*see* **Note 3**).
14. Incubate for 2 h at room temperature (20–25°C).
15. Desalt carrier protein as described in Steps 17–19. After desalting, set aside T3 carrier at room temperature (20–25°C) until the antibody conjugation has completed.
16. A control conjugation using T3 carrier should be performed alongside your monoclonal antibody to ensure conjugation occurs properly. For the control T3 carrier conjugation, add 5.8 μL of immobilizing attachment B to the carrier protein and incubate for 2 h at room temperature. Desalt as described in Step 20. After desalting, set aside at room temperature until the bovine IgG control antibody is ready.
17. After incubation, separately desalt the antibody and T3 carrier (*see* **Note 4**). Be sure to use a separate 2.5-mL desalting column for each sample (antibody and T3 carrier). A ring stand or other device may be used to hold the desalting column vertically.
18. Equilibrate each desalting column (2.5 mL size) with 1 × conjugation buffer. Fill the column with 1 × conjugation buffer and allow the buffer to enter the packed bed completely. Discard the flow through. Repeat as needed for a total of 25 mL of 1 × conjugation buffer.
19. Add 2.5 mL of the sample (either antibody or T3 carrier) slowly into the column bed. Discard the sample flow through. Add 3.5 mL of 1 × conjugation buffer to elute the sample. Collect the sample in a 15-mL conical tube.
20. The T3 carrier control should also be desalted with 1 × conjugation buffer. Equilibrate a new control PD-10 desalting column (1 mL size) with 1 × conjugation buffer. Fill the column with 1 × conjugation buffer and allow the buffer to enter the packed bed completely. Discard the flow through. Repeat as needed for a total of 15 mL of 1 × conjugation buffer. Add 1.0 mL of the T3 carrier control into the column bed. Discard the sample flow through. Add 1.5 mL of 1 × conjugation buffer to elute the sample. Collect 1.5 mL of the sample in a 15-mL conical tube. The bovine IgG control antibody must also be desalted. Equilibrate the control-desalting column (1 mL size) with 1 × conjugation buffer. Fill the column with 1 × conjugation buffer and allow the buffer to enter the packed bed completely. Discard the flow through. Repeat as needed for a total of 15 mL of 1 × conjugation buffer. Add 1.0 mL of bovine IgG antibody slowly into the column bed. Discard the sample flow through. Add 1.5 mL of 1 × conjugation buffer to elute the sample.

Collect 1.5 mL of the sample in a 15-mL conical tube (*see* **Note 5**).

21. Mix all column eluted, desalted carrier protein and desalted antibody together. Repeat with the controls. Incubate overnight at 4°C for 14–16 h (*see* **Note 6**).
22. After incubation, use the Centriprep® spin column to stop the reaction and to remove any excess carrier protein from your antibody conjugate sample. Remove the plastic top of Centriprep® spin column and the inner filtrate column and set aside. Be careful not to touch or tear the membrane on the filtrate column. Add your conjugated sample directly to the inside of the sample container and spin your sample at 1500 × *g* for 10 min. The conjugate will remain in the sample container while the carrier protein will move to the filtrate collector. After 1 spin, remove the seal cap and decant the solution in the filtrate collector.
23. Add 3 mL of 1 × conjugation buffer to remaining material in sample container and spin again.
24. Continue to remove the carrier protein and add conjugation buffer, spinning a total of 4–6 times (*see* **Note 7**). After spinning is complete, the remaining material in the sample container is the final antibody–carrier conjugation product. The antibody conjugate is now ready to be added to your RNAi molecule of choice (siRNA, miRNA). See **Sections 3.3** and **3.4** for in vitro and in vivo conjugated antibody delivery.
25. To determine conjugation efficiency, and compare the conjugation of your sample antibody to the control bovine IgG, run your conjugated sample, along with your unconjugated control samples, on an SDS-PAGE gel. Note that the conjugated samples run at a higher molecular weight compared to the unconjugated sample antibody (**Fig. 8.1**) (*see* **Note 8**).

3.3. SDS-PAGE

1. Visualizing conjugation efficiency by SDS-PAGE is recommended.
2. Prepare a 1.5-mm-thick, 12.5% separating gel by mixing 3.13 mL of 30% 29:1 acrylamide/bis-acrylamide, 1.9 mL of 4 × separating buffer, pH 8.8, 112 μL of 10% APS, 5 μL of TEMED, and 2.37 mL of distilled water. Pour the gel leaving a space of around 2.5 cm for a stacking gel. Overlay with water saturated butanol to create an even surface. The gel should polymerize in about 30 min.
3. Pour off butanol and rinse the top of the gel twice with water.
4. Prepare the stacking gel by mixing 450 μL of 30% 29:1 acrylamide/bis-acrylamide, 665 μL of 4 × stacking buffer,

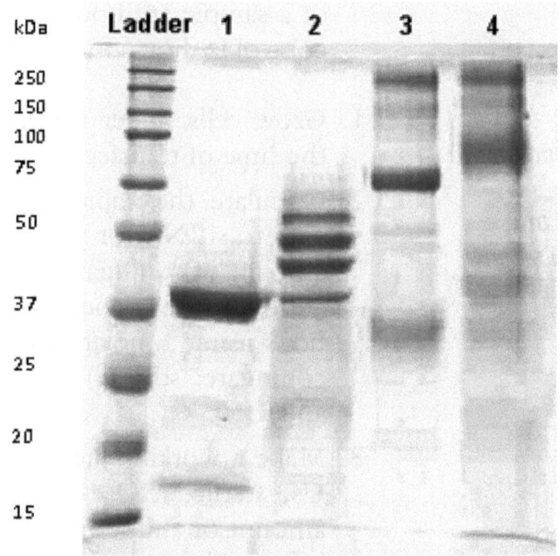

Fig. 8.1. An example of a scFv CD7 antibody (lane 1) and T3 conjugate (lane 2) along with the bovine IgG control (lane 3) and its respective conjugate (lane 4). The higher molecular weight shift in lanes 2 and 4 indicate that the conjugation has completed efficiently.

pH 6.8, 42 μL of 10% APS, 5 μL of TEMED, and 1.5 mL of distilled water. Pour the stacking buffer immediately into glass plates and insert comb before polymerization.

5. Allow the stacking gel to polymerize for at least 10–15 min before using.
6. Use a Bradford reagent kit to quantify the antibody conjugate concentration as well as the controls. Load approximately 20–40 μg of sample per lane. Adding less than 20 μg of antibody conjugate may not allow for proper visualization of the bands. Adding more than 40 μg of antibody conjugate may not allow for proper separation of the conjugated and un-conjugated bands.
7. Mix each sample with SDS sample buffer. Heat the samples at 95°C for at least 10 min.
8. Add ladder, individual samples, and antibody-conjugate to separate lanes in the gel.
9. Run the gel between 100 and 150 V.
10. Do not allow the dye front to run off the gel. Turn off when the dye reaches the bottom of the plate.
11. Stain with Coomassie brilliant blue R-250 staining solution and allow gel to rock in stain for at least 2 h. Pour off and add 5–10 volumes of Coomassie destain solution until the gel background is clear and only the molecular weight markers and samples are visible. An example of an SDS-PAGE run

of a sample antibody conjugate along with the bovine IgG control is shown in **Fig. 8.1**.

3.4. In Vitro siRNA–Antibody Delivery via Transfection of Adherent Cells (96-Well Plates)

1. Grow cells under optimal physiological conditions before the time of transfection.
2. Calculate the appropriate molar ratio of antibody conjugate to siRNA for your particular experiment. The optimal amount of conjugate and siRNA will vary for each cell line, antibody used, and experimental condition. Making titrations using a molar ratio between 1:5 and 1:100 (antibody conjugate: siRNA) is a good starting point (*see* **Notes 9** and **10**).
3. Make a working stock solution of the antibody conjugate. Use sterile, RNase-free 1 × PBS as the diluent. Titrate the amount of antibody conjugate needed for every new experiment. The recommended amount of antibody that should be added to each well (in a 96-well plate) may range between 100 and 500 ng.
4. Make an siRNA working stock solution, separate from the antibody conjugate, and use sterile, nuclease free 1 × PBS as the diluent.
5. Add the antibody conjugate to the 96-well plate (*see* **Note 11**).
6. Add siRNA to the respective wells, mix gently, and wait for the antibody conjugate and siRNA to complex for at least 30 min. Alternatively, to ensure no unbound siRNA is added to cells mix siRNA and antibody separately and use a Centricon® centrifugal filter device to remove unbound siRNA before adding to well.
7. Immediately make a stock of suspended cells. Cells should be in a concentration where each well (96-well plate) will receive approximately 6000–12,000 cells.
8. After the siRNA and antibody conjugate have finished complexing, immediately add the cells to the plate.
9. If wells contain a total volume less than 200 μL, bring the volume up with complete media to a total of 200 μL.
10. Cover the plate and place in a 37°C incubator with (5% CO_2) for 16–24 h.
11. After overnight incubation, aspirate the media with gentle vacuum suction.
12. Add 200 μL of DMEM to each well and incubate for an additional 48 h.
13. Aspirate the medium and wash the wells three times with 1 × PBS.

14. Process the cells and analyze using the desired method to screen for knockdown (i.e., qPCR, Western, or ELISA).

3.5. In Vivo siRNA–Antibody Delivery into Mouse

1. For optimal results, dissolve the RNAi agent (siRNA, miRNA, antisense oligonucleotide) in RNase-free, sterile 1 × PBS at a concentration between 5 and 10 mg/mL. The optimal amount of conjugate and siRNA will vary for each experiment. Titrate the amount of siRNA added to antibody to ensure that the right amount is used. Start titrations with a molar ratio between 1:1 and 1:50 (antibody:RNAi). Each oligonucleotide and antibody will require different conditions.
2. Incubate the complex at room temperature for 15 min.
3. Use a syringe to inject the animal in the desired location (**Figs. 8.2** and **8.3**).

3.5.1. Intravenous Injection

1. Warm the mouse tail to about 37°C (approximately 10 min) using a heat lamp.
2. The mouse should be restrained using a mouse-restraining device or hand.
3. Disinfect the tail using an alcohol swab and slightly rotate the tail to visualize the vein.
4. Once the vein has been located, disinfect the site of injection with ethanol and insert a 29-guage insulin needle at a slight angle. Inject 100 μL slowly (~20 μL/s) and watch for clearing of the blood in the vein. If a bulge appears in the tail, remove the needle and repeat the process proximal to the previous site.

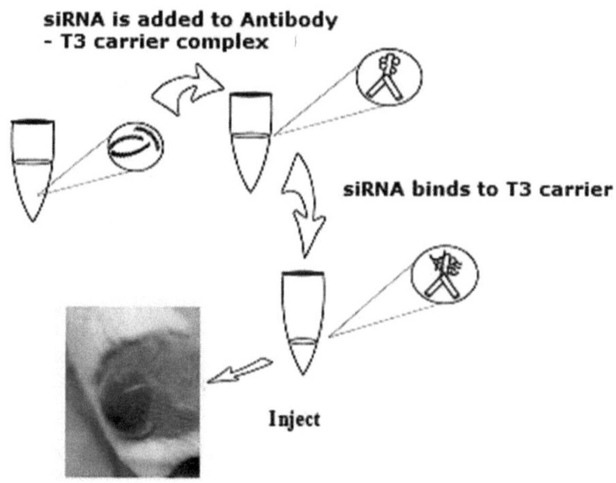

Fig. 8.2. Schematic of the treatment protocol for human xenograft tumors. siRNA is first mixed with the antibody delivery vehicle under sterile and RNase-free conditions. This is then injected directly into the tumor as shown or injected intravenously (**Fig. 8.8.3**).

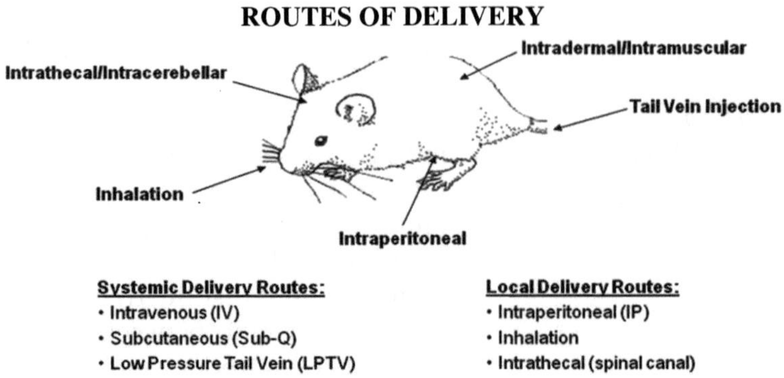

Fig. 8.3. In vivo injection of siRNA–antibody conjugated agents. The siRNA–antibody delivery vehicle can be administered using several methods. In addition to those described in the figure, the complex can be directly injected into a subcutaneous xenograft tumor or into a specific organ using surgical procedures.

5. Upon completion remove needle and apply pressure to the injection site.

3.5.2. Intraperitoneal Injection

1. Restrain the mouse using the fold of skin near the neck and expose the abdomen.
2. Disinfect the injection site by swabbing the area with an alcohol swab.
3. Insert a 29-guage needle, at an angle, into the peritoneal cavity of the abdomen, avoiding puncture of internal organs, and inject 50–100 μL slowly (20 μL/s).

3.5.3. Intranasal Administration

1. Mice should be anesthetized by injecting 0.2–0.3 mL of anesthetic, such as ketamine or phenobarbital, in the lower flank of the mouse.
2. Place the mouse on its back over a warm pad or under a heat lamp for about 10 min.
3. Inject 20–25 μL of the RNA agent formulation in each nostril using a narrow gel loading pipette tip. Inject slowly and wait about 1 min between injections to help recovery.

3.5.4. Intratumoral Injection

1. When the subcutaneous tumor areas reach at least 250–300 mm^3 (approx 10–12 days post injection) in size, the mice are ready for injection. Larger tumors may also be targeted. The volume of injection should be 50–200 μL/tumor, depending on the tumor size (**Fig. 8.2**).
2. The mouse should be restrained using a mouse-restraining device or by hand.
3. Use forceps to gently hold the tumor.

4. Disinfect the site of injection using an ethanol wipe and insert the needle directly into the tumor. Penetrate as deep as possible without passing through the tumor.
5. Slowly push approximately 20–50 μL into the tumor. Then gently retract the needle and shift to a different location within the tumor and inject an additional 20–50 μL of preparation. Repeat this process until all material has been injected.
6. After injection, leave the needle in the tumor for about 20 s, and then slowly pull it out and pinch the opening with fine forceps to avoid leakage.

3.5.5. Animal Harvest and Sample Preparation

After injection (one or more) of the RNAi agent, the recommended incubation time in the animal is 3–4 days. Sacrifice the animal using an approved method (vertebrae dislocation) and harvest desired organs or tumor. Samples may be stored at 2–4°C for no longer than 1–2 days. If samples need to be stored for longer than 2 days, freeze them at −20°C. Frozen samples may be thawed at room temperature (20–25°C/68–77°F) or in a refrigerator before use.

3.5.6. Serum

1. Blood should be collected without anticoagulant and left at room temperature for 3 h or at 4°C overnight to clot.
2. Centrifuge at 4°C for 10 min at 3000 × *g*.
3. Take the serum from the upper layer of the blood sample.
4. Use the serum directly in the desired assay.

3.5.7. Tissue-Protein Extraction

1. Add 500 μL of protein extraction buffer to 500 mg or less of tissue samples.
2. Grind the sample using a micro-homogenizer or a pestle.
3. Freeze the sample for 30 min in a −80°C freezer or in a dry ice/ethanol bath.
4. Thaw sample at room temperature (20–25°C) and spin in a microcentrifuge at 14,000 rpm for 10 min.
5. Remove supernatant and dispose of the pellet.
6. Dilute the sample in Bradford diluent. Recommended starting dilutions are between 1:50 and 1:100.
7. Quantify the total protein using a Bradford assay kit.
8. The total concentration of protein in the sample should be normalized using sample diluent to between 1.0 and 3.0 mg/mL.
9. Use the normalized sample in the desired assay.

4. Notes

1. The written procedure allows for a total of 3 mg of antibody to be conjugated. Attempts to conjugate different quantities of antibody may result in protein precipitation or inefficient conjugation.
2. Antibodies must be purified and not contain any auxiliary proteins. If antibodies are in a solution other than 1 × modification buffer, a buffer exchange or dialysis will be necessary. Antibodies in solution other than 1 × modification buffer may result in reduced conjugation efficiency.
3. Adding reconstituted immobilizing attachment B quickly to T3 carrier may cause precipitation. It is important that the immobilizing attachment B is added slowly while mixing.
4. The purpose of this step is to stop the reaction and remove any remaining un-conjugated immobilizing attachment A or immobilizing attachment B, which may interfere with siRNA binding.
5. To determine conjugation efficiency, set aside 50 μL of desalted T3 carrier, sample antibody and bovine IgG control antibody. These un-conjugated samples can be used as a reference when examining efficiency on an SDS-PAGE gel. Running your antibody on the gel is also recommended to ensure proteolysis has not occurred.
6. It is important that you do not exceed 16 h of incubation as over-conjugation may occur, resulting in precipitation and loss of antibody.
7. The purpose of this step is to stop the reaction and concentrate the antibody–carrier conjugate. Spinning the sample for greater than 10 min (15–20 min) will yield a more concentrated antibody–carrier conjugate.
8. Bands on a SDS-PAGE gel with higher molecular weight than the antibody indicate proper conjugation efficiency. Multiple bands indicate that several carrier groups are conjugated to the antibody. For every experiment, reaction conditions may be titrated to increase or decrease the number of bound carrier units. While conjugating many carrier units to the antibody domain increases the ability to bind more siRNA, this runs the risk of masking the paratope.
9. Not all of the siRNA added will bind to the T3 carrier. Use a spectrophotometer or fluorescently labeled siRNA to quantity the actual amount bound.
10. It is essential that your sample preparation is tested for RNase activity before in vitro or in vivo experimentation.

If your antibody is contaminated with nuclease, siRNA integrity may be compromised. Alternatively, the use of serum stable or nucleotide base modified siRNA may be used.

11. Use controls such as antibody conjugate alone or siRNA alone with cells to properly test gene knockdown.

References

1. Tuschl, T., Zamore, P. D., Lehmann, R., Bartel, D. P., and Sharp, P. A. (1999) Targeted mRNA degradation by double-stranded RNA in vitro. *Genes Dev* **13**, 3191–3197.
2. Elbashir, S. M., Harborth, J., Lendeckel, W., Yalcin, A., Weber, K., et al. (2001) Duplexes of 21-nucleotide RNAs mediate RNA interference in cultured mammalian cells. *Nature* **411**, 494–498.
3. Fire, A., Xu, S., Montgomery, M. K., Kostas, S. A., Driver, S. E., et al. (1998) Potent and specific genetic interference by double-stranded RNA in Caenorhabditis elegans. *Nature* **391**, 806–811.
4. Nguyen, T., Menocal, E. M., Harborth, J., and Fruehauf, J. H. (2008) RNAi therapeutics: an update on delivery. *Curr Opin Mol Ther* **10**, 158–167.
5. Rossi, J. J. (2005) Receptor-targeted siRNAs. *Nat Biotechnol* **23**, 682–684.
6. Dykxhoorn, D. M., and Lieberman, J. (2006) Silencing viral infection. *PLoS Med* **3**, e242.
7. Soutschek, J., Akinc, A., Bramlage, B., Charisse, K., Constien, R., et al. (2004) Therapeutic silencing of an endogenous gene by systemic administration of modified siRNAs. *Nature* **432**, 173–178.
8. McCaffrey, A. P., Nakai, H., Pandey, K., Huang, Z., Salazar, F. H., et al. (2003) Inhibition of hepatitis B virus in mice by RNA interference. *Nat Biotechnol* **21**, 639–644.
9. Lasic, D. D., Vallner, J. J., and Working, P. K. (1999) Sterically stabilized liposomes in cancer therapy and gene delivery. *Curr Opin Mol Ther* **1**, 177–185.
10. Cho, W. G., Albuquerque, R. J., Kleinman, M. E., Tarallo, V., Greco, A., et al. (2009) Small interfering RNA-induced TLR3 activation inhibits blood and lymphatic vessel growth. *Proc Natl Acad Sci USA* **106**, 7137–7142.
11. Kariko, K., Bhuyan, P., Capodici, J., Ni, H., Lubinski, J., et al. (2004) Exogenous siRNA mediates sequence-independent gene suppression by signaling through toll-like receptor 3. *Cells Tissues Organs* **177**, 132–138.
12. Song, E., Zhu, P., Lee, S. K., Chowdhury, D., Kussman, S., et al. (2005) Antibody mediated in vivo delivery of small interfering RNAs via cell-surface receptors. *Nat Biotechnol* **23**, 709–717.
13. Peer, D., Zhu, P., Carman, C. V., Lieberman, J., and Shimaoka, M. (2007) Selective gene silencing in activated leukocytes by targeting siRNAs to the integrin lymphocyte function-associated antigen-1. *Proc Natl Acad Sci USA* **104**, 4095–4100.

Chapter 9

Aptamer–Drug Conjugation for Targeted Tumor Cell Therapy

Michael J. Donovan, Ling Meng, Tao Chen, Yunfei Zhang, Kwame Sefah, and Weihong Tan

Abstract

Aptamers developed for applications in cancer therapy can improve the efficacy of drug treatment and enhance molecular imaging. Aptamers for these purposes are generated from SELEX (Systematic Evolution of Ligands by EXponential enrichment), more precisely cell-based SELEX, a process described in detail in this chapter. Experimental applications are also provided for aptamer-based drugs.

Key words: DNA aptamer, SELEX, cancer, therapy, therapeutic agents, targeting ligands.

1. Introduction

The greatest challenge to overcome when fighting cancer with chemotherapy is protecting healthy cells while still delivering a fatal toxicity level to cancer cells. To address this, a new procedure of linking cancer drugs to single-stranded oligonucleotides has been developed in recent years, promising the potential to greatly improve chemotherapy and therapeutic procedures. Specifically, these single-stranded oligonucleotides are synthetic probes known as aptamers that can recognize and bind to their target with high affinity and specificity. Aptamers are created to bind to specific targets, such as cell-surface proteins, through a process termed Systematic Evolution of Ligands by EXponential enrichment (SELEX). By linking, or conjugating, aptamers with drugs, such as doxorubicin, drug delivery to tumor cells can treat diseased cells with high specificity and reduced overall toxicity to healthy cells.

J. Goodchild (ed.), *Therapeutic Oligonucleotides*, Methods in Molecular Biology 764,
DOI 10.1007/978-1-61779-188-8_9, © Springer Science+Business Media, LLC 2011

1.1. SELEX

The method by which aptamers are selected, SELEX, was first independently introduced by the Gold (1) and Szostak (2) groups. In general, SELEX is a combination of in vitro evolution and combinatorial chemistry involving a series of steps including incubation, partitioning, and amplification.

The process starts with the design of a large nucleic acid library pool created by solid-phase technology. It is essential that this pool contains at least a few molecules having the unique conformations required to facilitate selective binding with the target. To accomplish this, the initial library sequence is randomized from 22 to 100 nucleotides in length (10^{15} different oligonucleotide molecules for 40 random nucleotides), flanked at either side by pre-defined primer binding sites for polymerase chain reaction amplification (PCR).

In a typical round of SELEX, the first step involves incubating the library with the target under a defined buffer condition. During incubation, some sequences of the library will bind to the target molecule tightly, but other sequences will only bind weakly, and a majority of the initial sequences do not bind to their target at all. A second step is therefore required to physically separate the binder: target complexes from unbound or weakly bound sequences, partitioning true binders from the others. The success of the entire process depends on this step, since separation results in the differentiation of different binders. Therefore, if the technique can eliminate most of the weak or non-binding aptamer molecules while retaining those tightly binding to their target, PCR will mainly amplify the tight binders. As a consequence, the enrichment process can be achieved rapidly with fewer rounds of selection. After separation, the high-affinity sequences are eluted from the target molecules and then enzymatically amplified by PCR to generate a new DNA pool for the next round of SELEX. To speed up the selection process and ensure that the successful aptamer sequences are actually high-affinity binders, the stringency of the binding conditions and/or elution conditions is generally increased during the later rounds. However, if the conditions are too harsh, there is a risk of losing binders at the earlier rounds, subsequently, resulting in the failure of selection. Typically, it takes approximately 20 rounds of SELEX to obtain aptamer sequences with good affinity (for protein selection, the process is generally around 10 rounds). After selection, the resulting oligonucleotides are subjected to DNA sequencing. The sequences corresponding to the initially variable region of the library are screened for conserved sequences and structural elements indicative of potential binding sites. Finally, binding potency of the aptamer candidates is verified. Even though aptamers were developed almost 20 years ago, most aptamers are still selected by this traditional method. However, other advanced technologies are now involved in the process, and they can vary from traditional capillary electrophoresis and flow cytometry to the more recently developed microfluidic channel (3). The overall goal for various selection approaches remains the same: quick and efficient selection of high-affinity ligands.

Based upon this process, a whole cell-SELEX strategy has been developed to generate a panel of aptamers for specific disease cells (4, 5). Moreover, the cell-SELEX process, which is illustrated in **Fig. 9.1**, can be performed without any prior knowledge about the target. Essentially, this means that an aptamer can be developed for a cell line without having to specifically target a unique cell membrane characteristic. In the context of this review, the cell-SELEX strategy is highly relevant because one blind selection process can yield an entire panel of aptamers to specifically deliver therapeutic reagents to the tumor mass. Similar to the SELEX method as outlined above, cell-SELEX involves the same series of steps: incubation, elution, and amplification. However, cell-based SELEX differs in that a counter selection is introduced after positive selection by performing similar incubation steps, but with a negative cell line. By doing so, the common binders for regular receptors on the cell membrane are removed from the resulting pools. Thus, the probability of recognizing unique molecules exclusively expressed in the target cancer cells is greatly enhanced. It is well documented that cancers originate from mutations of human genes and that these genetic alterations cause morphological and physiological changes in diseased cells. As such, the rational selection of aptamers must take into consideration the

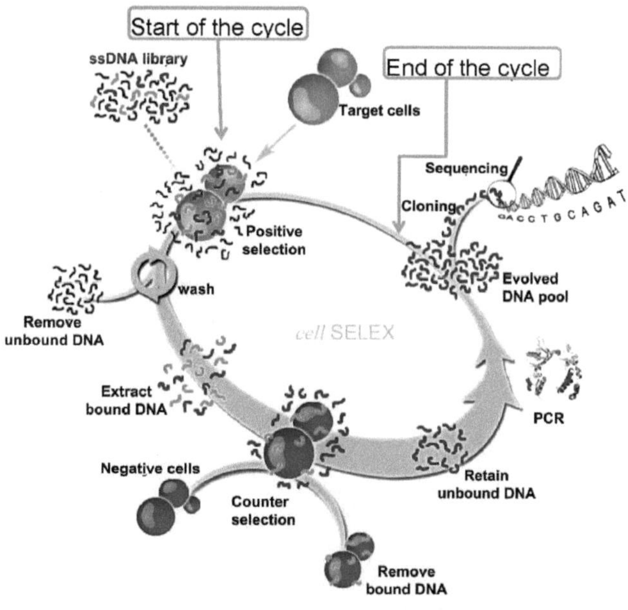

Fig. 9.1. Schematic representation of the whole cell-SELEX.

identification of molecular differences between normal and tumor cells as well as discriminate among tumor cells of different classifications at different disease stages or from different patients. Moreover, this blind selection strategy is an innovative method by which potential disease biomarkers can be discovered by identification of the aptamer's binding targets (6, 7).

1.2. Aptamer Applications

The first aptamer drug, Macugen® (pegaptanib) by Pfizer, a treatment for age-related macular degeneration, was approved by the US FDA in 2004. After that, aptamer technology greatly improved and expanded into many fields (8, 9). While emerging applications of aptamers have been treated in a variety of reviews over the last 15 years, incorporation of aptamers with a defined therapeutic function and recognition capability for cancer therapy is our focus. Thus far, aptamers and aptamer assemblies have been validated as essential molecular tools in the areas of anti-infectives, anticoagulation, anti-inflammation, antiangiogenesis, antiproliferation, and immune therapy.

1.2.1. Aptamer-Based Drug: Sgc8c-Doxorubicin Conjugate

The toxic side effects of chemotherapy are a continuing challenge in biomedicine (10). However, the advent of aptamer-based drugs brings the potential to significantly decrease toxicity to healthy

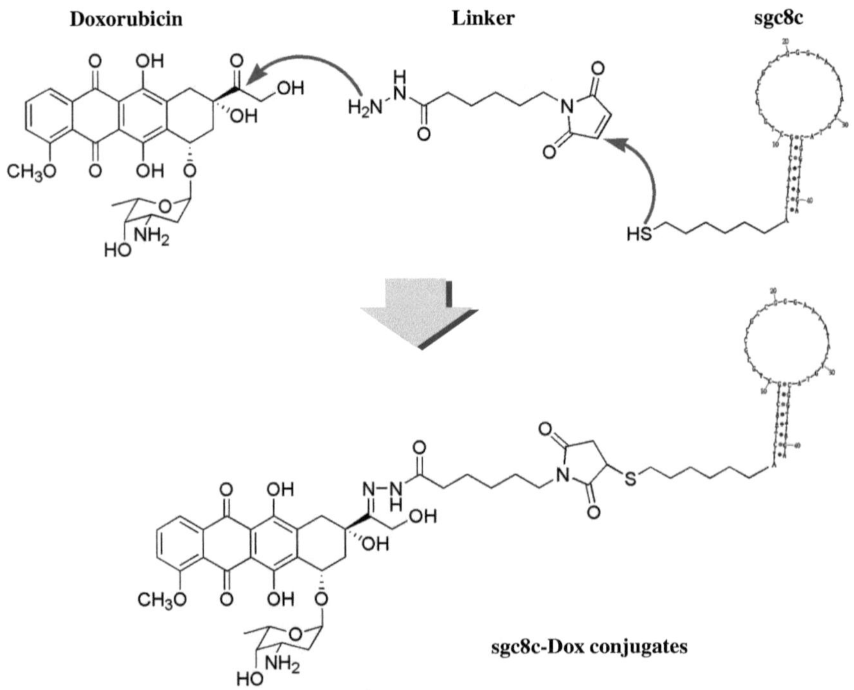

sgc8c: 5'-ATC TAA CTG CTG CGC CGC CGG GAA AAT ACT GTA CGG TTA GA

Scheme. 9.1. Conjugation of the drug doxorubicin (Dox) to aptamer sgc8c for targeted delivery to cancer cells.

cells while specifically delivering drugs to cancer cells. One such aptamer–drug conjugate with proven viability is a sgc8c-Dox conjugate. Sgc8c is a DNA-based aptamer, which was selected for human T-cell ALL CCRF-CEM cell lines. Sgc8c can recognize the protein tyrosine kinase 7 (PTK7), a transmembrane receptor highly expressed on CCRF-CEM cells, with high binding affinity ($K_d \sim 1$ nM). Because of its high binding affinity, sgc8c is able to bind to target leukemia cells while, at the same time, avoiding normal human bone marrow aspirate. Doxorubicin (Dox) is the most utilized anticancer drug against a range of neoplasms, including acute lymphoblastic and myeloblastic leukemias, as well as malignant lymphomas. Conjugation of sgc8c with Dox is fairly simple, as demonstrated in **Scheme 9.1**.

2. Materials

2.1. SELEX

1. Non-enzymatic cell dissociation solution
2. Medium: Appropriate medium depends on cell line cultured
3. Washing buffer (WB) 1 L Dulbecco's phosphate-buffered saline
   a. 4.5 g/L glucose
   b. 5 mL/L MgCl2 1 M
4. Binding buffer (BB) 500 mL Dulbecco's phosphate-buffered saline
   a. 4.5 g/L glucose: Weight: 2.25 g glucose
   b. 5 mL/L MgCl2 1 M: Add 2.5 mL MgCl2
   c. 1.0 g/L BSA: Weight 0.5 g BSA
   d. 100 mg/L tRNA: Weight: 50 mg tRNA
5. UV spectrophotometer
6. SpeedVac evaporating centrifuge
7. DNA library (FITC labeled)
8. Fetal bovine serum
9. DNA-free water
10. PCR buffer (TaKaRa – 10× PCR buffer)
11. dNTP – 2.5 mM
12. Taq (TaKaRa Taq HS)
13. Primers – forward primer labeled with FAM. Reverse primer labeled with biotin.
14. 3% agarose gel

   a. 1.35 g agarose
   b. 45 mL TBE
15. DNA ladder
   a. 1.5 μL 25 base pair DNA step ladder
   b. 2.0 μL 6× Blue/Orange loading dye
   c. 85 μL water
16. Streptavidin sepharose
17. DNA synthesis column
18. PBS (1× Dulbecco's PBS)
19. 200 mM NaOH
20. Desalting column

2.2. Sgc8c-Doxrubicin Conjugation

1. Dox hydrochloride
2. *N*-ϵ-Maleimidocaproic acid hydrazide (EMCH)
3. Methanol
4. Trifluoroacetic acid
5. Acetonitrile
6. Disulfided DNA
7. Aptamer sgc8c (5′-ATC TAA CTG CTG CGC CGC CGG GAA AAT ACT GTA CGG TTa gA-3′)
8. Tris-(2-carboxyethyl)phosphine (TCEP)
9. PBS (pH 7.4)
10. G-25 Sephadex size-exclusion column (NAP™-5, Amersham Pharmacia Biotech)
11. Dimethylformamide (DMF)
12. Tris–HCl buffer (10 mM, pH 7.4) and acetonitrile eluent for chromatography

3. Methods

3.1. Cell-SELEX

3.1.1. Prepare Cells

1. Treat cells with non-enzymatic buffer to break the adhesion interaction between them and suspend the cells in medium. Determine cell concentration and volume needed (*see* **Note 1**).
2. Pipette cells with medium into tube. Centrifuge cells (137×*g* for 3 min).
3. Remove supernatant.
4. Add 10 μL washing buffer and disperse.

5. Centrifuge cells (137×*g* for 3 min).
6. Repeat Steps 3–6 two times.

3.1.2. Prepare DNA Pool

1. Dissolve DNA library in water.
2. Get absorbance at 260 nm and calculate concentration (*see* **Note 2**).
3. Dry DNA using evaporating centrifuge.
4. Mix DNA library (5–20 nmoles) in specific volume of binding buffer to final volume of 500–700 μL, depending upon library concentration.
5. Add FBS – 10% volume (not necessary for first two rounds) (*see* **Note 3**).
6. Denature DNA at 95°C for 5 min (using heating block) and instantly cool on ice afterwards. This will force the ssDNA to fold by forming only the most favorable structure.

3.1.3. Incubation

1. Add binding buffer and DNA library solution to cells to make 500 μM library solution.
2. Shake for 1 h at 4°C. Check cells every 20 min for settling (*see* **Note 4**).
3. Centrifuge at 203×*g* for 3 min.
4. Remove supernatant (*see* **Note 5**).
5. Wash with 5 mL washing buffer.
6. Repeat Steps 3–5 two times.
7. Centrifuge at 203×*g* for 3 min.
8. Remove supernatant.
9. Add 500 μL of DNA-free water. (Use binding buffer for second round onwards in lieu of DNA free water.)

3.1.4. Retrieval of Bound DNA

1. Pipette cells to disperse.
2. Put solution into 1.5 mL test tube.
3. Heat at 95°C for 15 min.
4. Centrifuge for 5 min at 14,000 r.p.m. in a microfuge.
5. Obtain supernatant, which contains DNA.
6. Properly dispose of used cells.

3.1.5. Remove the Sequences That Bind to Other Cell Lines Using Negative Selection

Prepare negative cell line while denaturing DNA library. Use a higher cell concentration than the target cell line.

1. Get volume needed to have desired cell concentration.
2. Pellet the cells (centrifuge at 77–95×*g* for 3–5 min at 4°C).
3. Wash cells two times with 3–5 mL of washing buffer.

4. After centrifugation, remove medium.
5. Add 3–5 mL of washing buffer and re-suspend cells.
6. Centrifuge again to pellet cells.
7. Remove supernatant.
8. Repeat Steps 5–7.
9. Add library pool and incubate on ice for 50 min inside shaker.
10. Centrifuge to pellet cells with non-specific binder sequences (86×*g* for 3–5 min at 4°C).
11. Keep supernatant (contains sequences that bind to target cell line, but not to negative cell line).

3.1.6. Amplify Eluted Pool by PCR

The number of cycles depends upon the PCR optimization. Use about 10 cycles for initial amplification. Since library is about 500 μL, use a final volume of 1000 μL for PCR mixture.

1. Adjust library volume to 500 μL, adding water as needed.
2. Add the following reagents to library based upon concentration of library: 20% PCR buffer, 16% dNTP, 10% primer pool, 0.6% Taq, 54% DNA water (*see* **Note 6**).
3. Split 1000 μL mixture into 10 tubes (100 μL each) (*see* **Note 6**).
4. After PCR, pool samples together.

3.1.7. PCR Cycle Optimization

1. Since library concentration is unknown, use about 5–10% of total pool.
2. Prepare PCR mixture (50 μL samples). For 6 samples and 1 negative control, each having final volume of 50 μL, add 35 μL PCR buffer, 28 μL dNTP, 7.5 μL primers (pulled together), 234.5 μL water.
3. Transfer 45 μL to negative control tube and add 5 μL of water and 0.15 μL Taq.
4. Add 30 μL of library and 0.9 μL of Taq to sample pool. Transfer 50 μL of mixture into six sample Eppendorf tubes.
5. Run PCR using optimized annealing temperature. Choose six different cycles for samples.
6. Run 3% agarose gel to observe PCR product (*see* **Note** 7). Select the cycle with the best band (intense band, but no non-specific amplification). Use DNA ladder to check size of library.

3.1.8. Amplification of Pool

Amplify pool using optimized number of cycles. Use percentage to calculate volume of library needed.

3.1.9. Prepare ssDNA (Remove Biotin-Labeled Strand)

1. Prepare streptavidin beads. Take empty DNA synthesis column and place filter on one end. Place syringe on the other end. Add 200 μL of streptavidin beads solution onto syringe (first shaken; *see* **Note 8**). Beads will deposit on top of filter inside column while solution passes through. When all the solution is out, remove syringe from column and remove plunger (*see* **Note 9**). Then, put the syringe back onto the column.
2. Wash with PBS. Add 2–2.5 μL PBS to syringe and replace the plunger in its position. Push in a little in order to start dripping. When all PBS is out, remove syringe from column and remove the plunger from syringe. Put syringe back into column.
3. Load PCR product onto syringe and collect solution. The dsDNA library (one strand with biotin and one strand with FITC) will stay in the beads, as well as the reverse primer labeled with biotin. Repeat five times to make sure the entire pool stays on beads. Do not collect filtrate the third time.
4. Wash with 2–2.5 mL PBS to remove any forward primer trapped on beads (*see* **Note 10**).
5. Break hydrogen bonds on the dsDNA. Add 500 μL 200 mM NaOH to melt DNA. Do not put too much pressure on plunger; let it drip slowly. Collect the filtrate. It contains library labeled with FITC. Repeat three times in all.
6. Use desalting column to remove salt. Since library is in NaOH, salt needs to be removed. Prepare desalting column. Open column and pour out liquid. Add 15 mL water into column. Open tip and let water pass through column. Do not let it dry (*see* **Note 11**)! When last drop of water is out, load sample. Let sample penetrate into column.
7. Add 1 mL water and collect filtrate. Sample is here! When completed, wash column with 15–20 mL of water.

3.1.10. Obtain Library Concentration

1. Take absorbance at 260 nm and calculate the number of moles.
2. Dry sample in an evaporating centrifuge.
3. Re-suspend in binding buffer to get the desired concentration (200–300 nM).

Perform another round of selection using lower cell concentration with the same procedure.

3.2. Sgc8c-Doxrubicin Conjugation

1. Dox hydrochloride (5 mg, 8.62 μmol) and EMCH (10 mg, 44.4 μmol) were dissolved in methanol (4 mL).
2. Trifluoroacetic acid (3 μL) was added, and the solution was stirred at room temperature for 24 h while being protected from light. The methanolic solution was concentrated under reduced pressure at room temperature to a volume of 0.25 mL.
3. Acetonitrile (2.5 mL) was added, and the resulting suspension was allowed to stand at 4°C for 48 h for crystallization of the product.
4. The red solid hydrazone was isolated by centrifugation, washed with fresh methanol–acetonitrile (1:l0), and dried under vacuum to yield the (6-maleimidocaproyl) hydrazone of Dox (2.9 mg, 3.85 μmol, 44.6% yield).
5. A disulfided DNA (300 nmol) was reduced with TCEP in PBS (pH 7.4) for 2 h at room temperature. TCEP was removed by G-25 Sephadex size-exclusion column equilibrated with PBS.
6. The eluate was added to the (6-maleimidocaproyl) hydrazone of Dox (2 μmol) dissolved in DMF (50 μL) and incubated on ice for 12 h.
7. The product conjugates were purified by HPLC with Tris–HCl, pH 7.4, and acetonitrile as mobile phase and monitored with UV at 260 and 495 nm.
8. After concentration under vacuum to 0.1 mL, the amount of conjugate can be measured by UV scan. The amount of Dox can be determined by UV absorption at 495 nm ($\varepsilon_{495} = 12{,}000\ \text{cm}^{-1}\ \text{M}^{-1}$), while that of sgc8c was calculated from the absorption at 260 nm ($\varepsilon_{260} = 24{,}000\ \text{cm}^{-1}\ \text{M}^{-1}$), according to the formula:

$$sgc8c\,(M) = \frac{A_{260} - (\varepsilon_{260} \times A_{495}/\varepsilon_{495})}{397600}$$

From the absorption ratio, the Dox:sgc8c ratio can be calculated. One is able to control the number of Dox molecules conjugated to each sgc8c; however, adding more than one is relatively hard to achieve with current mAb-based immunoconjugation methodology.

The conjugation of sgc8c with Dox does not affect the ability of the sgc8c-Dox conjugate to bind to CCRF-CEM cells. Furthermore, sgc8c-Dox conjugates possess drug efficacy similar to unconjugated Dox. With these properties, aptamer–drug conjugates have the potential to kill cancer cells efficiently while reducing toxicity to healthy cells.

4. Notes

1. While using non-enzymatic buffer to break the adhesion forces between cells, it is imperative to obtain single cells with no clusters. As the cells adhere together, there is less surface area for the DNA to bind. The greater the surface area, the greater chance the library has to form complementary bonds with the cell's membrane.
2. The percentages given in Step 2 under **Section 3.1.6** are just recommendations. The actual concentrations will depend on the image of the agarose gel (Step 4 under **Section 3.1.7**). An ideal concentration will have a rich band with no shadows above or below. If the band has shadows below it, the number of rounds ran from PCR need to be decreased or the primer concentration needs to be decreased.
3. FBS is not necessary for the first two rounds. However, it can be used after the first two rounds in order to increase amplification and binding affinity.
4. While incubating, shake the cells every 20 min to keep the cells from settling. If the cells settle, the DNA on the top layer will only bind to the top layer of cells. This leads to inefficient incubation and loss of DNA.
5. Once centrifugation is complete, the supernatant needs to be removed. It is imperative to be prudent in the removal of all liquid, for this is where the unbound DNA remains. A Kim wipe may be used to remove the supernatant which cannot be obtained from pipetting.
6. When placing the tubes into PCR machine, make sure there are no cavities in the solution.
7. Make sure there are no cavities in the gel.
8. Streptavidin beads settle when being stored; therefore, shake the bottle to get beads into solution.
9. When removing plunger, first make sure the filter portion of the syringe is disconnected from the plunger and column. If it is still connected while removing the plunger, the vacuum created will suck the streptavidin beads off the filter tip.
10. If forward primer is not removed from the beads, a false concentration of DNA will be recorded. The primer will give a significantly higher reading.
11. Keeping the salt column wet is important. If the salt dries, undesired paths may be formed in the column due to uneven settling of the salt and salt will clump together.

Acknowledgments

The authors would like to extend a special thanks to our colleagues at the University of Florida for their excellent work reported here. This work is supported by NIH grants.

References

1. Tuerk, C., and Gold. L. (1990) Systematic evolution of ligands by exponential enrichment -RNA ligands to bacteriophage-T4 DNA-polymerase. *Science* **249**, 505–510.
2. Ellington, A. D., and Szostak, J. W. (1990) In vitro selection of RNA molecules that bind specific ligands. *Nature* **346**, 818–822.
3. Xiaohong, F., and Weihong T. (2010) Aptamers generated from Cell-SELEX for molecular medicine: a chemical biology approach. *Accounts Chem Res* **43**, 48–57.
4. Shangguan, D., Li, Y., Tang, Z. W., Cao, Z. H. C., Chen, H. W., Mallikaratchy, P., Sefah, K., Yang, C. Y. J., and Tan, W. H. (2006) Aptamers evolved from live cells as effective molecular probes for cancer study. *Proc Nat Acad Sci USA* **103**, 11838–11843.
5. Shamah, S. M., Healy, J. M., and Cload, S. T. (2008) Complex target SELEX. *Accounts Chem Res* **41**, 130–138.
6. Shangguan, D., Cao, Z. H., Meng, L., Mallikaratchy, P., Sefah, K., Wang, H., Li, Y., and Tan, W. H. (2008) Cell-specific aptamer probes for membrane protein elucidation in cancer cells. *J Proteome Res* 7, 2133–2139.
7. Mallikaratchy, P., Tang, Z. W., Kwame, S., Meng, L., Shangguan, D. H., and Tan, W. H. (2007) Aptamer directly evolved from live cells recognizes membrane bound immunoglobin heavy mu chain in Burkitt's lymphoma cells. *Mol Cell Proteomics* **6**, 2230–2238.
8. Lee, J. H., Canny, M. D., De Erkenez, A., Krilleke, D., Ng, Y. S., Shima, D. T., Pardi, A., and Jucker, F. (2005) A therapeutic aptamer inhibits angiogenesis by specifically targeting the heparin binding domain of VEGF165. *Proc Natl Acad Sci USA* **102**, 8902–18907.
9. Fraunfelder, F. W. (2005) Pegaptanib for wet macular degeneration. *Drugs Today* **41**, 703–709.
10. Huang, Y-. F., Shangguan, D., Liu, H., Phillips, J. A., Zhang, X., Chen, Y., and Tan, W. (2009) Molecular assembly of an aptamer-drug conjugate for targeted drug delivery to tumor cells. *ChemBiochem* **10**, 862–868.

Chapter 10

Five-Step Process for Screening Antisense Compounds for Efficacy: Gene Target IL-12Rb2

Nikki B. Marshall, Laura L. Hauck, and Dan V. Mourich

Abstract

Antisense technologies are widely used for the inhibition of gene expression. Although traditionally the AUG start codon of the open reading frame is targeted to disrupt ribosome assembly and initiation, an emerging approach is targeting sequences to disrupt pre-mRNA splicing. The primary advantage to using this approach is a positive read-out for an antisense effect through detection of a novel splice product, but additional benefit can be found in generating a novel splice product with altered functional properties. The antisense compounds used here are phosphorodiamidate morpholino oligomers conjugated to an arginine-rich cell penetrating peptide (P-PMO). We describe a five-step process for selecting the best candidate antisense compound for altering IL-12Rb2 expression including (1) detecting mRNA splice products by RT-PCR, (2) measuring protein expression, (3) evaluating protein function, (4) checking cellular viability, and (5) validating efficacy of the final candidate compound. The significance of targeting exons composed of a number of base pairs divisible by 3 is also discussed. The five steps described here for selecting the best candidate P-PMO to alter IL-12Rb2 expression should be applied for designing and screening antisense compounds for other gene targets.

Key words: Antisense, P-PMO, pre-mRNA, splicing, novel splice product, phosphorodiamidate morpholino oligomers, altered gene expression, altered gene function.

1. Introduction

Antisense technologies are widely used for altering gene expression to examine the role of specific genes in a given biological process. Antisense can be used to disrupt normal cellular gene expression, which arguably has more biological relevance than the use of gene-specific knock-out (KO) cells or tissues. The classic approach to antisense design is targeting on and around the AUG start codon to interfere with ribosome initiation and assembly. However, one drawback to this approach is that the assay read-out

J. Goodchild (ed.), *Therapeutic Oligonucleotides*, Methods in Molecular Biology 764,
DOI 10.1007/978-1-61779-188-8_10, © Springer Science+Business Media, LLC 2011

for effectiveness is solely a reduction in protein signal. This introduces a potential for a false positive and inappropriate interpretation of gene-specific antisense effectiveness due to confounding factors related to the treatment and viability of targeted cells. A preferred approach for measuring antisense efficacy is an assay that provides a positive read-out achieved by altering the normal splice pattern of pre-mRNA. The splicesome recognizes conserved sequences within the pre-mRNA as splice-acceptor (5′ end of exon) and splice donor (3′ end of the exon). The splicesome, in turn, excises intronic sequences as a lariat and ligates the exons to produce a mature mRNA transcript. If the splice-acceptor/donor or splicing factor recognition sequences are blocked due to the binding of a sequence-specific antisense molecule, the exon will be excised with the surrounding intronic sequence. The splicesome will proceed along the transcript and find subsequent splice recognition sites. This will often produce a unique mRNA transcript and can be designed to disrupt the open reading frame and thus effect expression of the normal gene product. The unique mRNA is then quantifiable by RT-PCR and can be correlated to a measured reduction in protein levels.

Here we used phosphorodiamidate morpholino oligomers conjugated to an arginine-rich delivery peptide (P-PMO), an antisense platform that has been repeatedly proven to be efficacious for altering gene expression, including splice-altering in leukocytes (1). Six P-PMO compounds were synthesized to target a subunit of the heterodimeric interleukin-12 receptor, IL-12Rb2 (Gene ID: NM_008354), which is found on activated T cells. The goal was to sufficiently disrupt IL-12Rb2 expression to inhibit IL-12R signaling, which normally results in phosphorylation of STAT4 and increased production of the cytokine IFN-γ. To determine the best compound to alter IL12rb2 expression, we deployed a five-step process that involved careful screening of mRNA splice products, protein expression and function, cellular viability, and sequencing of the altered splice product to choose the compound targeting the splice-acceptor site of exon 3 (SA3). Finally, SA3 was shown to sufficiently disrupt IL-12Rb2 mRNA expression in a dose-dependent manner by inducing excision of exon 3, which effectively inhibited downstream phosphorylation of STAT4.

2. Materials

2.1. Detecting Altered Splice Product by End-Point and Quantitative RT-PCR

1. DNA engine PCR apparatus (MJ Research)
2. iCycler with iCycler iQ realtime PCR Detection System or C1000 Thermal Cycler with the CFX96-Realtime System (BioRad)

3. Phosphorodiamidate morpholino oligomers conjugated to arginine-rich peptide (P-PMO) (AVI BioPharma Inc.)
4. RNeasy mini kit (Qiagen)
5. SuperScript^{TM} III One-Step RT-PCR System with Platinum® Taq DNA Polymerase (Invitrogen)
6. PCR Primers
7. 2× SYBR Green Supermix: dNTPs, 50 U/mL iTaq DNA polymerase, 6 mM $MgCl_2$, SYBR Green I, 20 nM fluorescein (Bio-Rad)
8. iScript Reverse Transcriptase for One-Step RT-PCR (Bio-Rad)
9. Mouse complete medium (MCM): Roswell Park Memorial Institute 1640 medium (RPMI 1640), 2.05 mM L-glutamine, 10% fetal bovine serum (FBS), 1% antibiotic-antimycotic solution (10,000 I.U./mL penicillin, 10,000 μg/mL streptomycin, 25 μg/mL amphotericin B), 50 μM β-mercaptoethanol
10. Suspension buffer: Hanks' balanced salt solution (HBSS), 2.5% FBS, 50 μg/mL gentamicin, 20 mM HEPES
11. Vector NTI software (Invitrogen)

2.2. Measuring the Effects of Antisense Treatment on IL-12Rb2 Protein Expression

1. Mouse complete medium: *see* **Section 2.1**
2. P-PMO: *see* **Section 2.1**
3. Rat IgG (Jackson ImmunoResearch)
4. Anti-mouse CD4 monoclonal antibody (clone RM4-5) conjugated to APC-Alexa Fluor® 750-CD4 (eBioscience)
5. Armenian hamster anti-mouse IL-12 Rβ2 monoclonal antibody (clone HAM10B9) (BD Pharmingen)
6. Anti-armenian hamster IgG secondary antibody conjugated to fluorescein isothiocyanate (FITC) (eBioscience)
7. Beckman Coulter FC-500 flow cytometer
8. WinList (Verity Software)
9. Flow cytometry buffer (FCB): PBS, 2% FBS, 0.1% sodium azide

2.3. Measuring the Effects of Antisense Treatment on IL-12rb2 Protein Function

1. Mouse complete medium: *see* **Section 2.1**
2. Functional grade anti-CD3: 1 μg/μL
3. Functional grade anti-CD28: 1 μg/μL
4. Functional grade recombinant IL-12: 10 μg/mL
5. Phosflow anti-mouse STAT4 monoclonal antibody (clone 38/p-Stat4) conjugated to R-Phycoerythrin (PE) (BD Biosciences)

2.4. Determining the Effects of Antisense Treatment on Cellular Viability

1. Mouse complete medium: *see* **Section 2.1**
2. P-PMO: *see* **Section 2.1**
3. 7-Amino-actinomycin D (7-AAD) (BD Biosciences)

2.5. Final Identification and Validation of the Best Candidate Compound and Measuring Efficacy of the Selected Candidate

1. QIAquick® Gel Extraction Kit (Qiagen)
2. TOPO TA Cloning Kit with pCR4-TOPO vector (Invitrogen)
3. QIAprep Spin Miniprep Kit (Qiagen)
4. P-PMO: *see* **Section 2.1**
5. Soluble anti-CD3: *see* **Section 2.3**
6. Soluble anti-CD28: *see* **Section 2.3**

3. Methods

1. C57Bl/6 mice were housed in microisolator cages under pathogen-free conditions and treated according to animal use protocols approved by the Institutional Animal Care and Use Committee of Oregon State University.
2. Phosphorodiamidate morpholino oligomers conjugated to arginine-rich peptide NH_2-RXRRXRRXRRXRXB-COOH (where X represents 6-aminohexanoic acid and B represents β-alanine) (P-PMO) were over 90% pure. P-PMO sequences (**Table 10.1**) were designed to target six areas within the gene's open reading frame including the AUG start codon and 5′ splice acceptor (SA) ends of exons 3, 5, 7, 9, and 11 (**Fig. 10.1a**).

Table 10.1
P-PMO sequences targeting IL-12Rb2

Name	Sequence (5′->3′)
Control	CCTCTTACCTCAGTTACA
AUG	CAGTCTGTGCCATGAGTCTTC
SA3	CAGTGCCAAGCTTGCACACAT
SA5	AGATTGTTTGGTCCACTTAAC
SA7	TTTGGCATTTGTAGCATTGAC
SA9	TGCCTCTGATGGATTCAGACT
SA11	AGATATAGGGCTTTATGTTCT

P-PMO sequences are shown in order from 5′ to 3′
Control sequence is non-specific for mouse

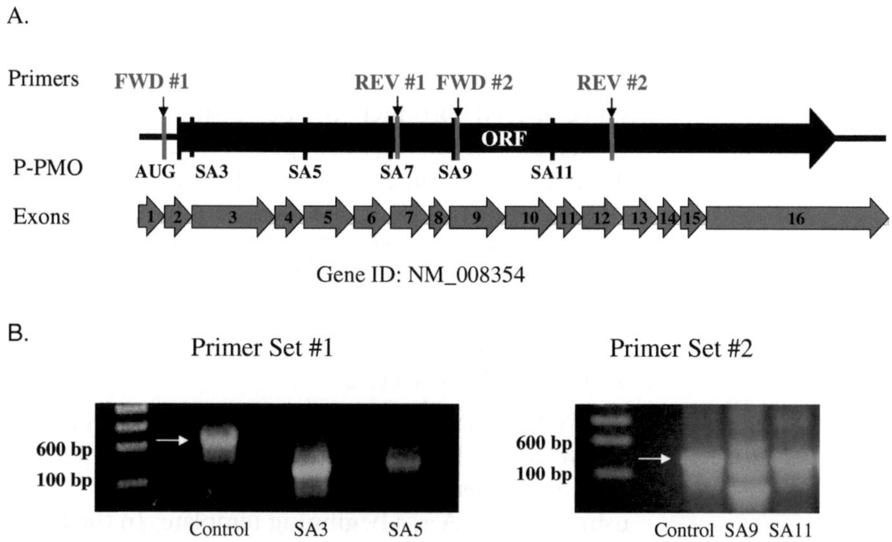

Fig. 10.1. P-PMOs targeting IL-12Rb2 alter normal pre-mRNA splicing. Locations of P-PMO designed to target the 5′ splice acceptor sites (SA) of exons 3, 5, 9, and 11 of the IL-12Rb2 gene (**a**). C57/Bl6 splenocytes were stimulated with 5 μg/mL concanvalin-A and treated with 5 μM P-PMO for 48 h. Whole RNA was isolated and used as template in one-step rt-PCR reactions containing primer pair #1 or #2 (**a**) to amplify the target region. The PCR products were visually examined by agarose gel electrophoresis and ethidium bromide staining alongside a reference ladder (**b**).

3. Lyophilized P-PMO were dissolved in DI water at 1–2 mM and stored at 4°C. P-PMOs were heated for 1 min at 55°C to release compound bound to the glass vial, followed by cooling to room temperature before use.

3.1. Detecting Altered Splice Product by End-Point and Quantitative RT-PCR

Endpoint RT-PCR is the quickest way to assess the presence of splicing but quantitative RT-PCR is a more accurate method for quantification of splice products (2). We will describe end-point RT-PCR as well as two quantitative methods that can be used to assess splicing activity of P-PMO. Both quantitative methods use measurements of fluorescence during a PCR reaction to quantitate the amplified product (*see* **Note 1** for additional general information about quantitative RT-PCR methods).

3.1.1. Endpoint RT-PCR

1. Splenocyte suspensions were prepared by dissociation of spleens between frosted microscope slides in suspension buffer followed by a 10-s hypotonic water lysis of red blood cells.
2. Splenocytes were cultured in MCM and stimulated with 5 μg/mL concanvalin-A. Cells were simultaneously treated for 48 h with 5 μM P-PMO.
3. Total RNA was extracted from the cultured cells using the Qiagen RNeasy mini-kit per the manufacturer's instructions and used as template material for RT-PCR using one of the two pairs of sequence specific primers described below.

4. Primer pairs were designed using Vector NTI software to amplify small regions of the target gene encompassing the splice junction. Smaller RT-PCR products allow a splice-altered product to be easily differentiated when visualized on an agarose gel. Thus, for IL-12Rb2, we designed two primer sets, which spanned exons 2–7 and 9–12, amplifying PCR products around 1000 and 500 base pairs (bp), respectively (**Fig. 10.1a**). Primer pair #1: FWD 5′-TCTGGAGAACCAGAGGTTGC-3′; REV 5′-CTCC AATTACTCCAACTTCCTC-3′; Primer pair #2: FWD 5′-GTCTGAATCCATCAGAGG-3′; REV 5′-TCTGCTG TCGAGTCTCGTTC-3′
5. RT-PCR reactions were performed using Superscript III according to the manufacturer's instructions. This kit allows cDNA synthesis and PCR to be conducted in the same tube using only RNA as a beginning template. In the first step, the RNA is converted into cDNA with the gene-specific primers and a 30-min incubation at 55°C. This is followed by denaturation and a standard PCR cycling protocol with an annealing temperature of 55°C to amplify the final product.
6. PCR products were visualized by agarose gel electrophoresis followed by ethidium bromide staining. As shown in **Fig. 10.1b**, smaller RT-PCR products were found for samples treated with SA3, SA5, SA9, and SA11 P-PMO compared to the full size product seen in the control. No PCR product was detected in the SA7 sample due to placement of the reverse primer in exon 7 (data not shown).

3.1.2. Detecting Altered Splice Product by Quantitative RT-PCR Using SYBR Green

1. To measure the presence of a spliced product using SYBR Green, the primers must be designed so that one hybridizes to the unique exon boundary created by the spliced product (**Fig. 10.2a**). Priming of the PCR reaction will only occur if the novel splice junction is present. A second reaction should also be set up with a second set of primers. These primers should not prime across the novel splice junction and must create a product close to the same size as the first set of primers. This secondary PCR reaction allows you to quantify the amount of template available, both spliced or unspliced, for the gene of interest. The ratio between the two splice forms allows you to determine the % of splicing as a function of P-PMO treatment.
2. Template should be prepared as described in **Section 3.1.1** Steps 1–4. RT-PCR reactions are set up according to manufacturer's instructions for SYBR Green Super Mix using RNA as the template and adding 0.25 μL of reverse transcriptase and the specially designed primers.

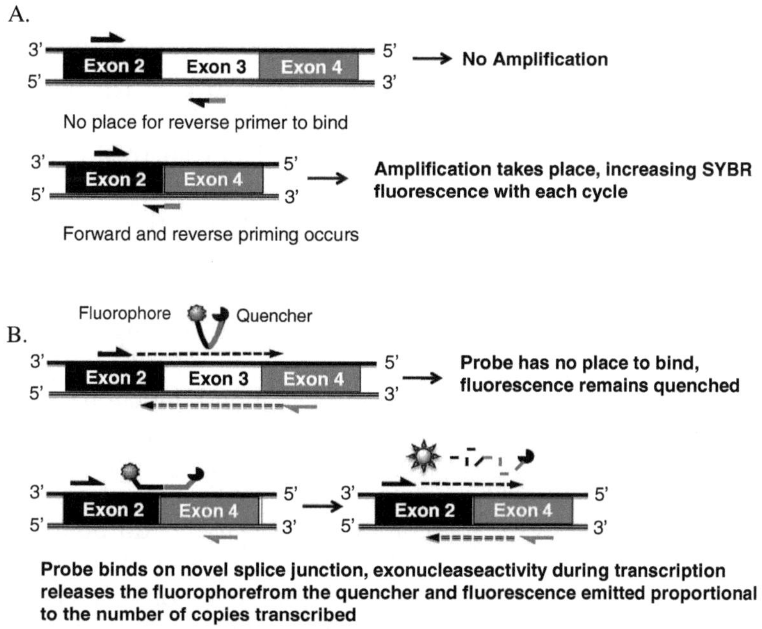

Fig. 10.2. Quantitative RT-PCR approaches to analyze alternative splice products. Designing one of the primers, in this case the reverse, to bind to a novel exon junction allows you to specifically amplify the novel splice product, which can then be measured by SYBR Green fluorescence (**a**). Alternatively a Taqman probe can be designed to bind to the novel splice junction. During elongation Taq polymerase exonuclease activity will release the fluorophore from the quencher and result in measurable fluorescence (**b**).

3. Perform quantitative PCR reactions on a machine capable of measuring fluorescence during amplification; we use the iCycler and C1000 Thermal Cycler (*see* **Section 2**). Cycling includes a 15-min RT step at 55°C, 2 min at 95°C to activate the taq, and then a three-step amplification cycle (×40) including an annealing step that will be specific to the primers being used.
4. Analyze the data according to manufacture's instructions.

3.1.3. Detecting Altered Splice Product by Quantitative RT-PCR Using a Molecular Probe

1. When designing the probe you must make sure that it binds to the novel splice junction (**Fig. 10.2b**). Primers flanking this junction are then used during a PCR reaction. Only when the template has the novel splice junction for the probe to bind to can exonuclease activity release the flourophore (**Fig. 10.2b**). Companies like Biosearch Technologies (biosearchtech.com) have on-line software programs that assist in the design of the primers and probe as you order them, making this technology very approachable.
2. Template should be prepared as described in **Section 3.1.1** Steps 1–4. RT-PCR reactions are set up according to manufacturer's instructions for Reaction Mix for Probes using

RNA as the template and adding 0.25 μL of reverse transcriptase, the probe, and the probe-based primers.

3. Perform quantitative PCR reactions on a machine capable of measuring fluorescence during amplification; we use the iCycler and C1000 Thermal Cycler (*see* **Section 2**). Cycling includes a 15-min RT step at 55°C, 2 min at 95°C to activate the taq, and then a three-step amplification cycle (×40) including an annealing step that will be specific to the primers and probe being used.
4. Analyze the data according to manufacture's instructions.

3.2. Measuring the Effects of Antisense Treatment on IL-12Rb2 Protein Expression

1. Plates with bound functional grade anti-CD3 are commercially available, but can also be prepared as follows. For a 24-well plate, make 7.5 mL of PBS containing 3.34 μg/mL of functional grade anti-CD3 (stock usually comes at 1 μg/μL). Add 300 μL of this solution to each well and incubate in a sealed bag overnight at 4°C. Aspirate each well and gently rinse twice with 500 μL PBS per well. The plate is then ready for culture.
2. To measure surface expression of IL-12Rb2 protein on treated cells, splenocyte suspensions were prepared as described in **Section 3.1.1** and cultured in MCM, stimulated with plate-bound anti-CD3, and treated 48 h with 5 μM P-PMO.
3. The treated splenocytes were then harvested, washed, and incubated with non-immunized rat IgG (200 μg/mL) 15 min for FC receptor blocking. FC receptors are cell surface receptors found on certain immune cells. They serve a protective role as the part of the immune system that binds to the FC (fragment crystallizable) region of antibodies carrying pathogens, ultimately resulting in destruction of the bound pathogen. To avoid non-specific binding of conjugated antibody FC domains, these receptors must be blocked prior to staining with specific antibodies. The FC blocked cells were then stained on ice in FCB with optimal concentration of monoclonal antibodies recognizing IL-12Rb2 and fluorescently tagged monoclonal antibodies recognizing CD4 for 30 min. This was followed by 15 min of staining with FITC-conjugated armenian hamster IgG secondary antibody (for visualization of the IL-12Rb2 antibody). Optimal staining concentrations are based on the number of cells being stained and the staining volume. This is generally determined empirically by the researcher for their specific application. After staining the cells were washed twice in FCB prior to flow cytometry. For all flow cytometry experiments, samples lacking one of the individual stains (fluorescence minus one [FMO]) were used as staining controls.

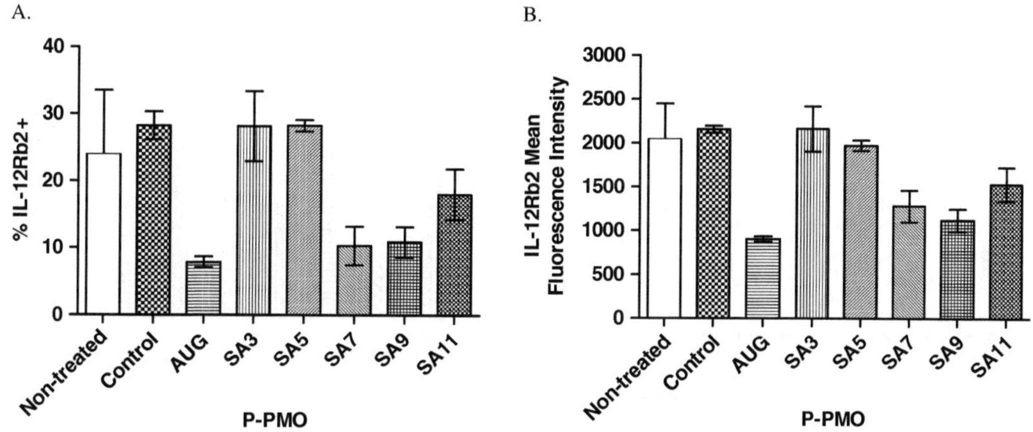

Fig. 10.3. Effects of P-PMO on IL-12Rb2 protein expression on CD4$^+$ T cells. Splenocytes were stimulated for 48 h with plate-bound anti-CD3 (5 μg/mL) and treated with 5 μM P-PMO. Cells were stained with anti-mouse IL-12Rb2 followed by FITC-anti-hamster IgG secondary and anti-mouse APC-Alexa Fluor® 750-CD4 and analyzed on a flow cytometer. The percent of cells that were FITC+ (IL-12Rb2+) (**a**) and the mean FITC fluorescence intensity (**b**) was measured on CD4$^+$ T cells (mean ± SEM).

4. A minimum of 10,000 CD4$^+$ T-cell events were collected per sample on a flow cytometer. All flow cytometry data analysis and software compensation were performed using WinList.
5. As shown in **Fig. 10.3**, decreased IL-12Rb2 expression was measured by flow cytometry on CD4$^+$ T cells for AUG-, SA7-, SA9-, and SA11-treated samples both in the percentage of cells (**Fig. 10.3a**) and the mean fluorescence intensity (**Fig. 10.3b**). This suggests that AUG, SA7, SA9, and SA11 were effectively disrupting translation of IL-12Rb2 mRNA. However, we did not yet exclude the other compounds, as it is possible for a partially translated protein to still be recognized by an antibody, but be non-functional.

3.3. Measuring the Effects of Antisense Treatment on IL-12rb2 Protein Function

Because a test compound may alter the function of a protein, without reducing the overall level of expression of that protein, we measured downstream activity of the targeted protein in addition to the expression of the targeted protein. For our example of IL-12Rb2, the binding of IL-12 to the IL-12Rb1 and IL-12-Rb2 heterodimer causes the dimerization and phosphorylation of STAT4. Thus, if IL-12Rb2 protein function is disrupted, we expect to see a decrease in STAT4 phosphorylation.

1. Splenocyte suspensions were prepared as described in **Section 3.1.1**.
2. Splenocytes were cultured in MCM and stimulated with soluble anti-CD3 (0.5 μg/mL) and soluble anti-CD28 (1 μg/mL) for 72 h with 5 μM P-PMO.
3. Cells were then pulsed with 5 ng/mL of recombinant IL-12 for 2 h.

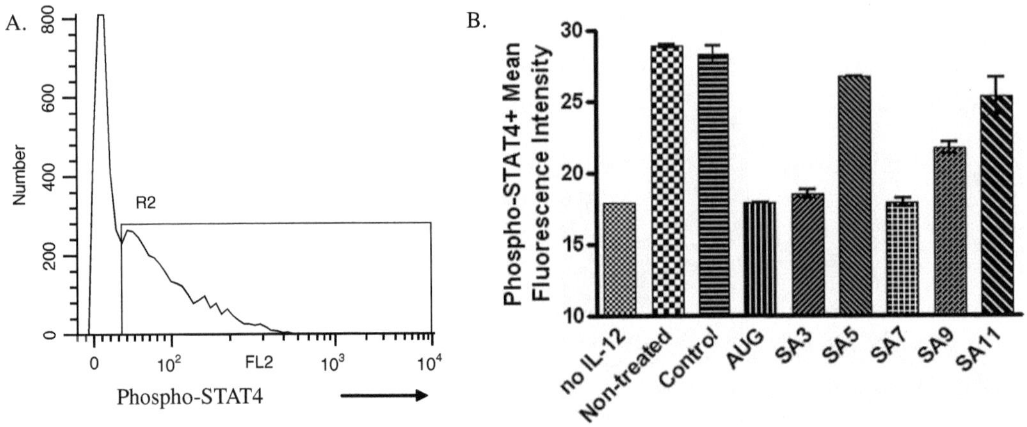

Fig. 10.4. Effects of P-PMO on phosphorylated-STAT4 protein expression in $CD4^+$ T cells. Splenocytes were stimulated with soluble anti-CD3 and anti-CD28 for 72 h with 5 μM P-PMO, then pulsed with 5 ng/mL IL-12 for 2 h. Phosphorylated STAT4 protein was measured intracellularly using BD PhosFlow reagents and analyzed by flow cytometry (**a**). The mean fluorescence intensity of phospho-STAT4 staining was measured in $CD4^+$ T cells (mean ± SEM) (**b**).

4. Phosphorylated STAT4 protein was measured intracellularly using a monoclonal antibody conjugated to phycoerythrin (BD Phosflow) per the manufacturer's instructions and analyzed by flow cytometry (**Fig. 10.4a**).
5. As shown in **Fig. 10.4b**, the mean fluorescence intensity of phosphorylated STAT4 staining was decreased in cells treated with AUG, SA3, SA7, and SA9. Thus, after examining mRNA splicing, protein expression, and protein function, AUG, SA3, SA7, and SA9 were our four leading candidate compounds.

3.4. Determining the Effects of Antisense Treatment on Cellular Viability

An alternate explanation for reduced protein expression in cells is a decrease in viability (*see* **Note 2**).

1. To determine if any of the compounds influenced cell viability, splenocyte suspensions were prepared as described in **Section 3.1.1** and were stimulated with plate-bound anti-CD3 (5 μg/mL; *see* **Section 3.2**) and treated with 5 μM P-PMO for 48 h.
2. The cells were then harvested, washed, and stained for 15 min with an optimal concentration of fluorescently tagged monoclonal antibodies recognizing CD4 as well as 7-amino-actinomycin D (7-AAD) per manufacturer's instructions. 7-AAD is a fluorescent compound that intercalates into nucleic acids but is excluded from intact cells making it useful as a measure of cellular viability. After staining the cells were washed twice in FCB prior to flow.
3. A minimum of 10,000 $CD4^+$ T-cell events were collected per sample on a flow cytometer. All flow cytometry data

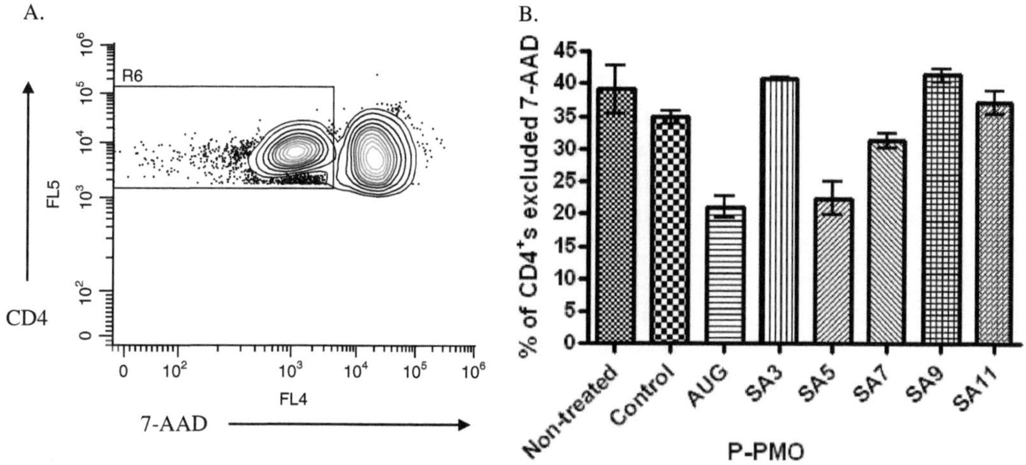

Fig. 10.5. Effects of P-PMO on $CD4^+$ T-cell viability (7-AAD exclusion). Splenocytes were stimulated for 48 h with plate-bound anti-CD3 (5 μg/mL) and treated with 5 μM P-PMO. Cells were harvested and stained with APC-Alexa Fluor® 750-CD4 and 2 μg/mL 7-AAD for 15 min, and analyzed by flow cytometry (**a**). The percent of $CD4^+$ cells which excluded 7-AAD (7-AAD negative) was measured (mean ± SEM) (**b**).

analysis and software compensation were performed using WinList.

4. As shown in **Fig. 10.5b**, there was a decrease in the percent of cells that excluded 7-AAD in the AUG-, SA5-, and SA7-treated samples. This likely explains why a decrease in IL-12Rb2 protein expression was measured for cells treated with these compounds. Thus, due to their effects on viability, AUG, SA5, and SA7 were omitted as candidate compounds.

3.5. Final Identification and Validation of the Best Candidate Compound and Measuring Efficacy of the Selected Candidate

1. To confirm the optimal candidate compound we then revisited the data collected from the first 4 steps and measured the efficacy of the final chosen compound (**Fig. 10.7a**). The initial screen for splice-altered mRNA products by RT-PCR showed every compound disrupted normal IL-12Rb2 mRNA splicing.
2. Compounds AUG, SA7, and SA9 were shown to inhibit IL-12rb2 protein expression on the cell surface, but SA3, SA5, and SA11 were not yet eliminated.
3. After looking at the phosphorylation of STAT4 we excluded SA5 and SA11 due to a lack of downstream effect, which all other compounds exhibited.
4. This left AUG, SA7, SA9, and SA3 as potential candidates. Of these four remaining candidates, two of the compounds were eliminated after a measure of viability suggested AUG and SA7 were impacting cellular viability.
5. Thus SA3 and SA9 are the two remaining candidate compounds.

6. Upon re-examination of the gel (**Fig. 10.1b**), SA9-treated cells had produced three visibly different sized PCR products, while SA3-treated cells produced only one. The ability of the SA3 compound to produce a cleaner and more consistent splice product makes it a more attractive compound. However, of the two, SA9 was the only compound that demonstrated decreased surface expression of the IL-12Rb2 protein. We decided to sequence the PCR product of the SA3-treated sample to determine why the expression of IL-12Rb2 protein was not affected when analyzed by flow cytometry.
7. Bands of the predicted size (for the SA3-treated splice product and full size product) were excised from the gel.
8. The DNA in the gel pieces was purified with the QIAquick® Gel Extraction Kit per manufacturer's instructions.
9. The clean fragments were cloned into pCR4-TOPO vector per manufacturer's instructions.
10. The cloned plasmid DNA was purified with Qiagen QIAprep Spin Miniprep Kit per manufacturer's instructions and then sequenced at the CGRB Core Laboratory at Oregon State University.
11. After comparing the sequence of the altered SA3 PCR product to the full-length IL-12Rb2 sequence, we found that exon 3 sequence was absent, and instead exons 2 and 4 were joined (**Fig. 10.6**). This indicated that the splicesome had continued onto the exon 4 SA site when the exon 3 SA site was blocked by the P-PMO. Interestingly, exon 3 is made up of 327 bp, which is divisible by 3, the number

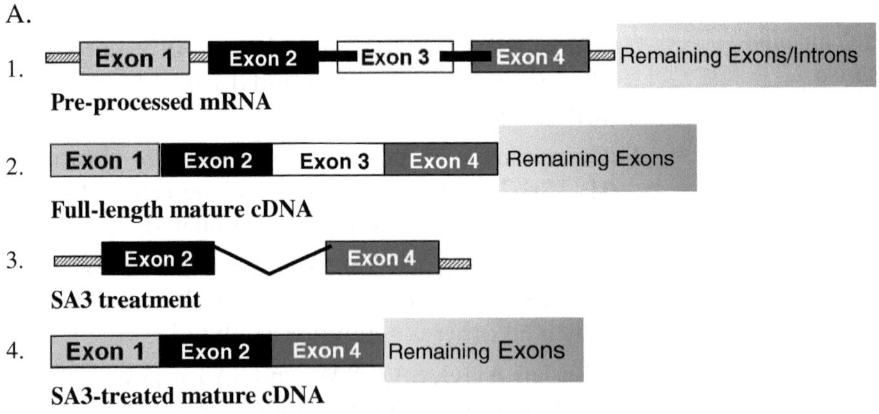

Fig. 10.6. Mechanism of action for the SA3 compound. The splice-altered PCR product produced from SA3-treated cells was excised from the agarose gel, cloned, and sequenced. The altered PCR product sequence showed removal of exon 3 and the joining of exons 2 and 4 as indicated graphically: 1. The pre-processed mRNA containing exons (*boxes*) and introns (*shaded bars*). 2. Once the introns are removed during processing the full-length mature cDNA is formed. 3. In the presence of the SA3 compound splicing machinery removes exon 3 with the surrounding introns. 4. This results in a mature cDNA lacking exon 3 that is expressed at normal levels.

of nucleotides in a codon which encodes for an amino acid. Thus, after excision of the 327 bp of exon 3, the remaining transcript was in frame and normal translation of subsequent exons was preserved, which explains why the IL-12Rb2 protein was still detectable by flow cytometry (*see* **Note 3**). Altering gene splicing within the coding frame has also shown therapeutic potential (*see* **Note 4**).

12. To further validate its efficacy, we measured a dose-dependent induction of altered splicing of IL-12Rb2 mRNA in SA3-treated cells. Splenocyte suspensions were prepared as described in **Section 3.1.1**.
13. They were then stimulated with 0.5 μg/mL soluble anti-CD3 and 1 μg/mL anti-CD28 and treated with 0, 0.5, 1, 2, or 4 μM SA3 or 4 μM control P-PMO for 48 h.
14. The PCR products were then analyzed by agarose gel electrophoresis confirming that the effect on splicing was dose dependent (**Fig. 10.7b**).

A.

	1. Altered mRNA expression	2. Decreased IL-12Rb2 protein	3. Decreased Phospho-STAT4	4. Preserved cellular viability	5. Review: Consistency of splice altered product	Attribute Totals:	Best candidate chosen & validated
AUG		→	++			++	
SA3	+		++	+	+	+++++	✓
SA5	+				+	+	
SA7	+	→	++		+	++++	
SA9	+	→	+	+		+++	
SA11	+	→		+		++	

B.

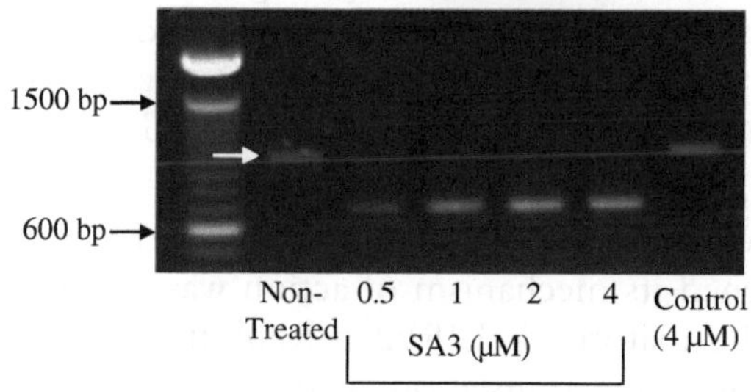

Fig. 10.7. Step-wise exclusion of IL-12Rb2-specific P-PMO for selection of optimal candidate compound, SA3. PPMO attributes based upon criteria of splicing, protein expression and activity, and cellular viability, and consistency of splice altered product. Number of "+" indicates level of activity, *arrows* and *grey boxes* indicate favorable or inconclusive results, respectively, *black boxes* indicate tests where compound was eliminated from consideration, and the check mark indicates the optimal compound (**a**). Splenocytes were stimulated with 0.5 μg/mL soluble anti-CD3 and 1 μg/mL anti-CD28 and treated with 0, 0.5, 1, 2, or 4 μM SA3 or 4 μM control P-PMO for 48 h. IL-12Rb2-specific PCR products were identified by agarose gel electrophoresis (**b**).

4. Notes

1. The simplest method for quantitative RT-PCR involves the incorporation of a double-stranded DNA binding molecule, such as SYBR Green, into an amplified product. Once bound to double-stranded DNA, SYBR Green absorbs blue light at 488 nm and emits a green fluorescent signal at 522 nm that can be measured during each cycle of PCR amplification. A more accurate method involves the deployment of a sequence-specific probe, such as Taqman, which is released from the template strand during amplification. The most common DNA-binding probe consists of a sequence-specific oligonucleotide with a flourophore on the 5′ end and a signal quencher at the 3′ end. When the quencher and the flourophore are in close proximity to one another, no fluorescence is emitted. During DNA amplification the 5′ to 3′ exonuclease activity of Taq polymerase degrades the probe from the 5′ end releasing the flourophore from the quencher resulting in measurable fluorescence that is directly proportional to the number of amplified products.
2. When using a measure of cellular viability to eliminate compounds it should be noted that in some cases inhibiting expression of a target gene could impact viability as the gene may play a role in preserving cell survival. Thus excluding compounds that impact cellular viability is not always an appropriate step. An interesting observation we have made while treating T cells with various AUG-specific P-PMO is an alteration in cellular activity. We have observed this in cells treated with P-PMO targeting the AUG of CTLA-4, a negative co-stimulatory molecule expressed by T cells, and the AUG of Foxp3, a transcription factor that confers suppressive regulatory function to T cells (data not shown). We again saw this effect when targeting the AUG of IL-12Rb2. The significance of this is not known; however, it presents another advantage to using a splice-altering approach to antisense design over targeting the AUG start codon, particularly in T cells.
3. After selection of the SA3 compound as our lead candidate, we learned its mechanism of action was removal of exon 3 from the mature IL-12Rb2 mRNA transcript, which preserved in-frame translation of the protein, but sufficiently disrupted IL-12Rb2 protein function. It should be noted that the skipping of exons made up of a number of bp divisible by 3 does not guarantee a frame-shift will not occur. Rather than find the next splice acceptor/donor site on the subsequent exon, the splicesome may find a cryptic splice

site within exonic or intronic sequence. There are natural protective mechanisms in place for when this occurs as aberrant mRNA and protein expression could be detrimental to the organism. Exon-junction complexes and the ribosome can recognize where aberrant-splicing occurs according to the position of other landmarks on the transcript. This results in nonsense-mediated decay of the transcript by the proteosome (3), causing the "nonsense" mRNA transcript to become degraded instead of translated. For antisense-targeting of exons not divisible by 3, this is likely a common mechanism for inhibition of normal gene expression.

4. The emerging approach for antisense is to target the 5′ and 3′ ends of exons to disrupt mRNA splicing. One advantage of this is a positive read-out for an antisense effect. Though most often the goal of antisense is to inhibit normal gene expression, it is estimated that at least 50% of genes are alternatively spliced, producing splice forms that play a specific role in biological processes. This means that antisense could be used to redirect splicing to induce naturally occurring splice forms as an approach to altering biological function. For example, targeting the SA site on exon 2 of the mouse CTLA-4 gene with P-PMO resulted in removal of exon 2, which encodes for the domain which interacts with B7 molecules on dendritic cells (2). This form can occur naturally in mice, though at disproportionally low levels in the non-obese diabetic (NOD) mice, which is linked to their spontaneous development of diabetes (4). Treating NOD mice with SA2 P-PMO resulted in protection from development of diabetes (2) through the increased production of a CTLA-4 splice form that does not need to bind to its cognate ligand to signal, thus a ligand-independent form. There are numerous possibilities for the use of gene-specific splice-altering antisense to modify biological processes. The critical step for their successful use remains careful screening for efficacious compounds as it is otherwise easy to overlook the best candidate.

Acknowledgments

We would like to acknowledge the technical assistance of Margaret Lubkin, and recognize the Cell and Tissue Analysis Facilities and Service Core of the Environmental Health Sciences Center and Center for Genome Research and Biocomputing at Oregon State University.

References

1. Marshall, N. B., Oda, S. K., London, C. A., Moulton, H. M., Iversen, P. L., Kerkvliet, N. I., and Mourich, D. V. (2007) Arginine-rich cell-penetrating peptides facilitate delivery of antisense oligomers into murine leukocytes and alter pre-mRNA splicing. *J Immunol Methods* **325**, 114–26.
2. Mourich, D. V., Oda, S. K., London, C. A., Marshall, N. B., Schnell, F. J., Hauck, L., Hinrichs, D. J., and Iversen, P. (2010, submitted) Ligand independent form of CTLA-4 induced by antisense EXON skipping in NOD mouse inhibits autoimmune diabetes *J. Immunol.*
3. Chang, Y. F., Imam, J. S., and Wilkinson, M. F. (2007) The nonsense-mediated decay RNA surveillance pathway. *Annu Rev Biochem* **76**, 51–74.
4. Vijayakrishnan, L., Slavik, J. M., Illes, Z., Greenwald, R. J., Rainbow, D., Greve, B., Peterson, L. B., Hafler, D.A., Freeman, G. J., Sharpe, A. H., Wicker, L. S., and Kuchroo, V. K. (2004) An autoimmune disease-associated CTLA-4 splice variant lacking the B7 binding domain signals negatively in T cells. *Immunity* **20**, 563–75.

Chapter 11

Diverse Small Non-coding RNAs in RNA Interference Pathways

Liande Li and Yi Liu

Abstract

Large numbers of diverse small non-coding RNAs have been discovered and characterized in eukaryotic RNA interference pathways. These small RNAs have distinctive characteristics and are associated with Argonaute family proteins to regulate gene expression and genomes at various levels. These small RNAs include the Dicer-dependent group such as microRNAs (miRNAs) and small interfering RNAs (siRNAs), and the Dicer-independent group such as Piwi-interacting RNAs (piRNAs). This review summarizes the various classes of eukaryotic small RNAs and the general knowledge of their characteristics, biogenesis, and functions, with emphasis on some of the recently identified small RNAs.

Key words: RNAi, miRNA, siRNA, piRNA, qiRNA.

1. Introduction

Small silencing RNAs (sRNAs) are 20–30 nt small non-coding RNA molecules, which function through the conserved RNA interference (RNAi) pathways to mediate both post-transcriptional and transcriptional gene silencing in eukaryotes (1–4). In general, the RNAi and its related pathways start with the generation of small RNA duplexes from double-stranded RNA (dsRNA) precursors by the RNase III protein Dicer-like enzymes, then the small RNA duplexes are loaded onto the RNA-induced silencing complex (RISC) containing an Argonaute family protein. After one strand (passenger strand) of the small RNA duplex is removed, the RISC is activated, which then uses the remaining single-stranded small RNA (guide strand) as a guide to find and silence the various RNA targets (5–7).

J. Goodchild (ed.), *Therapeutic Oligonucleotides*, Methods in Molecular Biology 764,
DOI 10.1007/978-1-61779-188-8_11,

Since the discovery of the first small RNA by Ambro's group and the establishment of the RNAi concept by Fire and Mello, various classes of small non-coding RNAs have been identified in eukaryotes with distinct characteristics (3, 8–11). The recent advancements of bioinformatics and deep sequencing technologies have greatly contributed to the explosive expansion of the small RNA world, with many novel types of small RNAs identified. The distinct characteristics of many types of sRNAs and their biogenesis pathways have also made the boundary between different classes of small RNAs vague (10). However, based on whether or not their biogenesis is dependent on Dicer, the dsRNA specific RNA III ribonuclease, all the known eukaryotic sRNAs can be simply classified into two groups: Dicer-dependent small RNAs and Dicer-independent small RNAs. The Dicer-dependent group includes microRNAs (miRNAs) and small interfering RNAs (siRNAs), which are the two well-studied classes. In this review, we will discuss the known sRNAs in three groups: miRNAs, siRNAs, and Dicer-independent small RNAs.

2. miRNAs

miRNAs arise from single-stranded RNA precursor transcripts with stem–loop structures produced from discrete genomic loci, which are processed by Dicer-like enzymes (12, 13). miRNAs have been reported in animals, plants, and algae (14–20), but not in the fungal kingdom. However, our recent discoveries have shown that miRNA and miRNA-like small RNAs exist in filamentous fungus *Neurospora crassa* (Since the submission of this review, miRNA like RNAs were discovered in the filamentous fungus *Neurospora* (21)), thus miRNAs exist in all major branches of eukaryotic organisms, indicating that miRNA-like gene regulatory mechanism evolved early in the eukaryotic lineage. The miRBase database (http://www.mirbase.org) (22) has more than 10,000 distinct miRNA genes listed from eukaryotes, including two from the slime molds besides those from animals, plants, and algae. However, further experimental validation may be needed for many of them.

2.1. miRNAs from Animals

The discoveries of the first miRNA *lin-4* and the second miRNA *let-7* in *Caenorhabditis elegans* set the foundation for the identification of miRNAs in animals (8, 23). In miRBase database (http://www.mirbase.org/) (22), there are more than 8600 registered animal miRNA genes.

Animal miRNA genes are often clustered and some are present within exons or introns (24, 25). In general, the miRNA genes are transcribed by RNA polymerase II (Pol II) into the primary transcripts called pri-miRNAs containing hairpin structures (26, 27). The pri-miRNA is cleaved in the nucleus at the stem of the hairpin into a 60–70-nt pre-miRNA, by the nuclear RNase III protein Drosha with a cofactor protein (Pasha in flies and nematodes and DGCR8 in mammals) (28–31). The pre-miRNAs are exported to the cytoplasm by exportin 5 and are then further processed into an approximately 22-nt miRNA/miRNA* duplex (where miRNA* is the passenger strand) by Dicer, with the assistance of a dsRNA-binding domain containing protein (TRBP in mammals or LOQS in flies) (3, 10). The small RNA duplex is then loaded onto an Argonaute protein to form the miRNA effector complex (miRISC), with the miRNA strand bound to the Argonaute as the guide and the miRNA* strand degraded, to regulate various biological processes (10). In animals, miRNAs are usually imperfectly complementary to multiple sites in the 3′ untranslated regions (UTRs) of their target mRNAs via their 5′ end sequences (seed regions), though a few have a nearly complete complementarity to their mRNA targets. As a result, animal miRNAs mostly regulate the targets by affecting mRNA stability and by repressing translation and only rarely by Argonaute-mediated slicing of mRNA targets (3, 25, 32). Animal miRNAs may play important roles in many aspects of biological functions, such as development and differentiation, by regulating gene expression or protein translation. Due to the relaxed base-pairing requirements between miRNAs and their targets, one miRNA can have many targets and it is predicted that a majority of animal genes may be regulated by one or more miRNAs except for some housekeeping genes (33).

2.2. miRNAs from Plants

miRNAs in plants have been extensively studied since their initial discovery in 2002, contributing significantly to our current understanding of miRNA function and biogenesis (11, 22, 25, 34). More than 2000 registered miRNA genes can be found in the miRBase database (http://www.mirbase.org/) (22). Plant miRNAs are generally different from those in animals as to their characteristics, biogenesis, and mode of action. Most plant miRNA genes are located between two ORFs, with more variable lengths and stem–loop structures, and are usually not arranged in tandem (25, 35). As animal miRNA genes, the plant miRNA genes are generally transcribed into pri-miRNAs by Pol II. In *Arabidopsis*, the Dicer-like protein DCL1 plays the function of both animal Drosha and Dicer to process the pri-miRNAs into pre-miRNAs and then to miRNA/miRNA* duplexes, assisted by the dsRNA binding protein HYL1 and a zinc finger protein SE. The miRNA duplexes are then exported to the cytoplasm by proteins including an Exportin 5 homologue HASTY (3, 11). In addition, the plant miRNAs are 2′-*O*-methylated at the 3′ termini by HEN1, which stabilizes sRNA by preventing their 3′ uridylation and degradation (36). The mature miRNAs are then loaded onto the miRISC containing the Argonaute AGO1 to perform their silencing functions.

Interestingly, many of the plant miRNAs are not conserved, even within closely related species such as *Arabidopsis thaliana* and *Arabidopsis lyrata*, demonstrating that miRNA genes evolve rapidly in plants (11, 37). This may also explain why no homologous miRNA species is present between animals and plants (38).

The miRNAs in plants are generally highly or perfectly complementary to their mRNA targets with mostly one single target sites in coding regions, and they direct the target cleavage at the site corresponding to the 10th and 11th nucleotides of the miRNA in the miRNA-mRNA duplex. However, the impact of translational repression mediated by plant miRNAs has been recognized recently as a widespread phenomenon (11, 17, 25, 38). Unlike the relaxed base-pairing requirements of animal miRNAs, which lead to many target mRNAs, most plant miRNAs have only limited number of predicted targets (3, 25).

Many mutants related to the plant miRNA biogenesis have developmental defects, indicating the important roles of miRNAs in plant development. miRNAs in plants have been reported functioning in phase transitions, hormone biosynthesis and signaling, pattern formation and morphogenesis, and other processes (11).

2.3. miRNAs from Green Algae

No miRNA was previously identified in unicellular organisms and thus miRNAs were previously considered to evolve independently in plants and animals as global regulators associated with multicellularity (39, 40). However, two independent studies in 2007 showed that the unicellular green alga *Chlamydomonas reinhardtii* contains many miRNAs (18, 19). These small RNAs share many characteristics with miRNAs in higher plants, such as the common features of miRNA precursors, miRNA directed target cleavage, 5′ phosphate, 5′uracil preference, and 3′-end modification. However, the biogenesis and function of these identified miRNAs are not yet clear in this organism (41).

3. Small Interfering RNAs (siRNAs)

siRNAs are processed from dsRNA precursors by Dicers, with multiple siRNA species produced from both strands of dsRNA (3). siRNAs can be classified into exogenous-siRNAs (exo-siRNA) and endogenously produced-siRNAs (endo-siRNAs) based on their origin (3). Here we discuss the different groups of siRNAs found in different eukaryotes.

3.1. Exo-siRNAs

Exo-siRNAs are produced from exogenous long dsRNAs from viruses, transgenes or exogenously supplied dsRNAs, which are 21–25 nt long and involved in the well-known RNA silencing phenomena such as co-suppression or PTGS (post-transcriptional gene silencing) in plants, quelling in *Neurospora*, and RNAi in animals (3, 9, 41–44). Dicer proteins cleave dsRNA into siRNA duplexes, which are then loaded onto the RISC with an Argonaute protein and the guide strand of the siRNA duplexes as the core components, similar to that of miRISC, to direct the gene silencing.

In animals, the exo-siRNAs processed by Dicers are generally 21 nt long, functioning in posttranscriptional regulation and antiviral defense; while the siRNAs can be 18–24 nt long in plants and mediate posttranscriptional gene silencing, antiviral defense, and transcriptional gene silencing (3, 25). In the filamentous fungus *N. crassa*, Dicer proteins cleave dsRNA into approximately 25 nt siRNAs, which are responsible for the transgene-induced post-transcriptional gene silencing phenomenon called quelling (45). In the fission yeast, siRNAs are predominantly 20–23 nt in length, which play a major role in heterochromatin formation and transcriptional silencing (2, 46–48).

Since introducing exo-siRNAs can provide specific and potent silencing of the target gene expression, this technology has been widely used in knocking-down individual genes for the identification of gene functions in almost all eukaryotes. Another very promising therapeutic application of RNAi is to use modified exo-siRNAs to silence gene expression that causes varieties of human diseases (49–51).

3.2. Endo-siRNAs

Endogenously produced-siRNAs (endo-siRNAs) have recently been found in plants, fungi, nematodes, flies, and mammals, indicating their wide presence in eukaryotes (3).

3.2.1. Plant Endo-siRNAs

Endo-siRNAs in plants include the casiRNAs (*cis*-acting siRNAs), tasiRNAs (*trans*-acting siRNAs), natsiRNAs (natural antisense transcript-derived siRNAs), and lsiRNAs (long siRNAs), which like other siRNAs, all require Dicers for their biogenesis (3). casiRNAs are 24 nt long, arise from transposons, repetitive elements, and tandem repeats, and function in chromatin modification by directing DNA methylation and histone modification (52). The natsiRNAs arise from dsRNA made by convergently transcribed RNAs, require DCL2 and/or DCL1 for their production with sizes of 21, 22, or 24 nt, and can regulate stress-response genes (3, 53, 54). lsiRNAs are also produced from natural antisense transcript pairs but are 30–40 nt long, and are induced by pathogen challenge or specific growth conditions (3, 55). lsiRNAs may regulate plant resistance to pathogens, since the induction of one of the lsiRNAs, AtlsiRNA-1, can silence the expression of AtRAP, a negative regulator of plant defense (55).

tasiRNAs are mostly 21 nt in size, and their biogenesis is a good example of the crosstalk between miRNA and siRNA biogenesis pathways (11, 56). tasiRNAs are produced by the dicer DCL4 from the RNA dependent RNA polymerase RDR6 converted dsRNA, which are generated from miRNA-cleaved transcripts from several *TAS* loci. These tasiRNAs, similar to plant miRNAs in both molecular and developmental functions, regulate targets in a posttranscriptional manner by target cleavage or translational inhibition. Since tasiRNAs may have the ability to move from cell to cell, they may function as signaling molecules (3, 11, 56).

3.2.2. Animal Endo-siRNAs

Endo-siRNAs in animals can be produced by Dicers from dsRNA precursors originated from various genomic regions (3). The dsRNA can come from structured loci where the transcripts can fold into long stem hairpins; from convergent or bidirectional transcription, or read-through transcription of the transposons in opposite direction; or from protein-coding genes which can pair with pseudogenes via trans-interaction and from regions with duplicated and inverted pseudogenes (3, 57–62). Though the general biological significance of most animal endo-siRNAs is not clear yet, the fly endogenous RNA pathway may play roles in silencing transposons (3, 60). Most recently, studies demonstrated that a novel type of endogenous 26 nt small RNAs with a 5′ guanine (26G-RNAs) in *C. elegans* may function as the primary small RNAs leading to the production of the transposon silencing and centromere function-related 22-nt RNAs (22G-RNAs; see below for details) (63–69). These 26G-RNAs are associated with the Piwi-clade ERGO-1 and are antisense with 5′ monophosphorylated. In addition, they also require the dicer DCR-1 for the their biogenesis (63–65).

3.2.3. DNA Damage-Induced qiRNA in Neurospora

Recently, a new class of Argonaute-binding siRNAs, which are induced after DNA damage, was discovered in the filamentous fungus *N. crassa* (70). This class of sRNA was called qiRNA due to its association with the *Neurospora* Argonaute protein QDE-2. qiRNA levels are very low under normal growth conditions but were greatly induced by treatment of DNA-damaging agents such as histidine, hydroxyurea, or EMS. These small RNAs originate from the highly repetitive rDNA locus, match both sense and antisense strands with approximately equal numbers and are about 20–21 nt long, several nucleotides shorter than *Neurospora* siRNAs. In addition, they exhibit a strong preference for uridine at the 5′ end. The biogenesis of qiRNAs requires the RNA-dependent RNA polymerase QDE-1, QDE-3 (a RecQ DNA helicase homologous to the Werner/Bloom Syndrome proteins), and

Dicers, but not the Argonaute QDE-2. qiRNAs may contribute to the DNA damage response by inhibiting protein translation. Consistently with this hypothesis, mutants that lack qiRNA production exhibit increased DNA damage sensitivity.

4. Dicer-Independent Small RNAs

The Dicer-independent group of small silencing RNAs were discovered recently, with the Piwi-interacting RNAs (piRNAs) being a major class, and were intensively studied in the past a few years (3, 10, 20, 71–80). This group of small RNAs, whose biogenesis does not require Dicers, includes piRNAs, 21U-RNAs which are actually piRNAs from *C. elegans*, and secondary siRNAs in *C. elegans* (3, 6, 63, 75, 76). Here we will mainly focus on our current understanding of piRNAs.

4.1. Characteristics and Functions of piRNAs

piRNAs are physically associated with the Piwi clade of Argonaute proteins, are highly enriched in germ line cells, and to date have only been found in animals, including mammals, flies, zebrafish, nematodes, cnidarian, and poriferan (20, 71–79). These small RNAs, with sizes ranging from 24 to 32 nt, are longer than miRNAs and siRNAs (80). piRNAs may function to protect germline integrity by repressing the activity of transposable elements, though the functions of many piRNAs and the mechanisms of the related silencing pathways are currently not clear (3, 81–83). Current evidences support the notion that piRNAs and PIWI proteins serve as the regulators of transposon activity and confer the protection of the eukaryotes from transposons (80, 82–86).

4.2. piRNAs in Mammals

piRNAs were first identified from mouse testes in 2006 by several independent labs and were named because of their binding to the PIWI subfamily of Argonaute proteins (3, 71–74, 80, 81, 87, 88). Later, however, the previously reported *Drosophila* repeat-associated small interfering RNAs (rasiRNAs) were shown to be piRNAs in flies (3, 71, 78, 80, 89). These studies found that in mouse, piRNAs were mostly clustered at a few hundred loci throughout the genome, and there are two types of piRNAs: pre-pachytene piRNAs and pachytene piRNAs (3, 10, 80). The pre-pachytene piRNAs, associated with the Piwi proteins MILI and MIWI2, are mostly from clusters rich in repeats and transposons and are present before meiotic pachytene. On the other hand, the pachytene piRNA's are highly expressed in the pachytene stage, and are mostly from unannotated regions of the genome, and are associated with MILI and MIWI (3, 10, 80). Though piRNA sequences are not conserved in mammals, the piRNA clusters in mice, rat and humans are syntenic, and there is a strong 5′U preference for the piRNAs (3, 10, 80).

4.3. piRNAs in Insects

As mentioned above, piRNAs were initially found in flies as rasiRNAs for they match to repetitive DNA elements (3, 71, 80, 87). piRNAs in *D. melanogaster* are associated with the PIWI subfamily proteins PIWI, Aubergine (AUB), and AGO3. The AGO3-associated piRNAs mostly originate from the sense strand of retrotransposons and have an adenosine preference at nucleotide 10 but not at the 5′ ends, while the AUB and PIWI-associated are largely from the antisense transcripts of retrotransposons, with a strong 5′ end uridine preference (6, 10). Large number of putative piRNAs were also isolated from the silkworm *Bombyx mori*, among which about one-third are repeat-associated and have similar uridine or adenosine preference as in *Drosophila*, indicating a similar piRNA biogenesis mechanism in insects (80, 90).

4.4. piRNAs in Zebrafish

piRNAs 26–28 nt in length are found associated with PIWI proteins ZIWI and ZILI in zebrafish male and female germlines (80, 91, 92). Similar to the fly piRNAs, ZIWI associated piRNAs prefer 5′U while ZILI associated piRNAs prefer A at the 10th nucleotide (80, 91, 92).

4.5. piRNAs in Nematodes

The initially identified 21U-RNAs in *C. elegans* are in fact piRNAs since they were found to interact with the PRG-1, a PIWI-subfamily protein, and their production requires PRG-1 but not the Dicer protein DCR-1 (3, 6, 63, 75, 76). Though the 21U-RNAs are shorter than the other reported piRNAs, they also have a 5′ U bias, and are required for the germ line maintenance and fertility (3, 6, 63, 75, 76). Unlike the *Drosophila* piRNAs, the *C. elegans* piRNAs lack a preference for A at the 10th nucleotide (63).

4.6. Biogenesis of piRNAs

piRNAs have very diverse sequences and originate mostly from repetitive elements, transposons, large piRNA clusters and mostly match only one strand of the DNA, indicating that piRNAs arise from single-stranded RNA precursors (10, 77, 80, 93). The biogenesis of piRNA requires Piwi-related proteins, which might function in cleaving the precursor into mature siRNA (10, 80). Based on the different sequence characteristics of piRNAs associated with different PIWI proteins, a feed-forward posttranscriptional amplification mechanism, the "ping-pong" model, was proposed to be involved in the production of secondary piRNAs from primary piRNAs via the slicer activity of Piwi-related proteins (84, 94–96). However, such an amplification mechanism does not seem to be present in the piRNA production in *C. elegans* (75, 76), and the primary piRNAs from the *flamenco* region in

Drosophila are independent of the ping-pong pathway (10, 81). The biogenesis mechanism for primary piRNA production and the nuclease(s) involved are currently unknown.

4.7. Secondary siRNAs in C. elegans

Secondary siRNAs are small RNAs processed from RNA dependent RNA polymerase (RdRP)-generated dsRNAs (6, 64, 65, 97). Secondary siRNAs in *C. elegans* are a type of small RNAs with only antisense polarity, 5′ ends mainly triphosphorylated and are mostly found upstream of the initial dsRNA trigger (6, 64, 65, 98–100). They are produced by RdRPs using the target mRNAs as the templates, but are not dependent on Dicers (6, 64, 65, 98–100) and are sometimes not considered as siRNAs technically (3). Secondary siRNAs are the majority of the siRNA populations in *C. elegans* (64, 99). It was proposed that an Argonaute protein is guided by low levels of primary siRNAs generated from long dsRNAs by the dicer DCR-1 to recruit a RdRP to the targeted transcripts, which leads to the production of secondary siRNAs directly from the target mRNA without the need for Dicer-mediated cleavage of dsRNA. These single-stranded secondary siRNAs then bind to another Argonaute protein to mediate downstream gene silencing (10, 99).

This model is supported by two studies published recently, which demonstrated that two different groups of RdRPs function sequentially in different stages of the RNAi in *C. elegans* (64, 65). The Argonaute ERGO-1 interacts with the primary 26G-RNAs, whose biogenesis requires the RdRP RRF-3 and DCR-1, which may recruit the RdRP RRF-1 (or EGO-1) to transcribe the more abundant secondary small RNAs 22G-RNAs. The 22G-RNAs then associate with the worm-specific AGOs (WAGOs) such as NRDE-3 to guide gene silencing (64, 65). The 22G-RNAs were identified recently as a new class of small RNAs with a strong preference for 5′ guanine, are antisense and triphosphorylated, and require the RRF-1 and WAGOs for their production (65, 67). By associating with the Argonaute proteins WAGOs and CSR-1, 22G-RNAs can regulate such functions as transposon/repetitive element silencing, chromosome segregation and centromere assembly (65–69).

5. Conclusions

miRNAs and siRNAs, including exo-siRNAs and endo-siRNAs, have been found in diverse eukaryotic organisms. The dicer-independent sRNAs, a recently discovered, major, small RNA group, are present in many different animal species. These results indicate that gene silencing pathways utilizing small non-coding

RNAs evolved very early in the eukaryotic evolution. Due to recent advancements of deep sequencing technologies and bioinformatics methods, more small RNAs and additional novel types of small RNAs will surely be discovered in many different and diverse organisms. Despite our current knowledge of RNAi pathways, the biogenesis mechanisms of many types of sRNA remain largely unknown. Perhaps more importantly, the biological functions for the vast majority of the currently identified sRNA are not known. The lack of evolutionary conservation for most of the sRNA sequences indicates that the sRNA universe is rapidly evolving during evolution. Detailed characterization and functional studies of different small RNAs in different organisms in the future will shed more light on these questions.

Acknowledgement

This work is supported by funds from the National Institute of Health and Welch Foundation (I-1560) to Y. Liu.

References

1. Ambros, V. (2004) The functions of animal microRNAs. *Nature 431*, 350–355.
2. Buhler, M., and Moazed, D. (2007) Transcription and RNAi in heterochromatic gene silencing. *Nat Struct Mol Biol 14*, 1041–1048.
3. Ghildiyal, M., and Zamore, P. D. (2009) Small silencing RNAs: an expanding universe. *Nat Rev Genet 10*, 94–108.
4. Hannon, G. J. (2002) RNA interference. *Nature 418*, 244–251.
5. Jinek, M., and Doudna, J. A. (2009) A three-dimensional view of the molecular machinery of RNA interference. *Nature 457*, 405–412.
6. Siomi, H., and Siomi, M. C. (2009) On the road to reading the RNA-interference code. *Nature 457*, 396–404.
7. Maiti, M., Lee, H. C., and Liu, Y. (2007) QIP, a putative exonuclease, interacts with the *Neurospora* argonaute protein and facilitates conversion of duplex siRNA into single strands. *Genes Dev 21*, 590–600.
8. Lee, R. C., Feinbaum, R. L., and Ambros, V. (1993) The C. elegans heterochronic gene lin-4 encodes small RNAs with antisense complementarity to *lin-14*. *Cell 75*, 843–854.
9. Fire, A., Xu, S., Montgomery, M. K., Kostas, S. A., Driver, S. E., and Mello, C. C. (1998) Potent and specific genetic interference by double-stranded RNA in *Caenorhabditis elegans*. *Nature 391*, 806–811.
10. Kim, V. N., Han, J., and Siomi, M. C. (2009) Biogenesis of small RNAs in animals. *Nat Rev Mol Cell Biol 10*, 126–139.
11. Chen, X. (2009) Small RNAs and their roles in plant development. *Annu Rev Cell Dev Biol 25*, 21–44.
12. Bartel, D. P. (2004) MicroRNAs: genomics, biogenesis, mechanism, and function. *Cell 116*, 281–297.
13. Ambros, V., Bartel, B., Bartel, D. P., Burge, C. B., Carrington, J. C., Chen, X., Dreyfuss, G., Eddy, S. R., Griffiths-Jones, S., Marshall, M., Matzke, M., Ruvkun, G., and Tuschl, T. (2003) A uniform system for microRNA annotation. *RNA 9*, 277–279.
14. Lee, R. C., Feinbaum, R. L., and Ambros, V. (1993) The *C. elegans* heterochromatic gene *lin-4* encodes small RNAs with antisense complementarity to *lin-14*. *Cell 75*, 843–854.
15. Lagos-Quintana, M., Rauhut, R., Lendeckel, W., and Tuschl, T. (2001) Identification of novel genes coding for small expressed RNAs. *Science 294*, 853–858.
16. Lee, R. C., and Ambros, V. (2001) An extensive class of small RNAs in *Caenorhabditis elegans*. *Science 294*, 862–864.

17. Llave, C., Kasschau, K. D., Rector, M. A., and Carrington, J. C. (2002) Endogenous and silencing-associated small RNAs in plants. *Plant Cell 14*, 1605–1619.
18. Molnar, A., Schwach, F., Studholme, D. J., Thuenemann, E. C., and Baulcombe, D. C. (2007) miRNAs control gene expression in the single-cell alga *Chlamydomonas reinhardtii. Nature 447*, 1126–1129.
19. Zhao, T., Li, G., Mi, S., Li, S., Hannon, G. J., Wang, X. J., and Qi, Y. (2007) A complex system of small RNAs in the unicellular green alga *Chlamydomonas reinhardtii. Genes Dev 21*, 1190–1203.
20. Grimson, A., Srivastava, M., Fahey, B., Woodcroft, B. J., Chiang, H. R., King, N., Degnan, B. M., Rokhsar, D. S., and Bartel, D. P. (2008) Early origins and evolution of microRNAs and Piwi-interacting RNAs in animals. *Nature 455*, 1193–1197.
21. Lee, H. C., Li, L., Gu, W., Xue, Z., Crosthwaite, S. K., Pertsemlidis, A., Lewis, Z. A., Freitag, M., Selker, E. U., Mello, C. C., and Liu, Y. (2010) Diverse pathways generate microRNA-like RNAs and Dicer-independent small interfering RNAs in fungi. *Mol Cell 38*, 803–814.
22. Griffiths-Jones, S., Saini, H. K., van Dongen, S., and Enright, A. J. (2008) miRBase: tools for microRNA genomics. *Nucl Acids Res 36*, D154–158.
23. Reinhart, B. J., Slack, F. J., Basson, M., Pasquinelli, A. E., Bettinger, J. C., Rougvie, A. E., Horvitz, H. R., and Ruvkun, G. (2000) The 21-nucleotide let-7 RNA regulates developmental timing in *Caenorhabditis elegans. Nature 403*, 901–906.
24. Kim, V. N. (2005) MicroRNA biogenesis: coordinated cropping and dicing. *Nat Rev Mol Cell Biol 6*, 376–385.
25. Voinnet, O. (2009) Origin, biogenesis, and activity of plant microRNAs. *Cell 136*, 669–687.
26. Lee, Y., Kim, M., Han, J., Yeom, K. H., Lee, S., Baek, S. H., and Kim, V. N. (2004) MicroRNA genes are transcribed by RNA polymerase II. *EMBO J 23*, 4051–4060.
27. Cai, X., Hagedorn, C. H., and Cullen, B. R. (2004) Human microRNAs are processed from capped, polyadenylated transcripts that can also function as mRNAs. *RNA 10*, 1957–1966.
28. Lee, Y., Ahn, C., Han, J., Choi, H., Kim, J., Yim, J., Lee, J., Provost, P., Radmark, O., Kim, S., and Kim, V. N. (2003) The nuclear RNase III Drosha initiates microRNA processing. *Nature 425*, 415–419.
29. Han, J., Lee, Y., Yeom, K. H., Kim, Y. K., Jin, H., and Kim, V. N. (2004) The Drosha-DGCR8 complex in primary microRNA processing. *Genes Dev 18*, 3016–3027.
30. Gregory, R. I., Yan, K. P., Amuthan, G., Chendrimada, T., Doratotaj, B., Cooch, N., and Shiekhattar, R. (2004) The microprocessor complex mediates the genesis of microRNAs. *Nature 432*, 235–240.
31. Landthaler, M., Yalcin, A., and Tuschl, T. (2004) The human DiGeorge syndrome critical region gene 8 and Its D. melanogaster homolog are required for miRNA biogenesis. *Curr Biol 14*, 2162–2167.
32. Carthew, R. W., and Sontheimer, E. J. (2009) Origins and mechanisms of miRNAs and siRNAs. *Cell 136*, 642–655.
33. Barringhaus, K. G., and Zamore, P. D. (2009) MicroRNAs: regulating a change of heart. *Circulation 119*, 2217–2224.
34. Reinhart, B. J., Weinstein, E. G., Rhoades, M. W., Bartel, B., and Bartel, D. P. (2002) MicroRNAs in plants. *Genes Dev 16*, 1616–1626.
35. Zhang, B., Pan, X., and Stellwag, E. J. (2008) Identification of soybean microRNAs and their targets. *Planta 229*, 161–182.
36. Yu, B., Yang, Z., Li, J., Minakhina, S., Yang, M., Padgett, R. W., Steward, R., and Chen, X. (2005) Methylation as a crucial step in plant microRNA biogenesis. *Science 307*, 932–935.
37. Felippes, F. F., Schneeberger, K., Dezulian, T., Huson, D. H., and Weigel, D. (2008) Evolution of *Arabidopsis thaliana* microRNAs from random sequences. *RNA 14*, 2455–2459.
38. Brodersen, P., Sakvarelidze-Achard, L., Bruun-Rasmussen, M., Dunoyer, P., Yamamoto, Y. Y., Sieburth, L., and Voinnet, O. (2008) Widespread translational inhibition by plant miRNAs and siRNAs. *Science 320*, 1185–1190.
39. Siomi, H., and Siomi, M. C. (2007) Expanding RNA physiology: microRNAs in a unicellular organism. *Genes Dev 21*, 1153–1156.
40. Odling-Smee, L. (2007) Complex set of RNAs found in simple green algae. *Nature 447*, 518.
41. Napoli, C., Lemieux, C., and Jorgensen, R. (1990) Introduction of a chimeric chalcone synthase gene into Petunia results in reversible co-suppression of homologous genes in trans. *Plant Cell 2*, 279–289.
42. Romano, N., and Macino, G. (1992) Quelling: transient inactivation of gene expression in *Neurospora crassa* by transformation with homologous sequences. *Mol Microbiol 6*, 3343–3353.
43. Carmell, M. A., and Hannon, G. J. (2004) RNase III enzymes and the initiation of

gene silencing. *Nat Struct Mol Biol* ***11***, 214–218.
44. Cecere, G., and Cogoni, C. (2009) Quelling targets the rDNA locus and functions in rDNA copy number control. *BMC Microbiol* ***9***, 44.
45. Catalanotto, C., Pallotta, M., ReFalo, P., Sachs, M. S., Vayssie, L., Macino, G., and Cogoni, C. (2004) Redundancy of the two dicer genes in transgene-induced post-transcriptional gene silencing in *Neurospora crassa*. *Mol Cell Biol* ***24***, 2536–2545.
46. Catalanotto, C., Nolan, T., and Cogoni, C. (2006) Homology effects in Neurospora crassa. *FEMS Microbiol Lett* ***254***, 182–189.
47. Freitag, M., Lee, D. W., Kothe, G. O., Pratt, R. J., Aramayo, R., and Selker, E. U. (2004) DNA methylation is independent of RNA interference in *Neurospora*. *Science* ***304***, 1939.
48. Chicas, A., Forrest, E. C., Sepich, S., Cogoni, C., and Macino, G. (2005) Small interfering RNAs that trigger posttranscriptional gene silencing are not required for the histone H3 Lys9 methylation necessary for transgenic tandem repeat stabilization in *Neurospora crassa*. *Mol Cell Biol* ***25***, 3793–3801.
49. Castanotto, D., and Rossi, J. J. (2009) The promises and pitfalls of RNA-interference-based therapeutics. *Nature* ***457***, 426–433.
50. Siomi, M. C. (2009) Short interfering RNA-mediated gene silencing; towards successful application in human patients. *Adv Drug Deliv Rev* ***61***, 668–671.
51. Jackson, A. L., and Linsley, P. S. (2010) Recognizing and avoiding siRNA off-target effects for target identification and therapeutic application. *Nat Rev Drug Discov* ***9***, 57–67.
52. Xie, Z., Johansen, L. K., Gustafson, A. M., Kasschau, K. D., Lellis, A. D., Zilberman, D., Jacobsen, S. E., and Carrington, J. C. (2004) Genetic and functional diversification of small RNA pathways in plants. *PLoS Biol* ***2***, E104.
53. Katiyar-Agarwal, S., Morgan, R., Dahlbeck, D., Borsani, O., Villegas, A., Jr., Zhu, J. K., Staskawicz, B. J., and Jin, H. (2006) A pathogen-inducible endogenous siRNA in plant immunity. *Proc Natl Acad Sci USA* ***103***, 18002–18007.
54. Borsani, O., Zhu, J., Verslues, P. E., Sunkar, R., and Zhu, J. K. (2005) Endogenous siRNAs derived from a pair of natural cis-antisense transcripts regulate salt tolerance in *Arabidopsis*. *Cell* ***123***, 1279–1291.
55. Katiyar-Agarwal, S., Gao, S., Vivian-Smith, A., and Jin, H. (2007) A novel class of bacteria-induced small RNAs in Arabidopsis. *Genes Dev* ***21***, 3123–3134.
56. Vazquez, F., Vaucheret, H., Rajagopalan, R., Lepers, C., Gasciolli, V., Mallory, A. C., Hilbert, J. L., Bartel, D. P., and Crete, P. (2004) Endogenous trans-acting siRNAs regulate the accumulation of *Arabidopsis* mRNAs. *Mol Cell* ***16***, 69–79.
57. Ghildiyal, M., Seitz, H., Horwich, M. D., Li, C., Du, T., Lee, S., Xu, J., Kittler, E. L., Zapp, M. L., Weng, Z., and Zamore, P. D. (2008) Endogenous siRNAs derived from transposons and mRNAs in *Drosophila* somatic cells. *Science* ***320***, 1077–1081.
58. Okamura, K., Balla, S., Martin, R., Liu, N., and Lai, E. C. (2008) Two distinct mechanisms generate endogenous siRNAs from bidirectional transcription in *Drosophila melanogaster*. *Nat Struct Mol Biol* ***15***, 581–590.
59. Okamura, K., Chung, W. J., Ruby, J. G., Guo, H., Bartel, D. P., and Lai, E. C. (2008) The *Drosophila* hairpin RNA pathway generates endogenous short interfering RNAs. *Nature* ***453***, 803–806.
60. Chung, W. J., Okamura, K., Martin, R., and Lai, E. C. (2008) Endogenous RNA interference provides a somatic defense against *Drosophila* transposons. *Curr Biol* ***18***, 795–802.
61. Tam, O. H., Aravin, A. A., Stein, P., Girard, A., Murchison, E. P., Cheloufi, S., Hodges, E., Anger, M., Sachidanandam, R., Schultz, R. M., and Hannon, G. J. (2008) Pseudogene-derived small interfering RNAs regulate gene expression in mouse oocytes. *Nature* ***453***, 534–538.
62. Czech, B., Malone, C. D., Zhou, R., Stark, A., Schlingeheyde, C., Dus, M., Perrimon, N., Kellis, M., Wohlschlegel, J. A., Sachidanandam, R., Hannon, G. J., and Brennecke, J. (2008) An endogenous small interfering RNA pathway in *Drosophila*. *Nature* ***453***, 798–802.
63. Ruby, J. G., Jan, C., Player, C., Axtell, M. J., Lee, W., Nusbaum, C., Ge, H., and Bartel, D. P. (2006) Large-scale sequencing reveals 21U-RNAs and additional microRNAs and endogenous siRNAs in *C. elegans*. *Cell* ***127***, 1193–1207.
64. Gent, J. I., Lamm, A. T., Pavelec, D. M., Maniar, J. M., Parameswaran, P., Tao, L., Kennedy, S., and Fire, A. Z. (2010) Distinct phases of siRNA synthesis in an endogenous RNAi pathway in *C. elegans* Soma. *Mol Cell* ***37***, 593–595.
65. Vasale, J. J., Gu, W., Thivierge, C., Batista, P. J., Claycomb, J. M., Youngman, E. M., Duchaine, T. F., Mello, C. C., and Conte, D., Jr. (2010) Sequential rounds of RNA-dependent RNA transcription drive

endogenous small-RNA biogenesis in the ERGO-1/Argonaute pathway. *Proc Natl Acad Sci USA* ***107***, 3582–3587.

66. Halic, M., and Moazed, D. (2009) 22G-RNAs in transposon silencing and centromere function. *Mol Cell* ***36***, 170–171.
67. Gu, W., Shirayama, M., Conte, D., Jr., Vasale, J., Batista, P. J., Claycomb, J. M., Moresco, J. J., Youngman, E. M., Keys, J., Stoltz, M. J., Chen, C. C., Chaves, D. A., Duan, S., Kasschau, K. D., Fahlgren, N., Yates, J. R., 3rd, Mitani, S., Carrington, J. C., and Mello, C. C. (2009) Distinct argonaute-mediated 22G-RNA pathways direct genome surveillance in the *C. elegans* germline. *Mol Cell* ***36***, 231–244.
68. Claycomb, J. M., Batista, P. J., Pang, K. M., Gu, W., Vasale, J. J., van Wolfswinkel, J. C., Chaves, D. A., Shirayama, M., Mitani, S., Ketting, R. F., Conte, D., Jr., and Mello, C. C. (2009) The argonaute CSR-1 and its 22G-RNA cofactors are required for holocentric chromosome segregation. *Cell* ***139***, 123–134.
69. van Wolfswinkel, J. C., Claycomb, J. M., Batista, P. J., Mello, C. C., Berezikov, E., and Ketting, R. F. (2009) CDE-1 affects chromosome segregation through uridylation of CSR-1-bound siRNAs. *Cell* ***139***, 135–148.
70. Lee, H. C., Chang, S. S., Choudhary, S., Aalto, A. P., Maiti, M., Bamford, D. H., and Liu, Y. (2009) qiRNA is a new type of small interfering RNA induced by DNA damage. *Nature* ***459***, 274–277.
71. Aravin, A. A., Naumova, N. M., Tulin, A. V., Vagin, V. V., Rozovsky, Y. M., and Gvozdev, V. A. (2001) Double-stranded RNA-mediated silencing of genomic tandem repeats and transposable elements in the *D. melanogaster* germline. *Curr Biol* ***11***, 1017–1027.
72. Aravin, A., Gaidatzis, D., Pfeffer, S., Lagos-Quintana, M., Landgraf, P., Iovino, N., Morris, P., Brownstein, M. J., Kuramochi-Miyagawa, S., Nakano, T., Chien, M., Russo, J. J., Ju, J., Sheridan, R., Sander, C., Zavolan, M., and Tuschl, T. (2006) A novel class of small RNAs bind to MILI protein in mouse testes. *Nature* ***442***, 203–207.
73. Girard, A., Sachidanandam, R., Hannon, G. J., and Carmell, M. A. (2006) A germline-specific class of small RNAs binds mammalian Piwi proteins. *Nature* ***442***, 199–202.
74. Grivna, S. T., Beyret, E., Wang, Z., and Lin, H. (2006) A novel class of small RNAs in mouse spermatogenic cells. *Genes Dev* ***20***, 1709–1714.
75. Batista, P. J., Ruby, J. G., Claycomb, J. M., Chiang, R., Fahlgren, N., Kasschau, K. D., Chaves, D. A., Gu, W., Vasale, J. J., Duan, S., Conte, D., Jr., Luo, S., Schroth, G. P., Carrington, J. C., Bartel, D. P., and Mello, C. C. (2008) PRG-1 and 21U-RNAs interact to form the piRNA complex required for fertility in *C. elegans*. *Mol Cell* ***31***, 67–78.
76. Das, P. P., Bagijn, M. P., Goldstein, L. D., Woolford, J. R., Lehrbach, N. J., Sapetschnig, A., Buhecha, H. R., Gilchrist, M. J., Howe, K. L., Stark, R., Matthews, N., Berezikov, E., Ketting, R. F., Tavare, S., and Miska, E. A. (2008) Piwi and piRNAs act upstream of an endogenous siRNA pathway to suppress Tc3 transposon mobility in the *Caenorhabditis elegans* germline. *Mol Cell* ***31***, 79–90.
77. Lau, N. C., Seto, A. G., Kim, J., Kuramochi-Miyagawa, S., Nakano, T., Bartel, D. P., and Kingston, R. E. (2006) Characterization of the piRNA complex from rat testes. *Science* ***313***, 363–367.
78. Vagin, V. V., Sigova, A., Li, C., Seitz, H., Gvozdev, V., and Zamore, P. D. (2006) A distinct small RNA pathway silences selfish genetic elements in the germline. *Science* ***313***, 320–324.
79. Lin, H. (2007) piRNAs in the germ line. *Science* ***316***, 397.
80. Thomson, T., and Lin, H. (2009) The biogenesis and function of PIWI proteins and piRNAs: progress and prospect. *Annu Rev Cell Dev Biol* ***25***, 355–376.
81. Aravin, A. A., Hannon, G. J., and Brennecke, J. (2007) The Piwi-piRNA pathway provides an adaptive defense in the transposon arms race. *Science* ***318***, 761–764.
82. Malone, C. D., and Hannon, G. J. (2009) Small RNAs as guardians of the genome. *Cell* ***136***, 656–668.
83. Girard, A., and Hannon, G. J. (2008) Conserved themes in small-RNA-mediated transposon control. *Trends Cell Biol* ***18***, 136–148.
84. Brennecke, J., Aravin, A. A., Stark, A., Dus, M., Kellis, M., Sachidanandam, R., and Hannon, G. J. (2007) Discrete small RNA-generating loci as master regulators of transposon activity in Drosophila. *Cell* ***128***, 1089–1103.
85. Brennecke, J., Malone, C. D., Aravin, A. A., Sachidanandam, R., Stark, A., and Hannon, G. J. (2008) An epigenetic role for maternally inherited piRNAs in transposon silencing. *Science* ***322***, 1387–1392.
86. Xu, M., You, Y., Hunsicker, P., Hori, T., Small, C., Griswold, M. D., and Hecht, N. B. (2008) Mice deficient for a small cluster of Piwi-interacting RNAs implicate Piwi-interacting RNAs in transposon control. *Biol Reprod* ***79***, 51–57.

87. Aravin, A. A., Lagos-Quintana, M., Yalcin, A., Zavolan, M., Marks, D., Snyder, B., Gaasterland, T., Meyer, J., and Tuschl, T. (2003) The small RNA profile during Drosophila melanogaster development. *Dev Cell 5*, 337–350.
88. Watanabe, T., Takeda, A., Tsukiyama, T., Mise, K., Okuno, T., Sasaki, H., Minami, N., and Imai, H. (2006) Identification and characterization of two novel classes of small RNAs in the mouse germline: retrotransposon-derived siRNAs in oocytes and germline small RNAs in testes. *Genes Dev 20*, 1732–1743.
89. Saito, K., Nishida, K. M., Mori, T., Kawamura, Y., Miyoshi, K., Nagami, T., Siomi, H., and Siomi, M. C. (2006) Specific association of Piwi with rasiRNAs derived from retrotransposon and heterochromatic regions in the *Drosophila* genome. *Genes Dev 20*, 2214–2222.
90. Kawaoka, S., Hayashi, N., Katsuma, S., Kishino, H., Kohara, Y., Mita, K., and Shimada, T. (2008) Bombyx small RNAs: genomic defense system against transposons in the silkworm, Bombyx mori. *Insect Biochem Mol Biol 38*, 1058–1065.
91. Houwing, S., Kamminga, L. M., Berezikov, E., Cronembold, D., Girard, A., van den Elst, H., Filippov, D. V., Blaser, H., Raz, E., Moens, C. B., Plasterk, R. H., Hannon, G. J., Draper, B. W., and Ketting, R. F. (2007) A role for Piwi and piRNAs in germ cell maintenance and transposon silencing in Zebrafish. *Cell 129*, 69–82.
92. Houwing, S., Berezikov, E., and Ketting, R. F. (2008) Zili is required for germ cell differentiation and meiosis in zebrafish. *EMBO J 27*, 2702–2711.
93. Ro, S., Park, C., Song, R., Nguyen, D., Jin, J., Sanders, K. M., McCarrey, J. R., and Yan, W. (2007) Cloning and expression profiling of testis-expressed piRNA-like RNAs. *RNA 13*, 1693–1702.
94. Gunawardane, L. S., Saito, K., Nishida, K. M., Miyoshi, K., Kawamura, Y., Nagami, T., Siomi, H., and Siomi, M. C. (2007) A slicer-mediated mechanism for repeat-associated siRNA 5′ end formation in *Drosophila*. *Science 315*, 1587–1590.
95. Malone, C. D., Brennecke, J., Dus, M., Stark, A., McCombie, W. R., Sachidanandam, R., and Hannon, G. J. (2009) Specialized piRNA pathways act in germline and somatic tissues of the *Drosophila* ovary. *Cell 137*, 522–535.
96. Li, C., Vagin, V. V., Lee, S., Xu, J., Ma, S., Xi, H., Seitz, H., Horwich, M. D., Syrzycka, M., Honda, B. M., Kittler, E. L., Zapp, M. L., Klattenhoff, C., Schulz, N., Theurkauf, W. E., Weng, Z., and Zamore, P. D. (2009) Collapse of germline piRNAs in the absence of argonaute3 reveals somatic piRNAs in flies. *Cell 137*, 509–521.
97. Sijen, T., Fleenor, J., Simmer, F., Thijssen, K. L., Parrish, S., Timmons, L., Plasterk, R. H., and Fire, A. (2001) On the role of RNA amplification in dsRNA-triggered gene silencing. *Cell 107*, 465–476.
98. Aoki, K., Moriguchi, H., Yoshioka, T., Okawa, K., and Tabara, H. (2007) In vitro analyses of the production and activity of secondary small interfering RNAs in *C. elegans*. *EMBO J 26*, 5007–5019.
99. Pak, J., and Fire, A. (2007) Distinct populations of primary and secondary effectors during RNAi in *C. elegans*. *Science 315*, 241–244.
100. Sijen, T., Steiner, F. A., Thijssen, K. L., and Plasterk, R. H. (2007) Secondary siRNAs result from unprimed RNA synthesis and form a distinct class. *Science 315*, 244–247.

Chapter 12

Quantification of siRNAs In Vitro and In Vivo

Angie Cheng, Alexander V. Vlassov, and Susan Magdaleno

Abstract

RNA interference (RNAi) is a regulatory mechanism of eukaryotic cells that uses small interfering RNAs (siRNA) to direct homology-dependent control of gene activity. Applications of RNAi include functional genomics, in vivo target validation, and gene-specific medicines. A key to in vivo application of siRNA is the advancement of efficient delivery to organs, tissues, or cell types of interest. There is a need to develop reliable and easy-to-use assays to evaluate siRNA delivery efficiency and distribution, study pathways, and stability of siRNAs in cells (post-transfection) and in animals (post- injection). We have adopted the Applied Biosystems TaqMan® based stem–loop RT-PCR technology, originally developed for quantification of endogenous microRNAs in cells, to fulfill these needs. In this chapter, application protocols are described, which enable robust quantification of siRNA, including chemically modified molecules, in vitro and in vivo.

Key words: siRNA, shRNA, miRNA, RNAi, quantification, PCR, TaqMan® siRNA assays.

1. Introduction

Three major categories of gene-silencing molecules exists: (i) antisense oligonucleotide derivatives that, depending on their type, recruit RNase H to cleave the target mRNA or inhibit translation by steric hindrance (1); (ii) ribozymes and deoxyribozymes – catalytically active oligonucleotides that cause RNA cleavage (2); (iii) small interfering double-stranded RNA molecules (siRNA) that induce degradation of the complementary mRNA through a natural gene-silencing pathway called RNA interference (RNAi) (3). RNAi is the latest addition to the family of antisense technologies and has rapidly become the most widely used approach for gene knockdown because of its potency. siRNAs have become the leading application to investigate gene functions, target identification, target validation, and therapeutic development (4, 5).

J. Goodchild (ed.), *Therapeutic Oligonucleotides*, Methods in Molecular Biology 764,
DOI 10.1007/978-1-61779-188-8_12, © Springer Science+Business Media, LLC 2011

A key to in vivo application of siRNA is efficient delivery to organs, tissues or cell types of interest. There is a need to develop reliable and easy-to-use assays to evaluate siRNA distribution, study pathways, stability, and delivery efficiency of siRNAs in cells (post-transfection) and in animals (post-injection). siRNAs with fluorescent labels are often used to obtain quick estimates of delivery efficiency and localization (6). However, this approach is neither quantitative nor indicative of siRNA integrity. Furthermore, interpretation of the obtained information might not be accurate in some cases due to (i) bulky conjugates altering siRNA properties and trafficking and (ii) free-floating fluorescent labels present (not completely removed post-synthesis or cleaved off the siRNA in vivo). For reliable quantification of siRNAs, miRNAs and other small RNAs, nuclease protection assay (7) and various RT-PCR approaches are typically used (8–12).

This chapter describes the detailed protocols for siRNA quantification using specially designed TaqMan® siRNA assays (**Fig. 12.1**). To quantify siRNAs in vitro (post-transfection) or in vivo (post-injection), we isolated total RNA from the cells or organs of interest, performed reverse transcription (RT) using specific stem–loop RT primer, and proceeded with real-time PCR with TaqMan® assays. While conventionally designed TaqMan real-time primer-probe sets cannot amplify very small targets, such as siRNAs and miRNAs, the stem–loop RT primer addresses this fundamental problem by producing a longer DNA template upon reverse transcription that is amenable to standard TaqMan® real-time PCR assays (13). We observed better specificity and

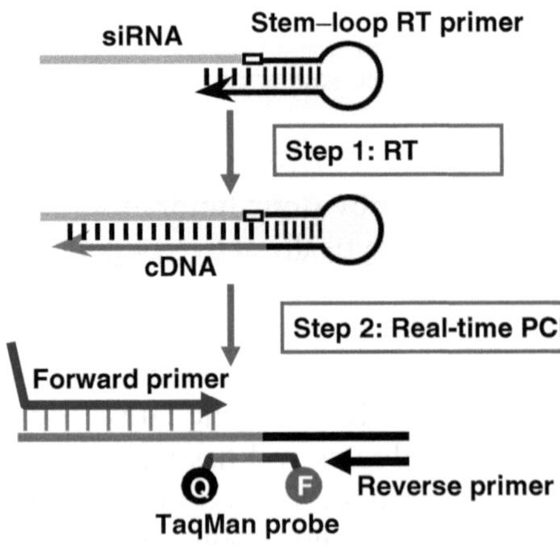

Fig. 12.1. Schematic description of TaqMan siRNA Assays. TaqMan-based real-time quantification of siRNAs includes two steps, stem–loop RT and real-time PCR. Stem–loop RT primers bind to the 3′-portion of siRNA molecules and are reverse transcribed. Then, the RT product is quantified using conventional TaqMan PCR that includes siRNA-specific forward primer, reverse primer and a dye-labeled TaqMan probes.

sensitivity with stem–loop RT primers than conventional linear ones presumably due to base stacking and spatial constraint of the stem–loop structure.

2. Materials

2.1. siRNA (see Note 1)

1. UQCR *Silencer*® Select siRNA (Ambion, Inc.).
   AS strand: 5′-UGUGGGUUUCCUCUUGUCCdAdT-3′ (5 nmols, for in vitro use). Store at −20°C.
2. GFP *Silencer*® siRNA (Ambion, Inc.).
   AS strand: 5′-UGCGUUCCUGUACAUAACCdTdT-3′ (250 nmols, for in vivo use). Store at −20°C.
3. Nuclease-Free water (Ambion, Inc.). Store at room temperature.

2.2. Cell Culture and Transfection

1. HeLa cells (or cell line of choice for transfections; ATCC)
2. Dulbecco's modified Eagle medium (DMEM) high glucose (Invitrogen) supplemented with 10% fetal bovine serum (FBS) (MediaTech, Inc.) and 1% penicillin (5000 Units)-streptomycin (5000 μg) (Pen/Strep) (Invitrogen). Remove 55 mL from a new DMEM bottle and dispose if not used. Add 50 mL of FBS and 5 mL of Pen/Strep. Invert the bottle 10 times to mix. Store at 4°C (*see* **Note 2**).
3. Opti-MEM® I reduced-serum medium (Invitrogen). Store in aliquots at 4°C.
4. PBS 10 ×, pH 7.4 (Ambion, Inc.). Store at room temperature.
5. 1000-mL 0.2-μM filter unit to filter sterilize PBS (Nalgene Labware).
6. Trypsin–EDTA (0.05% trypsin with EDTA) 1 × (Invitrogen). Store in aliquots at –20°C. Working aliquot can be stored at 4°C.
7. Lipofectamine™ 2000 Transfection Agent (Invitrogen). Store at 4°C. (*see* **Note 3**).
8. *Silencer*® Select siRNAs (Ambion, Inc.). Store at –20°C.

2.3. Cell Lysis

1. TaqMan® MicroRNA Cells-to-CT™ Kit (Ambion, Inc.)

2.4. Reverse Transcription

1. TaqMan® MicroRNA Reverse Transcription Kit (Applied Biosystems; included in the TaqMan® MicroRNA Cells-to-CT Kit)

2. siRNA-specific stem–loop primer (Applied Biosystems; component of Custom TaqMan® siRNA Assay)
3. 1 × TE buffer (Ambion, Inc.)
4. Nuclease-free Water (Ambion, Inc.)

2.5. Real-Time PCR: Quantification of siRNA

1. Real-Time PCR Instrument – Applied Biosystems 7900HT (Applied Biosystems)
2. Real-Time PCR plate (optically clear) specific to real-time PCR instrument
3. siRNA-specific forward and reverse primers (Applied Biosystems; components of Custom TaqMan® siRNA Assays)
4. siRNA-specific probe (Applied Biosystems; component of Custom TaqMan® siRNA Assays)
5. TaqMan® universal master mix, No AmpErase UNG (Applied Biosystems)
6. Nuclease-free water (Ambion)

2.6. Animals and Injections

1. Balb/c mice (6–8 weeks old)
2. Mouse restrainer (from Harvard Apparatus or analogous)
3. 5-mL syringes and 27G1/2 needles
4. Heat lamp
5. Cloth (gauze), ethanol

2.7. Sample Preparation and Total RNA Isolation

1. mirVana™ PARIS™ Kit (Applied Biosystems)
2. RNA*later* (Applied Biosystems)
3. PRO250 homogenizer (ProScientific)
4. Ethanol
5. NanoDrop™ 1000 Spectrophotometer

3. Methods

Multiple techniques exist for the introduction of siRNA into cultured cells and in animals. These include lipids, polymers, various nanoparticles, conjugates, cell-penetrating peptides, viruses, electroporation, and ultrasound (14). In some cases to improve in vivo performance, chemically modified siRNAs are used, with 2′-OMe, LNA, or other modifications placed at unique positions in order to enhance nuclease resistance and reduce interferon response (15–17). The capability to trace and quantify siRNA in different types of cells and various cell compartments provides invaluable information regarding the delivery efficiency by a technique/reagent of choice, siRNA distribution, and integrity, allowing one to study intracellular pathways, pharmacokinetics, and pharmacodynamics. Once siRNAs have been chosen and delivery conditions have been optimized, the quantification experiments can be performed with TaqMan® siRNA assays.

We first describe in **Section 3.1** how to prepare siRNA stocks for subsequent experiments. Then, we describe two versions of the quantification protocol: one for siRNA in vitro transfection, and another for siRNA in vivo injection (**Fig. 12.2**). **Sections 3.2** and **3.3** covers siRNA transfection into cultured cells with cationic lipid-based reagent, and subsequent cell lysis to recover total RNA for downstream analysis. **Sections 3.4** and **3.5** describe reverse transcription and qPCR (RT-PCR) for quantification of siRNA. To determine the number of siRNA copies delivered per cell upon transfection, Ct values from siRNA qPCR detection results are compared with a standard curve generated by adding known amounts of siRNA to untreated cell lysates and then processing samples using the same procedure. **Sections 3.6** and **3.7** describe siRNA injections into mice using a hydrodynamic technique, with subsequent organ harvesting, homogenization, and purification of total RNA for downstream analysis. qRT-PCR for quantification of siRNA in vivo is similar to siRNA in vitro protocol.

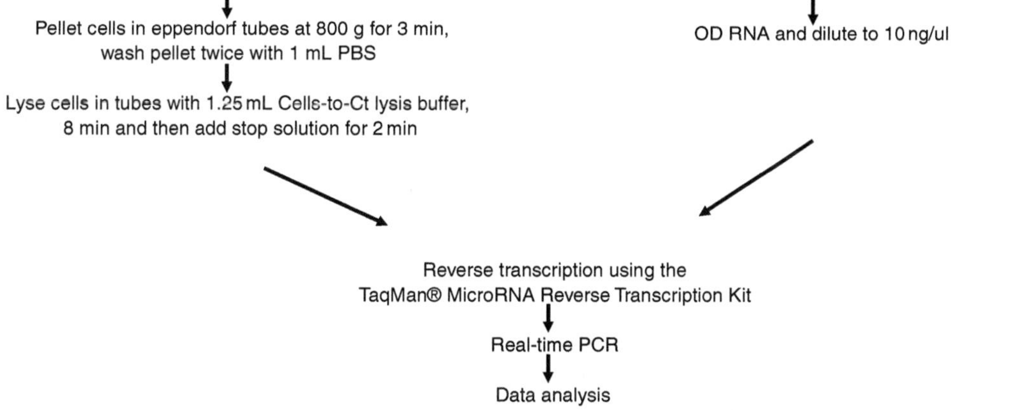

Fig. 12.2. Schematic depiction of the protocol for the quantification of siRNA in cells (post-transfection) and in animals (post- injection).

3.1. Assay Design

1. Submit your siRNA sequence to www4.appliedbiosystems.com/beta/smallrna for automated design of TaqMan® siRNA assays.
2. After assays are designed, order the desired reaction size. Assay includes specific stem–loop RT primer, forward and reverse PCR primers and probe specific to your siRNA of interest.

3.1.1. Prepare siRNAs

1. Resuspend the siRNAs with nuclease-free water. A convenient stock concentration is 100 μM, which can be diluted to meet downstream experimental needs (*see* **Note 4**).
2. Validate the concentration of the siRNA by measuring absorption at 260 nM using a spectrophotometer (e.g., Nanodrop) and adjust with water if necessary. Keep aliquots frozen at −20°C (*see* **Note 5**).

3.2. In Vitro Transfection of siRNAs

3.2.1. Prepare Reagents for Tissue Culture Use

1. Cell growth medium (DMEM, 10% FBS, 1% P/S) is prepared as described in **Section 2.2**.
2. 1 × PBS: Mix 100 mL of 10 × PBS and 900 mL of nuclease-free water in a 1000-mL graduated cylinder. Seal the top with parafilm and invert 10 times to mix. Then filter sterilize using a 1000-mL 0.2-μM filter unit and vacuum pump.

3.2.2. Prepare Cells for Transfections (in 6-Well Plates)

1. In a standard T75-cm^2 tissue culture flask after 2 days in culture, you can get 4–8 million cells depending on percent confluency (and cell type).
2. Make sure cells in the flask are healthy by observing both macroscopically and microscopically. There should not be obvious changes in media color, overt cell phenotype, or evidence of contamination.
3. After cells are determined healthy, warm up media and trypsin in a 37°C water bath.
4. Harvest cells for transfection according to standard tissue culture protocols. The most common method to harvest the cells is to expose the cells briefly to trypsin–EDTA at 0.05%, depending on the cell type, followed by 5- to 10-fold dilution in complete media to inactivate the trypsin and facilitate easy cell collection.
5. Collect the cells in a 50-mL polypropylene tube (*see* **Note 6**).
6. Count the cells with a cell counter of choice (e.g., hemocytometer or Invitrogen's Countess™ automated cell counter) according to standard tissue culture protocols.
7. Dilute HeLa cells to 100,000 cells/mL with cell growth media, plate 2.3 mL (250,000 cells) per well of a 6-well plate, and place into 37°C incubator for 24 h.

8. Include 14 extra wells of pre-plated cells for spike-in experiments to build a standard curve in order to determine the number of siRNA molecules that were delivered into cells upon transfection.

3.2.3. Warm Up the Reagents Before Transfections

1. Warm the Opti-MEM® I in a 37°C water bath or room temperature before use.
2. Warm Lipofectamine 2000 transfection reagent to room temperature before use.

3.2.4. Prepare siRNAs for Transfections (see ***Note 7****)*

1. Dilute stock siRNAs in 100 μL Opti-MEM® I per well to achieve a final concentration of 30 nM or your desired concentration (0.001–100 nM) in tubes or plates. Make master mixes when applicable for replicates to minimize variability (at least one well overage). For example, with a 100 μM siRNA stock, mix 0.75 μL of siRNA (75 pmols) + 99.25 μL Opti-MEM® I per well or 3 μL siRNA + 397 μL Opti-MEM® I for the master mix.

3.2.5. Prepare Transfection Complexes in a Sterile Environment

1. Dilute 5 μL/well of Lipofectamine 2000 in Opti-MEM® I for a total volume of 100 μL in a polystyrene 12 × 75 mm tube or conical tube. Make a master mix of sufficient volume to treat all wells to be transfected plus an extra 10% for pipetting variability. Mix by gently flicking bottom of the tube. Incubate mixture for 5 min at room temperature.
2. Combine 100 μL of the Lipofectamine 2000 mixture per 100 μL of diluted siRNA. Mix by tapping the tube. Incubate this mixture for another 20 min at room temperature.
3. After the incubation, add 200 μL of the siRNA/lipid complex to each well of a 6-well plate (final volume in wells 2.5 mL). Rock the plate back and forth to mix.
4. Place the plate in a 37°C incubator under normal cell culture conditions. Media can be changed after 24 h. Remove and assay siRNA or gene of interest expression levels at the desired time point. siRNA quantification can be done 1 h–5 days post-transfection.

3.3. Sample Preparation

From this stage downstream, take caution to prevent siRNA contamination. There is a risk of contamination if siRNA preparations and transfections are performed in the same environment as TaqMan® assay detection. For example, designate a "siRNA-free" area that is only used for setting up RT and PCR using PCR pipettes that were never used to pipette siRNAs.

3.3.1. Prepare TaqMan® Cells-to-CT™ Reagents at Time of Harvest

1. Prepare the cell lysis buffer by diluting DNase I (supplied with the kit) 1:100 into the lysis solution and make a master mix of sufficient volume to aliquot to all sample wells to be lysed. For example, prepare 1237.5 μL lysis solution +

12.5 μL DNase I and dispense 1.25 mL of prepared lysis solution per well of a 6-well dish.

2. Thaw enough stop solution for all wells (125 μL per well). Invert to mix (do not vortex) and put on ice.

3.3.2. Cell Washes and Lysis

1. Aspirate cell culture media and gently wash cells with 2 mL of 1 × PBS three times.
2. Aspirate the last PBS wash solution and add 400 μL of 0.05% trypsin–EDTA per well and incubate at 37°C for 3 min or until cells have visibly detached.
3. Inactivate the trypsin by adding 1 mL of complete growth medium (DMEM + 10% FBS). Swirl media around and triturate cells 3–5 times to dislodge cell clumps (*see* **Note 8**).
4. Transfer cells to 2-mL Eppendorf tubes, centrifuge the sample to pellet the cells (800 ×*g* for 3 min), and wash cell pellets twice with 1 mL of 1 × PBS by flicking the tube to dislodge the pellet for a complete wash.
5. Completely remove PBS then lyse cell pellets with 1.25 mL of TaqMan® MicroRNA Cells-to-CT™ lysis buffer. Mix lysates by inverting the tubes five times and then incubate at room temperature for 8 min.
6. Stop the lysis reactions by adding 125 μL stop solution followed by inverting the tubes to mix and incubate at room temperature for another 2 min. Then, proceed to reverse transcription or store samples. Lysates can be stored on ice for up to 2 h, or at –20°C or –80°C for up to 5 months.

3.3.3. Standard Curve

Make seven 5-fold serial dilutions by spiking siRNA (starting with 20 pmoles of siRNA) into non-transfected cells only lysates, which contain the same number of cells and process with the same method as the cell lysates that were transfected.

3.4. Reverse Transcription

1. A typical 10 μL RT reaction includes either 10 ng of total purified RNA or 1 μL lysate prepared using the TaqMan® MicroRNA Cells-to-CT™ kit and RT primer. Combine the following in the appropriate wells of a 96-well PCR plate or tubes that correspond to your PCR thermal cycler:

   1 μL Cells-to-CT lysate or total RNA (10 ng)

   2 μL 5 × stem–loop RT primer; included with the custom TaqMan® small RNA assay.

   Also include the proper controls to monitor reagent contamination such as no-RT enzyme and no-template controls.
2. Samples are then denatured at 85°C for 5 min, followed by 60°C incubation for 5 min, and then 4°C. This can be a program set on the thermal cycler.

3. Prepare the following enzyme mix in a PCR hood (numbers listed are for a single well, but a master mix can be prepared for numerous samples plus 10% for pipetting variability):

   0.1 μL 100 mM dNTPS

   0.7 μL MultiScribe™ RT enzyme 50 U/μL

   1 μL 10 × RT buffer

   0.13 μL RIP 20 U/μL

   5.07 μL Nuclease-free water

4. After the denaturation step is complete, add 7 μL of the enzyme master mix to the wells already containing RT primer and lysate. Seal the plate and briefly spin the plate down to collect sample at the bottom of each well. Place prepared plate into thermal cycler for cDNA synthesis.
5. Set the thermal cycler so samples are incubated at 16°C for 30 min, 42°C for 30 min, 85°C for 5 min, and then 4°C.
6. The cDNA may be stored on ice or frozen until real-time PCR is performed.

3.5. Real-Time PCR

1. Real-time PCR can be performed using a standard TaqMan® PCR assay protocol on an Applied Biosystems 7900HT Fast Real-Time PCR System.
2. Mix the following components for each gene being amplified in a PCR hood for a 10 μL reaction in a 384-well plate (numbers listed are for a single reaction, but a master mix can be prepared for numerous samples plus 10% extra for pipetting variability). Real-time PCR should be performed with at least technical duplicates.

   3.0 μL Nuclease-free water

   5.0 μL TaqMan® universal master mix

   1 μL 20 × Custom TaqMan small RNA assay primer/probe mix

3. Add 1 μL cDNA per 384-well.
4. Add 9 μL of the master mix.
5. Seal the plate with an optical cover and spin the plate to collect samples at bottom of the wells.
6. The thermocycling conditions using TaqMan® universal master mix and custom TaqMan® small RNA assays are 95°C for 10 min followed by 40 cycles of 95°C for 15 s and 60°C for 1 min.

3.5.1. Data Analysis

1. As the PCR amplification cycles proceeds, an increase in fluorescence is detected due to the increase in amount of DNA product formed. The starting amount of the target can be determined based on the number of cycles the fluorescent signal takes to reach a given "threshold" relative to a standard used to calculate the relative quantity of a product in a given sample. The cycle at which the sample reaches this level is called the cycle threshold, Ct. Once the real-time PCR is complete, calculate the average Ct and the standard deviation for the PCR duplicates for each sample.
2. There should be no DNA amplification in the no-RT enzyme and no-template controls. In case non-specific amplification is observed, that means samples were contaminated at one of the steps. It would be difficult to make conclusions regarding siRNA quantity inside cells, and we recommend repeating the experiment with fresh reagents.
3. To determine the number of siRNA copies delivered per cell upon transfection, compare Ct values from the treated samples to the standard curve generated using non-transfected cell lysate.
4. An example of results produced is shown in **Fig. 12.3**. In the figure for the standard curve, siRNA input (in pmols) is graphed on the *x*-axis while the corresponding Ct values are graphed on the *y*-axis.

Standard Curve Values

siRNA input for standard curve	Avg. Ct	SD
20 pmol	12.70	0.10
4 pmol	12.90	0.01
0.8 pmol	15.80	0.03
0.16 pmol	17.81	0.18
0.032 pmol	20.27	0.26
0.0064 pmol	22.31	0.09
0.00128 pmol	24.00	0.02

siRNA Transfection Values

siRNA transfected	Avg. Ct	SD
0.001nM	30.40	0.70
0.01nM	27.11	0.16
0.1nM	23.65	0.14
0.5nM	20.52	0.17
5nM	17.49	0.36
20nM	16.08	0.49
80nM	16.69	0.91

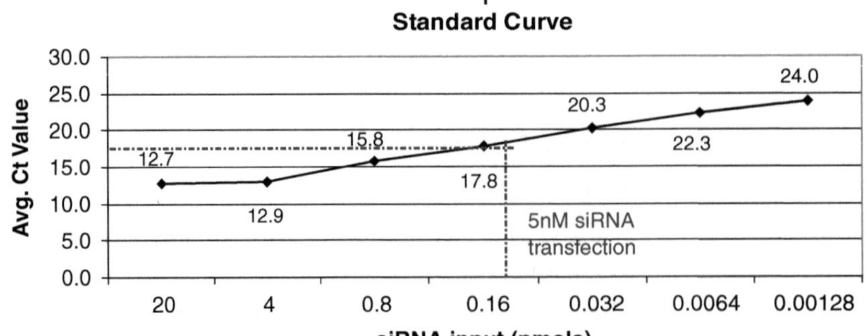

Fig. 12.3. siRNA in vitro quantification. HeLa cells were transfected with 0.001–80 nM of siRNA targeting UQCR. Cells were harvested 48 h later, and siRNA quantification was performed using TaqMan siRNA Assays. siRNA input (in pmols) is graphed on the *x*-axis while the corresponding Ct values are graphed on the *y*-axis. For example, for a 5-nM siRNA transfection the average of 17.5 Cts was obtained, which corresponds to approximately 0.1 pmols of siRNA detected.

5. For example, for a 5 nM siRNA transfection the average of 17.5 Cts was obtained, which corresponds to approximately 0.1 pmols of siRNA detected inside cells.
6. Don't use the straight-line equation, $y = mx + b$, to calculate the number of pmols detected because Ct values are not linear.

3.6. siRNA Quantification In Vivo

3.6.1. Prepare siRNA

1. Dilute 1 nmol GFP *Silencer*® siRNA in 2.5 mL 1 × PBS for each mouse to be injected (*see* **Note 9**).

3.6.2. Prepare Mice

1. Prepare Balb-c mice, 3 animals per siRNA to be quantified (per cage).
2. Prepare new labeled cages with fresh bedding, food and bottles of water.
3. Reserve one untreated animal for spike-in experiments to make a standard curve. Total RNA isolated from various organs and tissues of a single animal is sufficient for multiple spike-in experiments.

3.6.3. Injections

1. Put one mouse in the restrainer device.
2. Preheat the tail vein under the heat lamp for 10 s (*see* **Note 10**).
3. Rub the tail with a sterile cloth saturated with 75% ethanol.
4. Inject siRNA-1 × PBS mixture into the tail vein of the mouse within 5–8 s in a single bolus (hydrodynamic tail vein injection) (*see* **Note 11**).
5. Apply gentle pressure to injection site using gauze to stop bleeding (5 s).
6. Label the tail using a marker (one stripe for #1, etc.) and put mouse in the new cage (*see* **Note 12**).
7. Repeat procedure for 2 other mice.

3.7. Organ Harvesting and Total RNA Isolation

1. At desired time points post-injection (5 min–7 days), euthanize mice according to institution-approved procedures.
2. Harvest whole organs and tissues of interest (e.g., liver, spleen, kidney, and lung) by surgical dissection and immediately freeze on dry ice or liquid nitrogen. If RNA will not be isolated right away, store organs in RNA*later*® or other RNA stabilization solution.
3. Place whole organs into ice-cold cell disruption buffer (1–20 mL, depending on the size of the organ; supplied with the mirVana™ PARIS™ Kit) and homogenize samples using a PRO250 homogenizer at speed setting 5 for

10 s or until the tissue is completely blended. Place on ice immediately afterwards.

4. Transfer homogenized organ lysates (400 μL) into new Eppendorf tubes, add 400 μL of the denaturing solution to each tube, and mix well by vortexing at top speed for 30 s.
5. Perform extractions by adding 800 μL acid-phenol: chloroform (pH 4.5), vortexing for 5 min at top speed, and centrifuging in a microfuge at 13,000 rpm for 10 min. Transfer the upper phase (300 μL) into new tubes.
6. Mix 300 μL of the upper phase with 375 μL (1.25 vol) of 100% ethanol and pass through the supplied filter columns.
7. Add 700 μL wash 1 to each filter and centrifuge for 30 s. Discard flow-through.
8. Add 500 μL of wash 2/3 and centrifuge for 30 s. Discard flow-through and repeat.
9. Elute total RNA using 100 μL of elution buffer pre-heated to 95°C.
10. Measure total RNA concentration with a NanoDrop™ 1000 spectrophotometer and adjust to 10 ng/μL with water.
11. To the RNA samples isolated from an untreated mouse, add known amounts of siRNA (e.g., 500, 200, 100, 50, 10, 1, 0.1, 0.01 pmoles). This allows one to determine the amount of siRNA delivered based on comparison of Ct

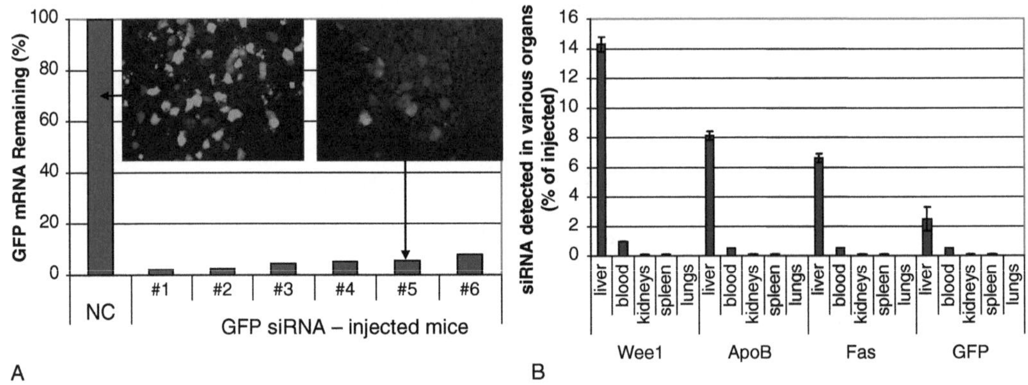

Fig. 12.4. siRNA delivery in vivo using hydrodynamic tail vein injection. (**a**) siRNA-induced knockdown of GFP in the mouse liver. GFP-encoding plasmid and siRNA targeting GFP were administered into mouse tail veins using hydrodynamic procedure. Mice were sacrificed 24 h later, their livers harvested, total RNA isolated, and mRNA levels determined by qRT-PCR and compared to the negative control (NC) group of five animals, which is normalized to 100%. #1–6 refer to six individual animals, all injected with the same GFP siRNA. At the *top*, representative liver sections are shown for both groups of mice displaying significant reduction of GFP protein levels. (**b**) Quantification of siRNA in the mouse organs. siRNAs targeting GFP, Fas, ApoB, Wee1 were administered into mouse tail veins using hydrodynamic procedure. Mice were sacrificed 5 min later, their organs harvested, total RNA isolated, and siRNA levels were determined by qRT-PCR. Percentage of the full-length siRNA (from injected dose), detected in various organs is shown.

values (from qPCR) between injected and spiked samples. Generate organ-specific standard curve samples for all organs and tissues (e.g., liver, spleen, kidney, lung) to be examined in the experiment.

3.7.1. Reverse Transcription and Real-Time PCR

Reverse transcription and PCR quantification are performed as described earlier for the in vitro samples (see **Sections 3.4** and **3.5**). Examples of siRNA quantification in vivo as well as siRNA-induced knockdown are shown in **Fig. 12.4**.

4. Notes

1. TaqMan siRNA assays can be used to detect and quantify unmodified siRNAs and siRNAs with limited modifications such as *Silencer*® siRNAs (two deoxyribonucleotide overhang at each 3′-terminus), and *Silencer*® Select siRNAs (LNA (locked nucleic acid) modifications in the passenger strand, and two deoxyribonucleotide overhangs at each 3′-terminus). The current analysis was focused on detection of the guide strand since it is the primary strand responsible for cleavage of the mRNA target, while the passenger strand is cleaved and expelled from the complex upon assembly with the RISC machinery (3).
2. For each cell line, use the recommended media (refer to ATCC's website). Supplementing the media with 1% Pen/Strep antibiotics is optional: High siRNA transfection efficiency can be achieved with or without antibiotics in the media.
3. Transfection conditions (type of transfection reagent and amount per well) are cell type-specific. Lipofectamine 2000 and RNAiMax are highly efficient transfection reagents enabling siRNA delivery into a wide variety of cells with minimal optimization required.
4. siRNA stocks are prepared in the same manner for in vitro and in vivo experiments. The major difference is that much larger amounts of siRNAs are needed for in vivo experiments, and the stock should be prepared at a 500 μM concentration. In addition to water, another typically used solvent is PBS.
5. If multiple subsequent experiments are planned, siRNAs can be stored at 4°C for up to 1 month with no observable degradation. The stock concentration should be no less than 10 μM (18).

6. Polypropylene tubes are important so that cells do not adhere to the plastic. Cells can be kept in such tubes for up to 1 h.
7. siRNA transfections could also be done in 24-well plates. We recommend larger-size wells because it is easier to manipulate with large number of cells >250,000/well (see **Section 3.3.2**).
8. Complete trypsinization and transfer of cells from plates to tubes is crucial for accurate quantification of siRNA molecules that are truly inside cells. We found that upon transfection, a large fraction of the siRNA/lipid complex is bound in the well even in the absence of cells and can variably influence quantitative measurements if the samples are processed in the cell culture plates. Such complexes can not be removed by multiple PBS washes, RNase1, EDTA, or 95% ethanol. However, siRNA is readily released into solution upon cell lysis. Taking into account that only a small fraction of the siRNA is delivered into cells (~1%), it is impossible to accurately quantify intracellular siRNA molecules without removing the extracellular molecules that are associated with the plastic. Overall, trypsinization was found to be the most efficient procedure that allows gentle removal of cells from the plates without releasing siRNA/lipid complexes bound to the wells. Also, trypsin treatment removes a fraction of nucleic acids that are tightly associated with cell surface components but are not inside cells.
9. GFP-encoding plasmid (e.g., pTRACER™-SV40, Invitrogen) can be co-injected with GFP-targeting siRNA to study in vivo efficacy of siRNA targeting exogenous gene.
10. By heating the tail, the vein becomes more visible. Another option is to use a warm water bath (e.g., 37°C).
11. Hydrodynamic tail vein injection was invented by scientists at Mirus Biocorporation (19). Any other reagent or technique of choice for siRNA in vivo delivery can be used, such as Invivofectamine 2.0 reagent (Life Technologies).
12. It is normal that mice are lethargic for the first 10 min following hydrodynamic injections.

Acknowledgments

The authors would like to acknowledge the work of C. Chen, Y. Liang, Y. Wang, and L. Wong on the development of TaqMan® siRNA assays and for helpful discussions. Hydrodynamic tail vein injections were conducted under a license grant from Mirus Bio Corporation.

References

1. Crooke, S. T. (2004) Antisense strategies. *Curr. Mol. Med.* **4**, 465–487.
2. Goodchild, J. (2000) Hammerhead ribozymes: biochemical and chemical considerations. *Curr. Opin. Mol. Ther.* **2**, 272–281.
3. Dorsett, Y., and Tuschl, T. (2004) siRNAs: applications in functional genomics and potential as therapeutics. *Nat. Rev. Drug Discov.* **3**, 318–329.
4. Castanotto, D., and Rossi, J. J. (2009) The promises and pitfalls of RNA-interference-based therapeutics. *Nature* **457**, 426–433.
5. Dallas, A., and Vlassov, A. (2006) RNAi: a novel antisense technology and its therapeutic potential. *Med. Sci. Monitor* **4**, 67–74.
6. Dunne, J., Drescher, B., Riehle, H., Hadwiger, P., Young, B. D., Krauter, J., and Heidenreich, O. (2003) The apparent uptake of fluorescently fabeled siRNAs by electroporated cells depends on the fluorochrome. *Oligonucleotides* **13**, 375–380.
7. Overhoff, M., Wunsche, W., and Sczakiel, G. (2004) Quantitative detection of siRNA and single-stranded oligonucleotides: relationship between uptake and biological activity of siRNA. *Nucleic Acids Res.* **32**, e170.
8. Varkonyi-Gasic, E., Wu, R., Wood, M., Walton, E. F., and Hellens, R. P. (2007) Protocol: a highly sensitive RT-PCR method for detection and quantification of microRNAs. *Plant Methods* **3**, 12.
9. Ro, S., Park, C., Jin, J., Sanders, K. M., and Yan, W. (2006) A PCR-based method for detection and quantification of small RNAs. *Biochem. Biophys. Res. Commun.* **351**, 756–63.
10. Jiang, M., Arzumanov, A. A., Gait, M. J., and Milner, J. (2005) A bi-functional siRNA construct induces RNA interference and also primes PCR amplification for its own quantification. *Nucleic Acids Res.* **33**, e151.
11. Sharbati-Tehrani, S., Kutz-Lohroff, B., Bergbauer, R., Scholven, J., and Einspanier, R. (2008) miR-Q: a novel quantitative RT-PCR approach for the expression profiling of small RNA molecules such as miRNAs in a complex sample. *BMC Mol. Biol.* **9**, 34.
12. Stratford, S., Stec, S., Jadhav, V., Seitzer, J., Abrams, M., and Beverly, M. (2008) Examination of real-time polymerase chain reaction methods for the detection and quantification of modified siRNA. *Anal. Biochem.* **379**, 96–104.
13. Chen, C., Ridzon, D. A., Broomer, A. J., Zhou, Z., Lee, D. H., Nguyen, J. T., Barbisin, M., Xu, N. L., Mahuvakar, V. R., Andersen, M. R., Lao, K. Q., Livak, K. J., and Guegler, K. J. (2005) Real-time quantification of microRNAs by stem-loop RT-PCR. *Nucleic Acids Res.* **33**, e179.
14. Whitehead, K. A., Langer, R., and Anderson, D. G. (2009) Knocking down barriers: advances in siRNA delivery. *Nat. Rev. Drug Discov.* **8**, 129–138.
15. Manoharan, M. (2004) RNA interference and chemically modified small interfering RNAs. *Curr. Opin. Chem. Biol.* **8**, 570–579.
16. Hickerson, R. P., Vlassov, A. V., Leake, D., Wang, Q., Contag, C. H., Johnston, B. H., and Kaspar, R. L. (2008) Stability study of unmodified siRNA and relevance to clinical use. *Oligonucleotides* **18**, 345–354.
17. Puri, N., Wang, X., Varma, R., Burnett, C., Beauchamp, L., Batten, D. M., Young, M., Sule, V., Latham, K., Sendera, T., Echeverri, C., Sachse, C., and Magdaleno, S. (2008) LNA incorporated siRNAs exhibit lower off-target effects compared to 2′-OMethoxy in cell phenotypic assays and microarray analysis. *Nucleic Acids Symp. Ser. (Oxf.)* **52**, 25–26.
18. Cheng, A., Varma, R., and Magdaleno, S. (2009) Your siRNA storage questions answered. *Ambion Tech. Notes* **16**(2), 9–10.
19. Lewis, D. L., and Wolff, J. A. (2005) Delivery of siRNA and siRNA expression constructs to adult mammals by hydrodynamic intravascular injection. *Methods Enzymol.* **392**, 336–350.

Chapter 13

Optimization of Transfection Conditions and Analysis of siRNA Potency Using Real-time PCR

Angie Cheng, Susan Magdaleno, and Alexander V. Vlassov

Abstract

RNA interference (RNAi) is a mechanism by which the introduction of small interfering RNAs (siRNAs) into cultured cells causes degradation of the complementary mRNA. Applications of RNAi include gene function analysis, pathway analysis, and target validation. While RNAi experiments have become common practice in research labs, multiple factors can influence the extent of siRNA-induced knockdown (and thus biological outcome). A properly designed and selected siRNA sequence, siRNA modification format, choice of transfection reagent/technique, optimized protocols of siRNA in vitro delivery, and an appropriate and optimized readout are all critical for ensuring a successful experiment. In this chapter, we describe a typical in vitro siRNA experiment with optimization of transfection conditions and analysis of siRNA potency, i.e., mRNA knockdown with quantitative real-time PCR.

Key words: siRNA, transfection, knock-down, reverse transcription, real-time PCR, qPCR, TaqMan® gene expression assays.

1. Introduction

RNA interference (RNAi) is a natural, powerful mechanism by which short double-stranded RNA (siRNA) silence the expression of complementary target mRNAs by inducing its cleavage with assistance of RISC protein complex. In the past few years RNAi has become the most widely used technology for gene knock-down. Applications of RNAi include gene and pathway analysis, target validation, and therapeutic development (1–5).

Many factors influence the success of an siRNA experiment. Before the experiment is initiated, some basic choices need to be made – what is the target(s) to be analyzed, what cell lines are

J. Goodchild (ed.), *Therapeutic Oligonucleotides*, Methods in Molecular Biology 764,
DOI 10.1007/978-1-61779-188-8_13, © Springer Science+Business Media, LLC 2011

going to be used, what type of siRNA should be chosen, what reagents should be used to deliver the siRNA into the cells, at what siRNA concentration, what method should be used to analyze knockdown of the target gene, what time should knockdown be analyzed, and what controls (positive and negative) should be used. **Figure 13.1** depicts a general workflow of an siRNA experiment and lists many of the choices that are required to make before starting out.

Typically well-characterized control siRNAs are utilized for optimizing transfection conditions for the cell type and growth conditions that are chosen. Several types of positive and negative controls are offered from commercial suppliers of siRNAs and can be easily obtained. We will describe the use of *Silencer*® Select siRNA targeting the glyceraldehyde 3-phosphate dehydrogenase (GAPDH) gene (Ambion positive control) as an example in this chapter. *Silencer*® Select siRNAs are designed with a novel algorithm and thus feature high potency (consistent >70% knockdown at 5 nM concentration) and incorporated locked nucleic acid (LNA) modifications ensure high specificity (6, 7).

Multiple techniques exist for the introduction of siRNA into cultured cells and in animals. These include lipids, polymers, various nanoparticles, cell-penetrating peptides, conjugates, viruses, ultrasound, and electroporation (8, 9). For some difficult-to-transfect cells (such as T and B cells in suspension) and neuronal cells (e.g., cortex and hippocampus) cationic lipid-based delivery is inefficient. However, for the vast majority of cells this approach is successful as well as time- and cost-effective. We will describe the use of one of the most efficient reagents in this

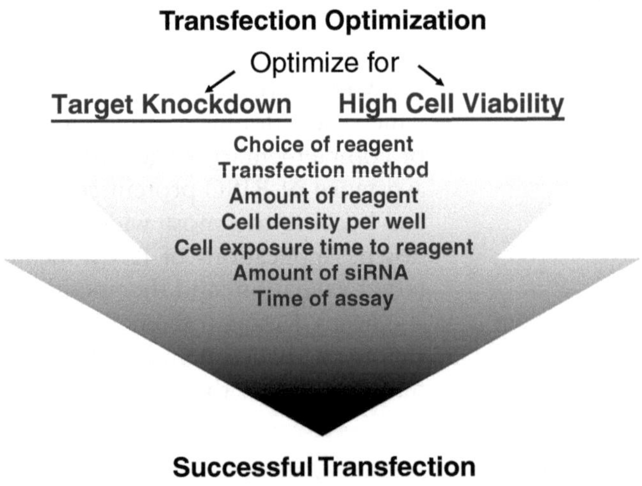

Fig. 13.1. Diagram of the siRNA experimental workflow and the variables that can be optimized to achieve effective siRNA-mediated knockdown of the target gene.

category – Lipofectamine™ RNAiMax. General optimization of transfection conditions is crucial for every reagent, which includes testing the transfection agent amount/well, cell number/well, siRNA concentration and assay time while balancing cellular toxicity and maximizing gene knockdown.

Assessing the activity of siRNA efficacy and optimization of delivery is most often performed by quantifying target mRNA levels using real-time PCR (10). This approach is the most quantitative, easily accessible and utilized in most labs versus quantification of protein levels by Western blot, which is time-consuming and incompatible with high throughput.

By following instructions described in this chapter, transfection conditions can be optimized for any cell type of interest, and subsequently siRNAs targeting gene(s) of interest can be evaluated under these optimized conditions. After quantification of the mRNA levels, protein and phenotypic analysis is often performed and correlation analyzed.

2. Materials

2.1. Tissue Culture and Preparation of Cells for Transfections

1. HeLa cells or cell line of choice for transfections.
2. HeLa cell growth medium: Dulbecco's modified Eagle medium (DMEM) high glucose (Life Technologies, Inc.) supplemented with 10% fetal bovine serum (FBS) and 1% penicillin (5000 Units)-streptomycin (5000 μg) (Pen-Strep) (Life Technologies, Inc.). Store at 4°C (*see* **Note 1**).
3. 10 × PBS: pH 7.4 (Life Technologies, Inc.). Store at room temperature.
4. Nuclease-free water.
5. Filter unit: 1000 mL, 0.2 μM.
6. T75-cm^2 tissue culture flasks.
7. 1 × Trypsin-EDTA: 0.05% trypsin with EDTA (Life Technologies, Inc.). Store in aliquots at −20°C. Working aliquot can be stored at 4°C.
8. Polypropylene tubes: 50 mL.
9. Hemocytometer (Invitrogen's Countess™ automated cell counter).

2.2. Preparation of siRNAs for Transfections

1. *Silencer*® Select GAPDH control siRNA (Life Technologies, Inc.). Store at –20°C.

2. *Silencer*® Select negative control siRNA #1 (Life Technologies, Inc.). Store at −20°C.
3. Nuclease-free water.
4. Spectrophotometer (e.g., Nanodrop).
5. Opti-MEM® I reduced-serum medium (Life Technologies, Inc.). Store in aliquots at 4°C.
6. Round bottom polystyrene tissue culture plate or polystyrene 12 × 75 mm tubes.

2.3. Reverse Transfection

1. Lipofectamine™ RNAiMax transfection agent (Life Technologies, Inc.). Store at 4°C (*see* **Note 2**).
2. Opti-MEM® I reduced-serum medium (Life Technologies, Inc.). Store in aliquots at 4°C.
3. Polystyrene 12 ×75 mm tube.

2.4. Cell Lysis

1. TaqMan® Gene Expression Cells-to-CT™ Kit (Life Technologies, Inc.)
2. PBS

2.5. Reverse Transcription

1. TaqMan® Gene Expression Cells-to-CT™ Kit (Life Technologies, Inc.)
2. 96-Well PCR plate or tubes that correspond to your PCR thermal cycler
3. PCR instrument such as 96-well GeneAmp® PCR System 9700 (Life Technologies, Inc.)

2.6. Real-Time PCR

1. Real-Time PCR instrument: 7900HT (Life Technologies, Inc.)
2. Real-Time PCR plate (optically clear) specific to real-time PCR instrument
3. TaqMan® gene expression master mix (Life Technologies, Inc.)
4. TaqMan® gene expression assay targeted to gene of interest (GOI) (Life Technologies, Inc.)
5. TaqMan® gene expression assay targeted to 18S and GAPDH (Life Technologies, Inc.)
6. Nuclease-free water

3. Methods

Since siRNAs (through sophisticated RNAi machinery) target and degrade their mRNA targets, assessing the potency of siRNAs and optimizing the intracellular delivery is most often performed by quantifying target mRNA levels using real-time PCR. Real-time PCR enables fast and highly accurate quantification of the amount of a specific gene in a sample using a fluorescent readout of the polymerase chain reaction (10, 11). As the PCR amplification cycles proceeds, an increase in fluorescence is detected due to the increase in amount of dsDNA product formed. The starting amount of the target can be determined based on the number of cycles the fluorescent signal takes to reach a given "threshold" relative to a standard used to calculate the relative quantity of a product in a given sample.

The extent of siRNA-induced gene knockdown is typically quantified using the ddCt method. This requires the following measurements to be taken: (i) target mRNA levels in siRNA-treated cells; (ii) mRNA levels in samples treated with negative control/non-targeting siRNA (assigned a value of 100% and used for calibration); (iii) mRNA levels for an endogenous control gene with consistent expression in each sample, e.g., 18S ribosomal RNA or the housekeeping gene β-actin. This should be measured in both the treated sample and control sample.

Here, we will describe siRNA in vitro transfection in the reverse format with cationic lipid-based reagent (**Sections 3.1**, **3.2**, and **3.3**, see **Fig. 13.2**), subsequent cell lysis to recover total RNA for downstream analysis (**Section 3.4**), reverse transcription (**Section 3.5**), and PCR (**Section 3.6**) followed by guidance on the data analysis – final step to quantification of siRNA-induced knockdown of the target gene (**Section 3.7**).

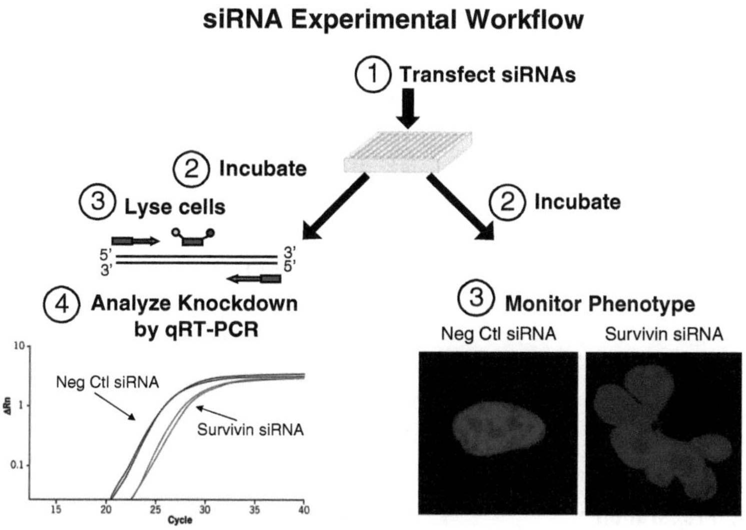

Fig. 13.2. Overview of the siRNA knockdown experiment.

3.1. Tissue Culture and Preparation of Cells for Transfections (96-Well Plate) (see Note 3)

1. To prepare HeLa cell growth medium, thaw a 50-mL aliquot of FBS and a 5-mL aliquot of Pen/Strep in a 37°C water bath at room temperature or overnight at 4°C. When thawed, remove 50 mL from a new DMEM bottle and dispose if not used. Then add the aliquots of FBS and Pen/Strep, and invert the bottle 5–10 times to mix.
2. Mix 100 mL of 10 × PBS and 900 mL of nuclease-free water in a 1000-mL graduated cylinder. Seal the top with parafilm and invert 5–10 times to mix. Then filter sterilize it with a 1000-mL 0.2-μm filter unit using a vacuum pump.
3. Warm cell growth medium and trypsin in a 37°C water bath or incubator before use.
4. In a standard T75-cm^2 tissue culture flask, you can get 4–8 million cells depending on percent confluency and cell type.
5. Make sure cells in the flask are healthy by observing both macroscopically and microscopically. There should not be obvious changes in medium color, overt cell phenotype, or evidence of contamination.
6. Trypsinize cells according to standard tissue culture protocols. The most common method for harvesting the cells is to expose them briefly to trypsin-EDTA at 0.05%, followed by 5- to 10-fold dilution in complete medium to inactivate the trypsin and facilitate easy cell collection.
7. Collect the cells in a 50-mL polypropylene tube (*see* **Note 4**).
8. Count the cells with a hemocytometer (or Invitrogen's Countess™ automated cell counter) according to standard tissue culture protocols.
9. Dilute cells to the appropriate cell density with cell growth medium in a total of 80 μL per well. Test several different cell densities. Cells in the well should be approximately 50–70% confluent depending on the cell type. Use a polypropylene tube. Optimized conditions for HeLa cells are 4000 cells per well. For our example, make enough for 10 wells (GAPDH control siRNA in triplicate, negative control siRNA in triplicate, non-transfected cells in triplicate and 10% overage). Place diluted cell suspension in a 37°C incubator or water bath until ready for use (*see* **Note 5**).

3.2. Preparation of siRNAs for Transfections

1. Resuspend the siRNAs with nuclease-free water (siRNAs are shipped lyophilized). A convenient stock concentration is 100 μM, which can be diluted to meet downstream

experimental needs (*see* **Note 6**). Validate the concentration of the siRNA by measuring absorption at 260 nm using a spectrophotometer (e.g., Nanodrop) and adjust with water if necessary. Keep aliquots frozen at –20°C (*see* **Note 7**).

2. Warm the Opti-MEM® I to room temperature before use.
3. Dilute siRNAs in 20 μL Opti-MEM® I per well to 5 nM (*see* **Note 8**) in a sterile round bottom polystyrene tissue culture plate or polystyrene 12 × 75 mm tubes. *Silencer®* Select siRNAs are up to 100 times more potent than other commercially available siRNAs thus a typical concentration upon transfection is 5 nM. Make master mixes when applicable for replicates to minimize variability. For example, make enough for four wells (40 μL). Transfections should be done in at least triplicates (*see* **Note 9**).
4. Dispense the 20 μL diluted siRNA into a 96-well plate in the desired plate format (*see* **Notes 10** and **11**).

3.3. Reverse Transfection in a 96-Well Plate – Preparation of Transfection Complexes

1. Warm the RNAiMax and Opti-MEM® I reagents to room temperature before use.
2. Dilute different amounts of RNAiMax in Opti-MEM® I for a total volume of 10 μL in a polystyrene 12 × 75 mm tube. Try 3–5 dilutions to find the right dose that balances good knockdown and low cell toxicity. Optimized dose for HeLa cells is 0.15 μL per well. Make a master mix for 10 wells. Mix by gently flicking the tube. It may be hard to collect all the liquid on the bottom (*see* **Note 12**.)
3. Combine 10 μL of the RNAiMax mixture per 20 μL of diluted siRNA for a total of 30 μL with a multichannel pipettor or repeater. Mix by tapping the four corners of the plate against the bench to ensure complete coverage of the whole well. Incubate this mixture for 10–20 min at room temperature.
4. After the incubation, add 80 μL of the diluted cell suspension with a multichannel pipettor or repeater to each well for a total of 110 μL. Be sure to mix the cells before adding them to the wells because cells may have settled to the bottom. Rock the plate back and forth to mix but do not swirl.
5. Place the plate in a 37°C incubator under normal cell culture conditions. Remove and assay at the desired time point. Maximal mRNA knockdown is usually observed 24–48 h after transfection (*see* **Note 13**).

3.4. Cell Lysis

Prepare TaqMan® Cells-to-CT™ reagents at time of harvest.

1. Prepare the cell lysis buffer by diluting DNase I (supplied with the kit) 1:100 into the lysis solution and make a master mix of sufficient volume to aliquot to all sample wells to be lysed (50 μL lysis mix per well in a 96-well plate).
2. Thaw enough stop solution for all wells (5 μL per well). Invert to mix (do not vortex) and put on ice.
3. Aspirate cell culture medium and gently wash cells with 100 μL of 1 × PBS (*see* **Note 14**).
4. Lyse cells with 50 μL of the TaqMan® Gene Expression Cells-to-CT™ lysis buffer per well and incubate at room temperature for 5 min with shaking.
5. Stop the lysis reactions by adding 5 μL stop solution; mix/shake at room temperature for 2 min. After the stop, proceed to reverse transcription or store plates on ice for up to 2 h, or at −20°C (or −80°C) for up to 5 months.

3.5. Reverse Transcription

1. Combine 6 μL lysate and 24 μL master mix in the appropriate wells of a 96-well PCR plate or tubes that correspond to your PCR thermal cycler for a 30-μL RT reaction (reaction volumes can be scaled up or down).
2. Mix 15 μL 2 × RT buffer, 1.5 μL 20 × RT enzyme, and 7.5 μL water (volumes listed are for a single well, but a master mix can be prepared for numerous samples plus 10% overage). For our example, make enough for 10 wells.
3. Add 24 μL of the above mixture to each well or tube to give a total of 30 μL, mix, and spin the plate or tube. Incubate at 37°C for 1 h, then 95°C for 5 min. The cDNA may be stored on ice or frozen until real-time PCR is performed.

3.6. Real-Time PCR

1. Real-time PCR can be performed using a standard TaqMan® PCR assay protocol on an Applied Biosystems 7900HT Fast Real-Time PCR System. Real-time PCR on each sample should be performed in duplicate at least.
2. Mix 2.5 μL nuclease-free water, 5.0 μL TaqMan® gene expression master mix, 0.5 μL 20 × gene expression assay (volumes are for a single 10 μL reaction in a 384-well plate, but a master mix can be prepared for numerous samples plus 10% extra for pipetting variability).
3. For each sample being examined, plate 2.0 μL of cDNA in four wells of a real-time PCR plate for gene detection and 18S detection in triplicate. We routinely use 18S as normalizer gene.
4. Add 8 μL of the GOI (gene of interest) mixture to 18 of the wells, and the endogenous control (18S) mixture to 18 wells

(biological triplicates + PCR duplicates on each loaded into six wells, siRNA treated in six wells, negative control treated and cells only in six wells).

5. Seal the plate with an optical cover and spin the plate.
6. The thermocycling conditions using TaqMan® gene expression master mix and a TaqMan® gene expression assay are 95°C for 10 min followed by 40 cycles of 95°C for 15 s and 60°C for 1 min.

3.7. Data Analysis

1. Once the PCR is complete, there are two settings that need to be adjusted for each gene: the threshold and the baseline.

   Baseline – this separates the background that occurs before the exponential phase of the amplification takes place. Set the upper limit just before the exponential phase begins. Set the lower limit such that it minimizes background.

   Threshold – this is the amount of fluorescence at which a cycle threshold (Ct) will be determined. This should be set such that it falls on the exponential amplification portion of the amplification plot for all of the samples for which that gene will be compared (see **Fig. 13.3** with the threshold line indicated).

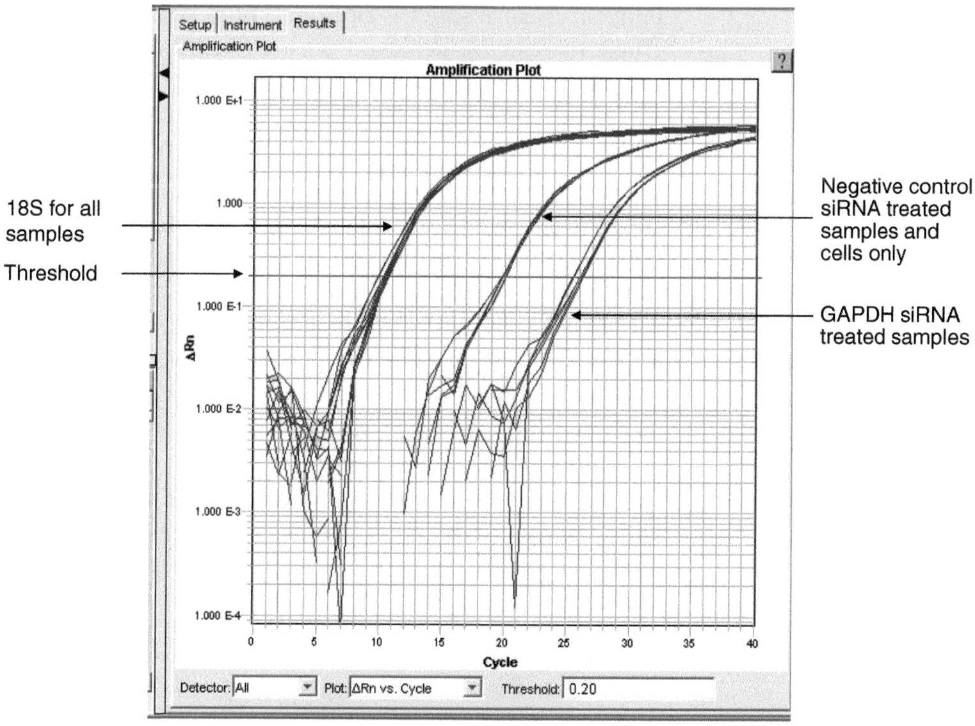

Fig. 13.3. Knockdown of GAPDH: real-time PCR amplification plot.

2. Once your baseline and threshold are set, export the file such that it can be edited in the spreadsheet program of your choice.
3. Calculate the average Ct and the standard deviation for the PCR triplicates for each sample and each gene. (For each comparison there should be four averages.)
4. Calculate the dCt for each sample: Subtract the Ct value of the endogenous control for the target sample from the target Ct. Then subtract the Ct value of the endogenous control for the calibrator sample from the calibrator Ct.
5. Calculate the ddCt for the two samples: Subtract the dCt for the target sample from the dCt of the calibrator sample. This value represents the fold difference between the levels of gene expression in the two samples.
6. To calculate the percent remaining mRNA use the following formula: Percent Remaining = $100*2^{-\Delta\Delta Ct}$ (*see* **Note 15**).

4. Notes

1. For each cell line, use the recommended medium (refer to ATCC's website). Supplementing the medium with 1% Pen/Strep antibiotics is optional: high siRNA transfection efficiency can be achieved with or without antibiotics in the medium. With new bottles make 50-ml aliquots of FBS and 5-ml aliquots of Pen/Strep upon arrival and store at –20°C.
2. Transfection conditions (type of transfection reagent and amount per well) are cell type-specific. Lipofectamine™ 2000 and Lipofectamine™ RNAiMax are highly efficient transfection reagents enabling siRNA delivery into a wide variety of cells with minimal optimization required. For co-transfection of siRNA and plasmid encoding target use Lipofectamine™ 2000.
3. Transfections can be done in reverse format (trypsinized cells added to siRNA-lipid complexes) or forward format (cells plated first, transfection performed 12–24 h later). Conditions need to be optimized for each cell line. The protocols described here are for the reverse format in 96-well plates. If other size plates are used, scale accordingly:

Culture vessel	Rel. surf. area[1]	Volume of plating medium	Cells plated per well		Dilution medium	RNAi duplex amount		Final RNAi duplex conc.		Lipofectamine™ RNAiMAX	
			Start point	Acceptable (Acc) Range	Reverse transf. (μL)	Start point (pmol)	Acc. Range (pmol)	Start point (nM)	Acc. Range (nM)	Start point (μL)	Acc. Range (μL)
96-well	0.2	80 μL	6,000	5000–15,000	2 × 10	3	0.12–6	30	1–50	0.15	0.1–0.4
48-well	0.4	160 μL	12,000	10,000–30,000	2 × 20	6	0.24–12	30	1–50	0.3	0.2–0.8
24-well	1	400 μL	30,000	25,000–75,000	2 × 50	15	0.6–30	30	1–50	0.75	0.5–2.0
6-well	5	2.0 mL	150,000	125,000–375,000	2 × 250	75	3–150	30	1–50	3.75	2.5–10.0

[1] Surface areas may vary depending on the manufacturer

4. Polypropylene tubes are important so that cells do not adhere to the plastic. Cells can be kept in such tubes for up to 1 h at 37°C.
5. Sample calculation to determine how many cells are needed:

   Cell count = 400,000 cells/mL

   Desired concentration = 4000 cells/well

   # of wells + overages = 10 wells

   volume per well = 80 μL

   Total volume needed = (# of wells)(volume per well)

   = (10)(80 μL) = 800 μL

   Total amount of cells needed = (# of wells)(desired concentration) = (10 wells)(4000 cells/well)

   = 40,000 cells

   Volume of cells needed = (amount of cells needed)/(cell count)

   = (40,000 cells)/(400,000 cells/mL) = 0.1 mL

   Mix the 0.1 mL cells with 0.7 mL growth medium to give the total volume needed of 0.8 mL or 800 μL.
6. To re-suspend siRNA, use water; another typically used solvent is PBS.
7. If multiple subsequent experiments are planned, siRNAs can be stored at 4°C for up to 1 month with no observable degradation. The stock concentration should be no less than 10 μM.
8. In most cases, 5 nM final concentration is sufficient to see robust mRNA knockdown. However, a dose curve should be performed to determine the best siRNA concentration to use for the desired cell line. In addition to the siRNA concentration, ideally test 2–3 negative control siRNAs to find the best one for your system that does not affect expression levels of gene of interest or induce cell toxicity.
9. Sample calculation to determine how much siRNA you need: $M_1 V_1 = M_2 V_2$

   Stock siRNA concentration = 10 μM

   Desired siRNA concentration = 5 nM

   Final transfection volume = 110 μL

   (10 μM)(×) = (5 nM)(110 μL)× = 0.055 μL/well

   Triplicate wells + 1 well overage = 0.055 μL × 4 = 0.22 μL

10. A repeater pipette may be useful for your replicates. Pipette up 60 μL of your master mix and multi-dispense 20 μL to your triplicate wells. There should be ~10 μL left in your master mix tube because you made enough for four wells.
11. To save time and if there are a lot of siRNAs to screen at once on the actual transfection day, pre-plating siRNAs is an option. Depending on how many replicates and plates, make a master mix and dispense the appropriate amount of siRNA in the wells. Freeze plates at –20°C if experiment will be performed next day or –80°C for long-term storage. Plates can then be thawed prior to use.
12. If the RNAiMax/Opti-MEM® I mixture does not fit in a polystyrene 12 × 75 mm tube, the RNAiMax can be diluted in a polypropylene tube such as a 15-mL conical tube.
13. After 12 h, the medium may be changed to reduce cellular toxicity due to transfection. Aspirate medium and replace with 100 μL fresh growth medium.
14. A vacuum pump aspirator with an 8-channel plastic adaptor for pipette tips (without filters) can be used to remove the medium cleaner and faster. After washing the plates with PBS, they can be frozen at –80°C until ready for lysis/RNA isolation.
15. Sample data analysis calculation: reverse transfected GAPDH control siRNA and Negl siRNA in triplicate. Duplicate wells for real-time PCR.

GAPDH knockdown with GAPDH siRNA

Ct	dCT	ddCT	%RE	Avg. %RE	SD
25.84	15.4283	6.24	1.32304	1.888728	0.62149
25.9	15.4883	6.3	1.26914		
24.75	14.3383	5.15	2.81641		
24.97	14.5583	5.37	2.41807		
25.57	15.1583	5.97	1.59533		
25.31	14.8983	5.71	1.91038		

dCT = Ct – Avg 18S target for GAPDH siRNA
ddCT = dCT – Avg GAPDH target for Negl
%RE = $100*2^{\wedge-ddCt}$

Normalization of GAPDH knockdown with Negl siRNA

Negl siRNA	Ct	dCT	
	19.82	9.27333	9.18833
	19.75	9.20333	Avg. Negl
	19.63	9.08333	
	19.64	9.09333	
	19.77	9.22333	
	19.8	9.25333	

dCT = Ct – Avg 18S target for Negl

18S RNA levels for Negl siRNA-treated wells

Ct	Avg.
10.35	10.5467
10.37	
10.53	
10.47	
10.84	
10.72	

18S RNA levels for GAPDH siRNA-treated wells

Ct	Avg.
10.49	10.4117
10.09	
10.43	
10.17	
10.7	
10.59	

References

1. Meister, G., and Tuschl, T. (2004) Mechanisms of gene silencing by double-stranded RNA. *Nature* **431**, 343–349.
2. Castanotto, D., and Rossi, J. J. (2009) The promises and pitfalls of RNA-interference-based therapeutics. *Nature* **457**, 426–433.
3. Pappas, T. C., Bader, A. G., Andruss, B. F., Brown, D., and Ford, L. P. (2008) Applying small RNA molecules to the directed treatment of human diseases: realizing the potential. *Expert Opin. Ther. Targets* **12**, 115–127.
4. Hickerson, R. P., Vlassov, A. V., Leake, D., Wang, Q., Contag, C. H., Johnston, B. H., and Kaspar, R. L. (2008) Stability study of unmodified siRNA and

relevance to clinical use. *Oligonucleotides* **18**, 345–354.
5. Dallas, A., and Vlassov, A. (2006) RNAi: a novel antisense technology and its therapeutic potential. *Med. Sci. Monitor* **4**, 67–74.
6. Puri, N., Wang, X., Varma, R., Burnett, C., Beauchamp, L., Batten, D. M., Young, M., Sule, V., Latham, K., Sendera, T., Echeverri, C., Sachse, C., and Magdaleno, S. (2008) LNA incorporated siRNAs exhibit lower off-target effects compared to 2′-OMethoxy in cell phenotypic assays and microarray analysis. *Nucl. Acids Symp. Ser. (Oxf.)* **52**, 25–26.
7. Wang., X., Wang, X., Varma, R. K., Beauchamp, L., Magdaleno, S., and Sendera, T. J. (2009) Selection of hyperfunctional siRNAs with improved potency and specificity. *Nucl. Acids Res.* **37**, e152.
8. Whitehead, K. A., Langer, R., and Anderson, D. G. (2009) Knocking down barriers: advances in siRNA delivery. *Nat. Rev. Drug Discov.* **8**, 129–138.
9. Vlassov, V. V., Balakireva, L. A., and Yakubov, L. A. (1994) Transport of oligonucleotides across natural and model membranes. *Biochim. Biophys. Acta* **1197**, 95–108.
10. Bustin, S. A. (2002) Quantification of mRNA using real-time reverse transcription PCR (RT-PCR): trends and problems. *J. Mol. Endocrinol.* **29**, 23–39.
11. Peters, I. R., Helps, C. R., Hall, E. J., and Day, M. J. (2004) Real-time RT-PCR: considerations for efficient and sensitive assay design. *J. Immunol. Methods* **286**, 203–217.

Chapter 14

siRNA Knockdown of Gene Expression in Endothelial Cells

Emily Dennstedt and Brad Bryan

Abstract

Cultured endothelial cells are renowned for being difficult to transfect, whether for the purpose of exogenous over-expression of plasmid DNA or for genetic knockdown via silencing RNA. Therefore, optimal conditions are absolutely necessary for achieving relatively high transfection efficiency coupled with low cellular toxicity in endothelial cells. This chapter will detail an optimized protocol that has been shown to knockdown gene expression using siRNA in primary cultures of human umbilical vein endothelial cells (HUVECs) – perhaps the most widely utilized endothelial cell line for vascular research. While developed for optimal siRNA transfection of HUVECs, aspects of this protocol can be empirically modified to yield efficient siRNA transfection in most other cell lines.

Key words: RNA interference, siRNA transfection, gene knockdown, human umbilical vein endothelial cells.

1. Introduction

Earlier this decade, disruption of targeted gene expression could only be achieved by genetically engineering knock-out mice. The potential for genetic background influence, functional genetic redundancy, embryonic lethality, and altered offspring fertility rates from such knock-outs, coupled with the expense and prolonged time to generate model organisms were some of the disadvantages faced by researchers when producing knock-outs (1–3). Moreover, any molecular analysis of the gene of interest would require isolating primary cell lines from knockout mice. Therefore developing a knock-out model became an overly cumbersome method to accurately determine the function of a particular gene. siRNA technology, which uses small interfering RNA (siRNA) to inactivate a target gene's expression before it can

J. Goodchild (ed.), *Therapeutic Oligonucleotides*, Methods in Molecular Biology 764,
DOI 10.1007/978-1-61779-188-8_14, © Springer Science+Business Media, LLC 2011

be translated into protein, has emerged as one of the premiere knockdown technologies for in vitro research (4), and has exhibited great success in vivo using viral vectors, electroporation, and in vivo transfection reagents (5, 6). This technology allows for the generation of a knock-down line that can be compared with wild type lines to assess the cellular effects rendered after disrupting expression of the gene of interest. Additionally, siRNA technology provides a significantly cheaper, more efficient means of creating knock-down samples compared to knockouts, requiring only the most basic of lab equipment that is provided to most research laboratories. Perhaps the greatest advantages of siRNA technology at present are that many pre-existing cell lines can be successfully utilized, and commercially available reagents and siRNA to match almost any defined RNA sequence can be cost effectively purchased.

Knockdown of gene expression using siRNA requires effective transfection into cell lines. Ultimately two factors must be taken into consideration for this to occur – RNA silencing and cell viability. An efficient protocol will lead to good reduction in RNA expression coupled with low cellular toxicity. As one would observe with transfection of plasmid DNA, different cell lines exhibit differing affinities for siRNA uptake; therefore optimized experimental protocols are absolutely necessary for effective knock-down of gene expression. Inadequate or partial transfection of siRNA into cells, which results in less than 70–75% reduction of mRNA levels, will ultimately lead to inadequate gene silencing and inaccurate conclusions regarding the biological function of the gene.

Three factors can be varied to generate a protocol for high-efficiency siRNA transfection – quality/quantity of transfection reagent, cell confluency, and quality of siRNA.

1. Selection of a suitable commercially available cationic liposomal or polyamine transfection reagent is essential for efficient siRNA transfection. Many companies that produce these agents will have already conducted optimization tests for commonly utilized cell lines and will provide their findings for consumer reference. It is advised to base your protocol around their suggested method and empirically tweak your method as necessary for each application (*see* **Note 1**).

2. The cell density can greatly affect the efficiency of siRNA transfection. The optimal confluency for transfection is between 30 and 70% depending on cell type and the amount of transfection reagent utilized (*see* **Note 2**).

3. The choice of siRNA can significantly influence the experimental outcome. While siRNA can be custom designed for experiments in non-standard species or with unique siRNA design requirements, we strongly recommend purchasing

siRNA pre-designed and validated by the company of purchase for general use in human, mouse, or rat cell lines (*see* **Note 3**).

In this chapter, we describe a protocol utilized in our laboratory for siRNA transfection of primary lines of human umbilical vein endothelial cells (HUVECs). HUVECs are perhaps the most commonly utilized endothelial cell line in scientific research due to their ease of culture, ability to respond to vascular growth factors, and capacity to form three-dimensional lumen-bearing capillary tubes upon defined culture conditions (7, 8). Because of these properties, HUVECs have been applied as a basic model system to understand vascular tissue engineering as well as vascular pathologies. Unfortunately, human umbilical vein endothelial cells (HUVEC) tend to be among the more difficult cell lines to transfect, characteristically exhibiting low uptake levels of transfection agent/nucleic acid conjugate. Additionally, we explain in detail a rapid, cost effective method to quantify transfection efficiency so that this protocol can be empirically modified to yield efficient siRNA transfection in most other cell lines.

1.1. siRNA Transfection of HUVECs

A number of commercial reagents and siRNAs are conveniently available for researchers, and our laboratory has quantified the efficacy of a number of these formulations. In our hands, Thermo Scientific DharmaFECT1 resulted in the highest transfection efficiency coupled with the lowest cytotoxicity for HUVECs, as well as a number of other primary and immortalized endothelial cell lines. Therefore, the following protocols are optimized to use DharmaFECT1 as the transfection reagent; however, empirical optimization of this protocol can be utilized for most other siRNA transfection systems and cell lines. This protocol is an optimized method for siRNA transfection of HUVECs. All steps should be conducted using sterile technique under a laminar flow cell culture hood. Included in this procedure are the volume calculations of each reagent necessary for transfection of *three* 96-wells (as experiments should be done at least in triplicate) (*see* **Note 4**).

We have observed that peak knockdown of steady-state mRNA occurs within 24–48 h, while peak knockdown of steady-state protein levels occurs within 48–72 h. An experimental example of the timing of siRNA-mediated mRNA and protein knockdown is demonstrated in **Fig. 14.1** using siRNA specifically targeted against mouse Rho Kinase 1, a Rho-GTPase effector protein involved in signal transduction controlling cytoskeletal dynamics (9). In some situations, efficiently transfected siRNAs that are accurately targeted for the gene of interest may result in little to no knockdown at the protein level, given the fact that steady-state protein levels largely depend on the half-life of the target protein. In other words, proteins with a short half-life may

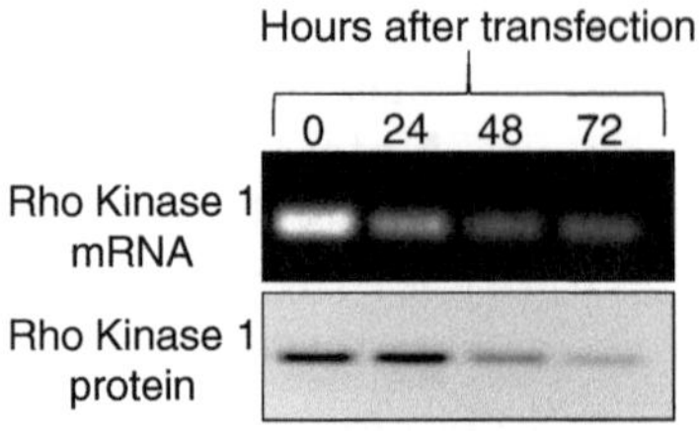

Fig. 14.1. Timecourse of siRNA-mediated mRNA and protein knockdown. HUVECs were transfected as indicated in **Section 3.1** utilizing siRNA specifically targeted against Rho Kinase 1. Reverse transcriptase polymerase chain reaction and Western blotting analysis was performed on cell lysates collected at 24, 48, and 78 h post-transfection. A 0-h timepoint (non-transfected cells) was used as a baseline control.

be more effectively knocked down than proteins with a very long half-life.

HUVECs are considered one of the more difficult cell lines to transfect, with a maximal target gene expression knockdown of around 70–80% and relatively high cytotoxicity in response to most transfection reagents. While HUVECs are perhaps the most commonly utilized endothelial cell lines, many laboratories also utilize primary endothelial lines isolated from human dermal microvasculature (HMVEC), human lung microvasculature (HMVEC-Ls), bovine aorta (BAE), bovine retina (BREC), porcine aorta (PAE), and many others. Moreover, a number of immortalized, hemangioma, and tumor microvascular endothelial cells have been used in many publications, each with their own unique characteristics that lend them for use in various applications. To investigate the biological relevance of target genes in these cell lines, siRNA protocols must be optimized to address the unique characteristics of these cells (i.e., high cytotoxicity, slow recovery, etc.).

1.2. Fluorescent Staining

Historically, siRNA targeted against the housekeeping gene GAPDH has been utilized as a measure of efficient siRNA transfection. However, detecting steady-state mRNA or protein levels using reverse transcriptase polymerase chain reaction (RT-PCR) or Western blot analysis can be expensive and time-consuming considering the relatively large number of samples needed in order to optimize siRNA transfection in a particular cell line. A much more efficient method for developing an optimal transfection procedure is through the use of non-targeting scrambled siRNA indicators covalently bound to fluorescent tags such as fluorescein (FITC), Cy-3, or their analogs such as 5-carboxy-fluorescein or DY-547 (10). Following transfection, immunoflourescence microscopy can be utilized to determine the number of cells exhibiting subcellular localization of the tagged siRNA molecules as compared to the total number of cells, thus quantifying transfection efficiency **(Fig. 14.2)**.

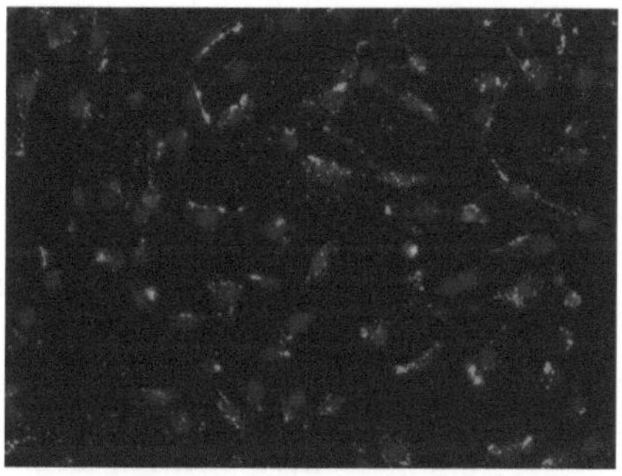

Fig. 14.2. Transfection of HUVECs with fluorescently labeled siRNA. HUVECs were transfected with Cy-3 conjugated siRNA as indicated in **Section 3.2.** After 24 h, transfection efficiency was determined using an inverted fluorescent microscope. Transfection efficiency for this sample was approximately 80%. Fluorescence is due to Cy-3 conjugated siRNA (the brighter, granular features) and DAPI nuclear marker.

The protocol detailed below utilizes transfection of FITC-conjugated siRNA indicators coupled with propidium iodide staining (which labels the nuclei of only dead cells) as a viability indicator, and Hoechst 33342 nuclear staining (which labels all nuclei) as a measure of total cell count. Healthy cells (fluorescing blue) take up only the cell-permeable Hoechst dye, but exclude the non-permeable propidium iodide. Dead cells (fluorescing both blue and red) take up both the cell permeable Hoechst dye and the non-permeable propidium iodide. Cells that are transfected with the FITC-labeled fluorescent siRNA indicator will fluoresce green.

2. Materials

1. siRNA Buffer (1×): 60 mM KCl, 6 mM HEPES (pH 7.5), 0.2 mM $MgCl_2$ (prepared in nuclease free water)
2. DharmaFECT1 transfection reagent
3. RPMI medium without supplements (*see* **Note 5**)
4. RPMI medium with supplements: 500 mL RPMI, 100 mL bovine calf serum, 0.1 g endothelial cell growth supplement (ECGS), 50 mg heparin sodium
5. FITC-conjugated scrambled siRNA (*see* **Note 6**)
6. 1× Hank's balanced salt solution (HBS)
7. Working propidium iodide solution: 5 μg/mL in cell culture medium (with supplements)
8. Working Hoechst 33342 solution: 10 μM in cell culture medium (with supplements)

3. Methods

Each experiment should include the following siRNA samples (*see* **Note 6**):

1. Untreated cells – as a reference to evaluate cytotoxicity
2. Mock-transfection (no siRNA) – compare to untreated cells to evaluate cytotoxicity
3. Positive control siRNA – siRNA which targets a housekeeping gene (usually GAPDH)
4. Negative control siRNA – non-targeting scrambled siRNA to evaluate non-specific activity
5. Experimental siRNA – targeting the gene of interest

3.1. siRNA Transfection of HUVECs

1. Using sterile technique, seed HUVECs to a density of 2.5×10^4 cells into each well of a 96-well plate. Incubate the cells overnight in RPMI media (with supplements) in a 5% CO_2 incubator.
2. Prepare a 2-μM siRNA solution in 1× siRNA Buffer. The total volume of 2 μM siRNA solution needed per triplicate reaction is 17.5 μL (*see* **Note 7**).
3. Per reaction, prepare the following solutions in sterile tubes:
   A. siRNA working solution – 17.5 μL of 2 μM siRNA solution plus 17.5 μL RPMI (without supplements)
   B. 0.05% DharmaFECT1 solution – 0.175 μL of DharmaFECT1 to 34.825 μL RPMI (without supplements).
4. Slowly pipette up and down to mix thoroughly and incubate at room temperature for 5 min.
5. Combine 35 μL of siRNA working solution with 35 μL of 0.05% DharmaFECT1 solution to make the siRNA master mix (70 μL total reaction volume). Mix contents by gently pipetting, and incubate at room temperature for 20 min.
6. Add 280 μL pre-warmed RPMI (with supplements) to the 70 μL siRNA master mix to make the siRNA final solution (*see* **Note 8**).
7. Remove culture media from cells and wash cells twice with RPMI (without supplements) to remove residual growth supplements or antibiotics.
8. Add 100 μL of the siRNA final solution to each well.
9. Incubate cells at 37°C in a 5% CO_2 cell culture incubator.
10. Using HUVECs, the abovementioned protocol leads to approximately 15–20% cytotoxicity. If increased cytotoxicity occurs, the researcher should measure cytotoxicity

using a viability assay such as (3-(4,5-dimethylthiazol-2-yl)-2,5-diphenyltetrazolium bromide) (MTT) or (3-(4,5-dimethylthiazol-2-yl)-5-(3-carboxymethoxyphenyl)-2-(4-sulfophenyl)-2H-tetrazolium) (MTS) colorimetric assays. Compare the untransfected cell number to the mock or negative control transfected cell number to determine the amount of cell death induced by the transfection procedure itself. Any cell death in the experimental siRNA condition that exceeds this baseline level can be attributed to cellular effects due to knockdown of the target gene.

3.2. Fluorescent Cell Staining

1. Follow the protocol in **Section 2** to transfect the cell line of interest with fluorescently labeled siRNA. Optimal cell density and transfection reagent volume should be empirically determined. It is recommended to seed cells (in a 96-well plate) between 5×10^3 and 2.5×10^4 cells/well. Transfection reagent volume should range between 0.05 and 0.5 μL per well.
2. At approximately 24 h post-transfection, cells should be briefly washed two times with HBS.
3. For each well, combine 50 μL of the propidium iodide solution with 50 μL of the Hoechst 33342 solution. Add the mixture to each well. Incubate for 30 min at 37°C.
4. Wash cells two times with HBS for 2 min for each wash.
5. Suspend cells in 100 μL of HBS and observe immediately using an inverted fluorescent microscope to determine transfection efficiency and cytotoxicity.

4. Notes

1. Upon choosing the best reagent, optimum concentration is absolutely essential as a condition of efficacy since an inadequate amount of transfection reagent can limit the siRNA uptake, while an excess often results in low survival rates that are detrimental to the experiment.
2. Suboptimal cell density can result in poor uptake of the siRNA transfection reagent and therefore insufficient silencing of the target gene.
3. If not validated, we suggest at minimum the purchase of at least four siRNA reagents to identify one that efficiently silences the target gene and reduces off-target effects.
4. If different plate formats or replications are required, scale the volumes of reagents and number of cells per well according to altered surface area of the dish.

5. Use of antibiotics will severely interfere with transfection efficiency.
6. siRNA should be stored in small, single use aliquots at −80°C.
7. RNA is very unstable. Once thawed, keep the siRNA solution on ice and discard after use.
8. The 350 μL total volume of the siRNA final solution is 50 μL more than necessary for triplicate reactions, but compensates for potential liquid loss during pipetting.

Acknowledgments

We would like to thank Dr. Magali Saint-Geniez of the Schepens Eye Research Institute (Boston, Massachusetts) for her contribution of images to this manuscript. We would also like to thank Dr. Dianne C. Mitchell of Acceleron Pharma (Cambridge, Massachusetts) for valuable insights and editing of this manuscript.

References

1. Rudmann, D. G., and Durham, S. K. (1999) Utilization of genetically altered animals in the pharmaceutical industry. *Toxicol Pathol* **27**, 111–114.
2. Lewandoski, M. (2001) Conditional control of gene expression in the mouse. *Nat Rev Genet* **2**, 743–755.
3. Doetschman, T. (2009) Influence of genetic background on genetically engineered mouse phenotypes. *Methods Mol Biol* **530**, 423–433.
4. Gopalakrishnan, B., and Wolff, J. (2009) siRNA and DNA transfer to cultured cells. *Methods Mol Biol* **480**, 31–52.
5. Draghia-Akli, R., and Khan, A. S. (2007) In vivo electroporation of gene sequences for therapeutic and vaccination applications. *Recent Pat DNA Gene Seq* **1**, 207–213.
6. de Fougerolles, A. R. (2008) Delivery vehicles for small interfering RNA in vivo. *Hum Gene Ther* **19**, 125–132.
7. Nakatsu, M. N., Sainson, R. C., Aoto, J. N., Taylor, K. L., Aitkenhead, M., Pérez-del-Pulgar, S., Carpenter, P. M., and Hughes, C. C. (2006) Angiogenic sprouting and capillary lumen formation modeled by human umbilical vein endothelial cells (HUVEC) in fibrin gels: the role of fibroblasts and Angiopoietin-1. *Microvasc Res* **66**, 102–112.
8. Bryan, B. A., and D'Amore, P. A. (2008) Pericyte isolation and use in endothelial/pericyte coculture models. *Methods Enzymol* **443**, 315–331.
9. Maruta, H., Nheu, T. V., He, H., and Hirokawa, Y. (2003) Rho family-associated kinases PAKl and rock. *Prog Cell Cycle Res* **5**, 203–210.
10. Surowiak, P. (2003) Evaluation of transfection effectiveness using fluorescein-labelled oligonucleotides and various liposomes. *Folia Morphol (Warsz)* **62**, 397–399.

Chapter 15

Using RNA Interference in *Schistosoma mansoni*

Rita Bhardwaj, Greice Krautz-Peterson, and Patrick J. Skelly

Abstract

Schistosomes are parasitic worms that infect over 200 million people and constitute an enormous public health problem worldwide. Molecular tools are being developed for use with these parasites in order to increase our understanding of their unique molecular and cell biology. Among the more promising methodologies is RNA interference (RNAi, or gene silencing), a mechanism by which gene-specific double-stranded RNA (dsRNA) triggers degradation of homologous mRNA transcripts. In this work we describe methods for applying RNAi to suppress gene expression in the intra-mammalian life stages of *Schistosoma mansoni*. These methods include isolating and culturing the parasites, preparing and delivering dsRNA targeting a specific gene and monitoring the outcome. Given the abundance of schistosome transcriptome and genome sequences now available, RNAi technology has the potential to rapidly expand analysis of the roles and importance of the genes of this globally important parasite.

Key words: Gene silencing, RNAi, double-stranded RNA, platyhelminth, parasite, trematode, schistosome, short interfering RNA.

1. Introduction

Schistosomes are platyhelminth intravascular parasites. Adults of the species *S. mansoni* reside in the mesenteric venules of the intestine, where they may cause the chronic debilitating disease schistosomiasis. The worms infect ~200 million people and another ~600 million are at risk of infection. The parasites have a two-host life cycle. They replicate asexually in specific freshwater snail species. Life cycle stages called cercariae (singular: cercaria) emerge from the snails and are infectious for humans. These parasites penetrate the skin and undergo a complex morphological and biochemical adaptation to their new environment that is called cercarial transformation. Following this, the parasites

J. Goodchild (ed.), *Therapeutic Oligonucleotides*, Methods in Molecular Biology 764,
DOI 10.1007/978-1-61779-188-8_15,

are called schistosomula (singular: schistosomulum) and these migrate through the dermal tissue and invade a blood vessel. Parasites migrate in the blood through the lungs to the hepatic portal vasculature where males and females pair. *S. mansoni* couples then migrate to the mesenteric venous plexus where eggs are laid. Eggs can pass into the intestinal lumen and into the environment. Juvenile forms emerging from the eggs may infect snails to complete the life cycle.

Molecular tools are being developed for use with these parasites in order to increase our understanding of their unique molecular and cell biology. Among the more promising methodologies is RNA interference (RNAi, or gene silencing), a mechanism by which gene-specific double-stranded RNA (dsRNA) triggers degradation of homologous mRNA transcripts. This technology currently represents the only approach for experimentally manipulating the expression of targeted endogenous schistosome genes, thereby providing insight into putative gene function (1, 2). Given the abundance of schistosome transcriptome and genome sequences now available (3–9), RNAi has the potential to revolutionize investigation of the roles and importance of the genes of this globally important parasite. The protocol used in this laboratory for RNAi in schistosome intravascular life forms is described here. Most emphasis is placed on our preferred method which involves electroporating the parasites in the presence of short interfering RNAs (siRNAs) targeting the gene of interest. We also describe our method for generating long double-stranded RNA that can be delivered to the parasites by electroporation or, in the absence of a suitable electroporator, by simple soaking.

2. Materials

2.1. Stock Solutions for Parasite Culture (Modified Basch) Medium

1. D-(+)-Glucose
2. Lactalbumin hydrolysate solubilized (MP Biomedicals LLC)
3. Basal Medium Eagle without Glutamine (BME)
4. HEPES buffer solution: 1 M, pK_a 7.55
5. Schnider's Drosophila Medium with L-glutamine (1×) (Invitrogen)
6. MEM Vitamin solution, 100× (Sigma-Aldrich)
7. Hypoxanthine: 1 mM stock solution in distilled water stored at 4°C
8. Serotonin hydrochloride: 1 mM stock solution in distilled water stored at 4°C

9. 3, 3′, 5-Triiodo-L-thyronine; 0.2-mM stock solution in 0.2 N NaOH stored at 4°C
10. Hydrocortisone solution: 50 μM
11. Insulin from bovine pancreas: 10 mg/mL in 25 mM HEPES, pH 8.2
12. Fetal bovine serum inactivated by heating in a water-bath at 56°C for 1 h. Aliquots stored at –20°C
13. Pen/Strep: penicillin, 10,000 U/mL and streptomycin, 10,000 μg/mL
14. Liquid filter system (Corning Incorporated)

2.2. Electroporating Adult Parasites with Short Inhibitory RNAs (siRNAs)

1. siRNAs targeting your gene of interest may be synthesized commercially, (e.g., from IDT). Our laboratory uses the on-line siRNA design tool: http://www.idtdna.com/Scitools/Applications/RNAi/RNAi.aspx (*see* **Notes 1** and **2**).
2. siRNA resuspension buffer, called RNase-free duplex buffer by the supplier: 100 mM potassium acetate, 30 mM HEPES, pH 7.5 (IDT).
3. Sterile transfer pipettes.
4. 0.4 cm Gene Pulser electroporation cuvettes (Bio-Rad).
5. siPORT electroporation buffer (Ambion) (*see* **Note 3**).
6. Gene Pulser Xcell Electroporator (Bio-Rad).
7. 12-well tissue culture plates.

2.3. Isolating RNA from Adult Parasites

1. Trizol reagent (Invitrogen)
2. RNase*Zap* Wipes (Ambion)
3. Pellet Pestle Motor and RNase free Pellet Pestle (Kontes)
4. Chloroform
5. Isopropyl alcohol
6. DEPC-treated water (Ambion)
7. TURBO DNA-free kit: 10× TURBO DNase buffer, DNA-free TURBO DNase, and DNase inactivation reagent (Ambion)
8. UV spectrophotometer

2.4. cDNA Synthesis

1. Oligo $(dT)_{12-18}$ Primer: 0.5 μg/μL (Invitrogen)
2. RNaseOUT recombinant ribonuclease inhibitor: 40 U/μL (Invitrogen)
3. RNase H, 2 U/μL
4. SuperScript III reverse transcriptase, 200 U/μL (Invitrogen)

5. DTT: 0.1 M
6. 10× Reverse transcriptase buffer (10× RT Buffer): Tris–HCl 200 mM, KCl 500 mM, pH 8.4
7. $MgCl_2$: 25 mM
8. dNTP 10 mM mix: dATP 100 mM 25 μL, dCTP 100 mM 25 μL, dGTP 100 mM 25 μL, dTTP 100 mM 25 μL, nuclease-free water 150 μL
9. RNase-Free 8-Strip 0.2 mL PCR tubes (Ambion)
10. Thermal cycler

2.5. Quantitative Real-Time-PCR (qRT-PCR)

1. TaqMan gene expression assay 20× mix: forward primer, reverse primer, reporter probe labeled with 6-carboxy-fluorescein (FAM) (Applied Biosystems) (*see* **Note 1**)
2. Taq Man Universal PCR Master Mix (2××) (Applied Biosystems)
3. Nuclease-free water
4. Fast optical 96-well reaction plates (Applied Biosystems)
5. Micro Amp optical adhesive film (Applied Biosystems)
6. Step One Plus Real-Time PCR System with Step One software v2.0 (Applied Biosystems)

2.6. Isolating RNA and Protein from the Same Parasite Sample

1. Phosphate buffered saline (PBS) (1×): 137 mM NaCl, 2.7 mM KCl, 10 mM Na_2HPO_4, 2 mM KH_2PO_4, pH 7.1
2. Protein And RNA Isolation System (PARIS kit): cell disruption buffer 2×, lysis/binding solution, wash solution 1, wash solution 2/3, filter cartridges, elution solution (Ambion)
3. BCA protein assay kit: protein assay reagents A & B, albumin standard (Thermo Fisher Scientific)

2.7. Working with Schistosomula

1. RPMI: RPMI 1640 (Invitrogen) supplemented with 10 mM HEPES, 2 mM glutamine, and antibiotics (100 U/mL penicillin and 100 μg/mL streptomycin)
2. Percoll: 35% sterile solution
3. Microslides: 25 × 75 × 1 mm
4. Cover-slips: 25 mm sq
5. Falcon conical centrifuge tubes: 15 mL and 50 mL BD
6. Trypan blue: 0.4%
7. Sterile transfer pipettes
8. 48-Well tissue culture plates
9. siPORT electroporation buffer (Ambion)
10. Trizol reagent (Invitrogen)

2.8. Preparing Long dsRNA

1. 10× PCR buffer: 200 mM Tris–HCl, 500 mM KCl, 15 mM $MgCl_2$, pH 8.4.
2. dNTP mix: 10 mM (Invitrogen)
3. Forward and reverse primers: 0.2 μM/reaction (Invitrogen)
4. Taq DNA polymerase: 2 U/reaction (Invitrogen)
5. QIAquick gel extraction kit: QIAquick spin column, buffer QG, buffer PE concentrate, nuclease-free water, ethanol, isopropanol (Qiagen)
6. MEGAscript, High Yield Transcription T7 and T3 Kits: T7 or T3 enzyme mix, RNA polymerase, 10× reaction buffer, ATP 75 mM, CTP 75 mM, GTP 75 mM, UTP 75 mM, Turbo DNAse 2 U/μL, nuclease-free water (Ambion)
7. Agarose
8. Ethidium bromide
9. 10× TBE (Sigma-Aldrich)
10. Agarose gel loading buffer (Sigma-Aldrich)
11. Wide-range DNA ladder (Sigma-Aldrich)
12. RNeasy MinElute Cleanup Kit: RNeasy MinElute columns, RLT and RPE buffers, ethanol, RNase-free water (Qiagen)
13. DEPC-treated water (Ambion)
14. RPMI is described in **Section 2.7**.

3. Methods

Schistosome adults can be recovered from infected experimental animals by vascular perfusion (10). All parasites are cultured in modified Basch medium at 37°C in an atmosphere of 5% CO_2.

Unlike complete Basch medium (11), the modified Basch medium used here lacks erythrocytes. Parasites may be cultured in an even simpler medium such as RPMI 1640 (Invitrogen) supplemented with 10 mM HEPES, 2 mM glutamine, 5% fetal calf serum, and antibiotics (100 U/mL penicillin and 100 μg/mL streptomycin) (RPMI) at 37°C, in an atmosphere of 5% CO_2. However, parasites cultured in this manner have decreased viability compared to those cultured in the modified Basch medium described. Parasites may be subjected to RNAi treatment immediately or at any time following culture.

3.1. Preparing Modified Basch Medium

1. Dissolve 1 g glucose and 0.5 g lactalbumin hydrolysate in 200 mL BME in a sterile 500-mL beaker using a magnetic stirrer. Add each of the remaining components listed in **Table 15.1**.

Table 15.1
Composition of modified Basch medium

	Component	Add stock	Stock concentration	Working concentration
1	Glucose	1 g		11.1 mM
2	Lactalbumin hydrolysate	0.5 g		1 g/L
3	HEPES buffer	5 mL	1 M	10 mM
4	Schnider's medium	25 mL	1×	5%
5	MEM vitamins	2.5 mL	100×	0.5×
6	Hypoxanthine	0.25 mL	1 mM	5×10^{-7} M
7	Serotonin	0.5 mL	1 mM	1 μM
8	Triiodothyronine	0.5 mL	0.2 mM	2×10^{-4} mM
9	Hydrocoritsone	10 mL	50 μM	1 μM
10	Insulin	0.4 mL	10 mg/mL	8 μg/mL
11	FCS (heat inactivated)	50 mL		10%
12	BME	405 mL	1×	
Total		500 mL		

2. Adjust the pH to 7.4 using 5 N NaOH and make the final volume up to 500 mL using BME.
3. Sterilize media by filtration under vacuum and store at 4°C.
4. Before use, add 1 mL Pen/Strep solution per 100 mL medium. Use fresh (within 2 weeks of preparation).

3.2. Electroporating Adult Parasites with siRNAs

1. Lyophilized, synthetic siRNAs are dissolved in siRNA resuspension buffer to a concentration of 10 μM (stock solution). siRNAs and electroporation buffer are kept on ice.
2. 12–15 adult worms in ~500 μL culture medium are transferred to a 1.5-mL tube for each experimental group/condition.
3. To gently pellet the worms, tubes are centrifuged for 1 min at 100×*g* at room temperature. Using a sterile transfer pipette, as much medium as possible is removed.
4. 100 μL of electroporation buffer is added to each tube and parasites are incubated on ice for 1 min. Incubating adult parasites briefly on ice in this manner immobilizes the worms and facilitates their transfer to the electroporation cuvettes in the next step.
5. Using a sterile transfer pipette, parasites and buffer are transferred to the bottom of an electroporation cuvette. siRNA is dispensed to the cuvette and mixed gently by tapping. We

recommend using 5 μg siRNA to test for suppression of a new target gene (*see* **Note 4**).

6. The parasites are electroporated by applying a square wave with a single 20 ms pulse, at 125 V in 4 mm cuvettes at room temperature (*see* **Note 5**).
7. Cuvettes are incubated at 37°C for 15 min.
8. 300 μL of modified Basch medium is added to the cuvettes and the parasites plus medium are transferred using a sterile transfer pipette to a well of a 12-well tissue culture plate containing 1 mL pre-warmed (37°C) modified Basch medium.
9. After overnight culture at 37°C in an atmosphere of 5% CO_2, medium is replaced with 2 mL of fresh medium. Thereafter, medium is changed every 2 days.

3.3. Isolating RNA Using Trizol Reagent from the Adult Parasites

The protocol below describes the methodology for RNA isolation using the Trizol reagent. This protocol is used when gene knockdown at the RNA level only is to be assessed. If knockdown at the protein level is also required, RNA is isolated using the protocol described in **Section 3.6** (*see* **Note 6**).

1. Parasites are transferred from culture wells to 1.5-mL tubes. As much medium as possible is removed from each tube.
2. 50 μL of Trizol reagent is added to each tube on ice (*see* **Notes 7** and **8**).
3. Parasites are homogenized on ice using a pellet pestle mortar for ~1 min.
4. An additional 950 μL Trizol is added and the samples are incubated on ice for 10 min.
5. Add 200 μL chloroform to each sample. Mix gently by inverting the tubes 15 times.
6. Incubate for 2 min at room temperature and then centrifuge at 12,000×*g* for 15 min at 4°C in a microfuge.
7. Collect 500 μL of the upper aqueous layer in a new tube and add 500 μL isopropanol. Vortex briefly and incubate for 10 min at room temperature.
8. Centrifuge at 12,000×*g* at 4°C for 10 min. It is helpful to align all the tubes in the same orientation in the microfuge since RNA forms a transparent pellet that is difficult to see. Knowing the location of the RNA pellet can help prevent dislodging it when removing the supernatant in the next step.
9. Using a vacuum pump, carefully remove the supernatant by placing the extraction tip on the side of the tube opposite the location of the RNA pellet.

10. Wash the pellet by adding 500 μL 75% ethanol along the wall of the tube without disturbing the RNA pellet.
11. Centrifuge at 7500×*g* for 5 min at 4°C, again aligning all the tubes in the same and known direction, as before.
12. Using a vacuum pump, carefully remove the ethanol.
13. Air-dry the pellet for not more than 10 min at RT. It is important not to let the RNA pellet dry completely as this will greatly decrease its solubility.
14. Add 50 μL of DEPC-treated water to the pellet and keep on ice for 30 min to help dissolve the pellet completely.
15. To remove any trace DNA that may contaminate the preparation, add 5 μL 10× TURBO DNase buffer and 1 μL of TURBO DNase. Mix well by tapping gently and incubate for 20 min at 37°C.
16. Add 5 μL of DNase Inactivation Reagent. Mix well by gently tapping the tube. Incubate 2 min at room temperature, mixing occasionally.
17. Centrifuge at 10,000×*g* for 1.5 min at room temperature and transfer the supernatant (40–45 μL) to a new tube.
18. Measure the RNA concentration using a spectrophotometer at OD 260 nm. The minimum expected RNA recovery is 5 ng/μL.
19. Keep RNA on ice. Immediately prepare cDNA or store at –80°C.

3.4. cDNA Synthesis

1. RNA samples, dNTP mix, 10× RT buffer, $MgCl_2$ and DTT are thawed on ice. Centrifuge briefly and keep on ice.
2. RNA/primer mixtures are combined in a nuclease-free tube as follows:

Component	Volume
RNA (≥5 ng/μL)	8 μL
dNTP mix	1 μL
Oligo(dT)	1 μL

3. Incubate each sample at 65°C for 5 min, and then cool on ice for ≥1 min.
4. Prepare the following reaction mixture, adding each component in the indicated order (*see* **Note 9**).

Component	Volume
10× RT buffer	2 μL
$MgCl_2$	4 μL
DTT	2 μL
RNaseOUT (RNase Inhibitor)	1 μL

5. Add 9 μL of reaction mixture to each RNA/primer mixture, mix gently and centrifuge briefly.
6. Incubate at 42°C for 2 min.
7. Add 1 μL (200 U) SuperScript III reverse transcriptase to each tube. Mix and incubate at 42°C for 50 min.
8. Terminate the reactions at 70°C for 15 min. Chill on ice.
9. Collect the reactions by brief centrifugation. Add 1 μL of RNase H to remove traces of RNA and incubate for 20 min at 37°C.
10. Store cDNA at –20°C.

3.5. Quantitative Real-Time-PCR (qRT-PCR)

1. Before starting the experiment, clean the bench, pipettors, and the required lab equipment with 20% bleach and dry with paper towel.
2. Thaw reagents and place on ice. Mark RT-PCR plate, or template, for orientation.
3. Mix buffer, probe, and nuclease-free water mixtures for each target probe in a tube as follows. Samples are routinely prepared and run in triplicate (*see* **Note 9**).

Component	Volume
TaqMan Universal PCR Master Mix (2×)	12.50 μL
Target or endogenous control (20×)	
Primers/probe mix	1.25 μL
Nuclease-free water	10.25 μL

4. Dispense 24 μL probe–buffer mixture into the Fast optical plate well marked for each reaction.
5. Dispense 1 μL of cDNA in the well marked for each sample. Because of the small volumes involved, careful pipetting is essential to minimize errors. Include triplicate blank wells (without cDNA) as one set of controls.
6. Cover the plate with adhesive film to avoid evaporation during amplification.

7. Place the plate in the StepOne PCR machine. Choose the ΔΔCt method (12). Open the template window and specify control wells and target wells. Control wells will contain reagents to amplify the internal endogenous control gene (often the alpha-tubulin gene) and target wells will contain reagents to amplify the cDNA of interest. Specify also the wells containing the reference samples (e.g., material from untreated parasites or parasites treated with an irrelevant siRNA) which will be set as 100% gene expression for the relative quantification of gene knockdown. Run the qRT-PCR reaction using the universal cycling conditions (40 cycles of the following: 95°C, 15 s, and 60°C, 1 min) from Applied Biosystems.
8. When the reaction is complete, Step One software can display the relative gene expression data. To most easily represent the data graphically, export them into an Excel file and plot the ΔΔCt data using MS Excel. An example of the results produced is shown in **Fig. 15.1.**

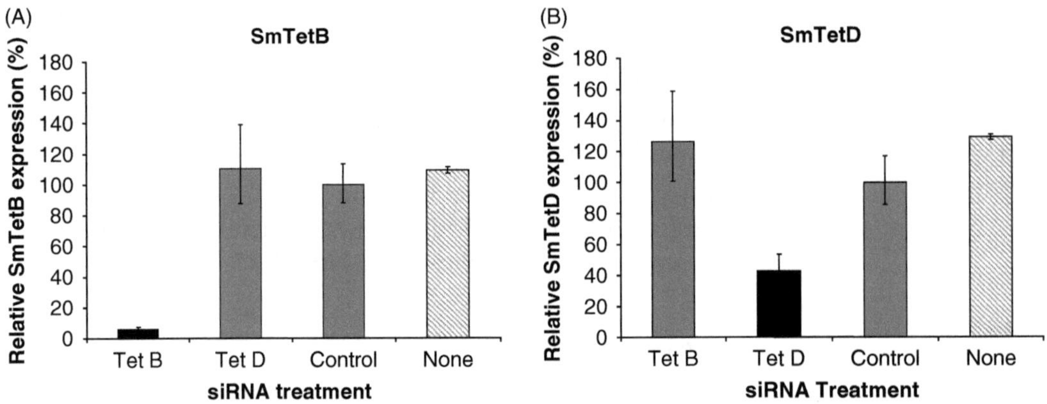

Fig. 15.1. Suppression of schistosome target genes encoding (**A**) tetraspanin B (SmTet B) and (**B**) tetraspanin D (SmTet D). Adult male schistosomes were treated with 5.0 μg of siRNA targeting SmTet B or SmTet D, or a control siRNA or no siRNA. Six days after treatment the relative expression (mean ± SE) of both genes in all groups was assessed by quantitative real-time PCR. The schistosome alpha-tubulin gene was used as the endogenous control and the expression level in parasites treated with control siRNA was set at 100%. In both cases, specific and robust suppression of the targeted gene is observed. The following siRNAs were used: SmTetD siRNA: 5′-ATTGAACCCCCACTACAATCTCTTA-3′, SmTetB siRNA: 5′-CAGGCAGCACCTATCGCTATTATTG-3′, Control siRNA: 5′-CTTCCTCTCTTTCTCTCCCTTGTGA-3′. The following primers/probe were used to detect SmTetB: *SmTetBfw*: 5′-GCCTCGTGCAAAGAAGAGAATTT-3′, *SmTetBRv*: 5′-GTTGCGTTTTAAGAATGCTCCAAAG-3′ and *SmTetB probe*: 5′-FAM-ACCGAAGGCTGTGTATCT-3′. The following primers/probe were used to detect SmTetD: *SmTetDfw*: 5′-TCCACCGACATCATGTTCAAAAGAT-3′, *SmTetBRv*: 5′-GCTTTGCTTAGATCGCTGACCTT-3′ and *SmTetB probe*: 5′-FAM-CATCCCTCTGTAAATTGT-3′. FAM is 6-carboxyfluorescein.

3.6. Isolating RNA and Protein from the Same Parasite Sample

When the suppression/knockdown is to be analyzed at both the RNA and protein level from the same parasite sample, the PARIS kit is used (*see* **Note 6**) as follows:

1. Incubate 2× lysis/binding buffer and wash solution 2/3 at 37°C for ~30 min to ensure complete reagent dissolution. Incubate elution solution at 94°C.
2. Prepare the reagents and standard protein dilutions required for the protein estimation protocol, following the manufacturer's instructions.
3. Transfer parasites from culture wells to microfuge tubes. Gently wash the parasites with PBS three times. After the last wash, remove as much PBS as possible.
4. Add 50 μL cell disruption buffer to each parasite sample. Homogenize samples on ice using an RNase free pestle for ~1 min.
5. Split the parasite homogenate into two halves. Use one for isolating RNA and the other for protein analysis. Incubate the fraction to be used for protein analysis (25 μL) on ice and proceed with RNA isolation using the remaining 25 μL parasite homogenate, as follows.
6. Mix the 25 μL parasite homogenate with 25 μL 2× lysis/binding solution by pipetting 3–4 times. Add 25 μL ethanol to the mixture. Mix gently by pipetting 3–4 times.
7. Apply the RNA sample mixture to a filter cartridge assembled in a collection tube. Centrifuge for 1 min at 10,000×*g* at room temperature. The filter retains the RNA. Discard the flow-through.
8. Apply 700 μL wash solution 1 to the filter cartridge. Centrifuge at 10,000×*g* for 1 min. Discard the flow-through.
9. Apply 500 μL wash solution 2/3 to the filter cartridge. Centrifuge at 10,000×*g* for 1 min. Discard the flow-through. Repeat this step once.
10. Centrifuge at 10,000×*g* for 30 s and transfer the filter cartridge to a fresh collection tube.
11. Elute RNA with 40 μL heated elution solution. Apply this solution to the center of the filter. Recover the RNA by centrifugation for 30 s.
12. Add a further 10 μL of hot elution solution to the center of the filter and spin for 30 s to collect residual RNA. Pool both RNA samples.
13. Remove any contaminating DNA by treating the pooled sample with DNase as described earlier (**Section 3.3,** Steps 15–17).

14. Measure the RNA concentration at OD 260 nm using a spectrophotometer. The minimum expected recovery is 5 ng/μL.
15. Recovered RNA is used for cDNA synthesis and qRT-PCR as described in **Sections 3.4** and **3.5**, respectively.
16. From the 25 μL parasite homogenate incubated on ice at Step 5, 10 μL may be used to determine the protein concentration of the extract using the BCA protein assay kit, following the manufacturer's instructions. The remaining 15 μL of parasite homogenate is used to test for RNAi-mediated suppression at the protein level (*see* **Note 10**).

3.7. Using RNAi in Schistosomula

RNAi can also be used to knockdown the expression of target genes in the larval parasites called schistosomula. The protocol is similar to that used for adult worms but has some modifications.

3.7.1. Preparing Schistosomula for Electroporation

1. Cercariae, recovered from infected snails, are vortexed for 2 min in water in order to shear their tails.
2. The preparation is then pelletted by centrifugation for 3 min at 200×*g*.
3. The pellet is resuspended in ~3 mL of RPMI medium without the serum, loaded on top of a sterile 35% Percoll solution and centrifuged at 1100×*g* at 15°C for 30 min.
4. Cercarial tails remain at the top of the gradient and are discarded while bodies are pelletted. Percoll solution is removed and the parasite pellet is washed in a large volume of sterile water (~50 mL) to remove residual Percoll.
5. Parasites are pelletted by centrifugation as in Step 2 and are then resuspended in modified Basch medium. In this medium the parasites transform into the juvenile intravascular forms called schistosomula. Parasites can be cultured for many weeks at ~2000 schistosomula/mL with medium changes every 2–3 days. Routinely schistosomula are not subject to RNAi treatment until ≥ 3 h after transfer to rich medium.
6. To prepare schistosomula for electroporation, transfer them from culture into 15-mL tubes. Rinse the culture wells with additional medium to detach and recover any parasites sticking to the bottom of the plate.
7. Centrifuge the schistosomula at 200×*g* for 3 min at RT. Gently remove all but 1 mL of medium from above the parasites.
8. Gently disperse the parasite pellet. Transfer 10 μL to a glass slide. For vital staining, add 5 μL trypan blue in PBS to the slide and mix well. Cover the parasites with a cover slip.

9. Count the number of live and dead schistosomula on the slide using light microscopy at 10× magnification. Live schistosomula can be distinguished from dead parasites as they are bright, translucent, and mobile, whereas the dead schistosomula are immobile, take up trypan blue dye to appear dark blue in color, and are often granular in appearance (*see* **Notes 11** and **12**).
10. Calculate the total number and percent live parasites. Routinely ~1000 live schistosomula are used per experimental condition. More parasites are used (~2000/group) if both RNA and protein levels are to be monitored.
11. Pellet parasites in a microfuge tube at 200×*g* for 3 min at room temperature.
12. Remove as much medium as possible and resuspend the parasites in electroporation buffer to a final concentration of 1000 parasites per 50 μL of electroporation buffer (*see* **Note 3**).
13. Transfer 50 μL of electroporation buffer containing schistosomula to the bottom of an electroporation cuvette and add siRNA. Follow the procedure described for adult worms (**Section 3.2**, Steps 6–9, *see* **Notes 4** and **5**), except that schistosomula are resuspended in total 500 μL of modified Basch media in a 48-well plate.

3.7.2. Collecting Schistosomula for RNA Isolation Using Trizol

1. Remove as much medium as possible (≥80%) from the culture wells, taking care not to disturb the schistosomula.
2. Add 1 mL Trizol to the wells. Mix by pipetting 3–4 times and incubate the plate at room temperature for 10 min (*see* **Note 7**).
   The protocol for RNA extraction from schistosomula using Trizol reagent is similar to that for the adult worms with minor modifications to adjust for the smaller amount of biological material usually available. The schistosomula RNA pellet is routinely resuspended in 20 μL DEPC-treated water. The amount of other reagents used in subsequent steps is then adjusted proportionally. For instance, for DNase digestion 2 μL 10× digestion buffer and inactivating reagent and 0.4 μL of DNase are used.

3.8. Using Long Double-Stranded RNA for Gene Suppression

As an alternative to the use of siRNA, schistosomula and adult worms may be treated with long double-stranded RNA (dsRNA). The efficiency of gene silencing in schistosomes using long dsRNA is similar to that observed with siRNA (13), but preparing long dsRNA is labor intensive and time consuming. In addition, the costs of RNA synthesis kits used to generate the long dsRNA may equal or exceed the cost for commercial siRNA synthesis. Nevertheless, laboratories lacking an electroporator that can generate a square wave, can effect suppression by simply soaking parasites in long dsRNA. Delivery of long dsRNA can be carried out by electroporation (using the same protocol described in **Section 3.2**) or by soaking. Here, we provide general guidelines for preparing long dsRNA and delivering it to the parasites by the soaking method.

3.8.1. Preparing Long dsRNA

1. Design a set of target-specific primers with the help of Primer Design software such as OligoPerfect Designer from Invitrogen (*see* **Note 1**). Include T7 and T3 RNA polymerase promoter sequences at the 5′end of the forward (Fw) and reverse (Rv) primers, respectively. A PCR product of about 400–500 bp in length is suitable (*see* **Note 13**).
2. Using the primers and a suitable template (e.g., cDNA from adult parasites, prepared as described in **Section 3.4**), amplify the fragment of interest by conventional PCR. Running multiple reactions at the same time can be helpful in generating maximal amounts of product.
3. PCR reactions are loaded onto a 0.7% agarose gel and the amplified DNAs are excised from the gel, purified and eluted with 30 μL of nuclease-free water using the QIAquick gel extraction kit.
4. Measure the amount of recovered DNA at OD 260 nm and synthesize RNA using MEGAscript, High Yield Transcription Kit. Use one RNA kit (T7) to make sense RNA and a second kit (T3) for anti-sense RNA. Combine the reagents below on ice. Label one tube for T7-derived RNA and another for T3-derived RNA and follow the kit instructions.

Component	Volume
ATP	2 μL
CTP	2 μL
GTP	2 μL
UTP	2 μL
10× Reaction buffer	2 μL
Purified PCR product	1 μg (max 8 μL)
T3 or T7 RNA polymerase	2 μL
Water to	20 μL

Incubate the reaction for 16 h at 37°C. Add 1 μL TURBO DNase, mix well, and incubate for 15 min at 37°C. Purify synthesized sense and anti-sense RNAs using RNeasy mini elute cleanup system following the kit instructions.

5. Measure the amount of RNA synthesized by T7 and T3 RNA polymerase using a spectrophotometer at OD 260 nm and proceed with the annealing reaction, as follows:

Component	Volume
10× Tris buffer (annealing buffer)	3 μL
T7 polymerase derived RNA	50 μg (up to 13 μL)
T3 polymerase derived RNA	50 μg (up to 13 μL)
DEPC water to	30 μL

Incubate the sample for 45 min at 68°C. Let the sample cool to room temperature for 10 min, before incubating on ice. Measure the amount of dsRNA at OD 260 nm and check the quality of the dsRNA generated by 1% agarose gel electrophoresis, applying single and double-stranded RNA for comparison. dsRNA migrates more slowly than single stranded RNA.

3.8.2. Soaking Parasites with dsRNA

1. Add 50 μg long dsRNA diluted in 50 μL RPMI media without serum to 1000 schistosomula, or 10–15 adult worms, in 150 μL RPMI media (*see* **Note 14**).
2. After overnight culture at 37 °C, remove 80% of the RPMI medium and replace with 300 μL of modified Basch medium for schistosomula and 1 mL for adults. Change the medium every 2 days thereafter, until parasites are collected for RNA and/or protein analysis as described in **Sections 3.3** and **3.6** (*see* **Note 14**).

4. Notes

1. Ensure that the sequence targeted by the siRNA is distinct from sequence that will be used in the later detection of mRNA. For instance, the first portion of a target sequence could be targeted by siRNAs, leaving the remainder of the sequence available for the development of assays (e.g., using Taq Man probes and primers) to measure RNA levels.
2. Use the available *S. mansoni* genome sequence (8) to target the TaqMan probe to an exon/exon junction of the target gene. This ensures that any residual genomic DNA contaminating the preparation will not be amplified during qRT-PCR.

3. RPMI medium has been used to replace electroporation buffer to good effect (1).
4. The empirical siRNA amount is 2.5 μg. Concentrations as low as 0.02–0.6 μg have been effective in schistosomula and adult worms, respectively (1, 14). Concentrations as high as 15 μg have been used in adult parasites without apparently compromising target specificity (1, 14).
5. Routinely ~15 adults or ~1000 schistosomula are treated in 100 and 50 μL buffer respectively containing 2.5 μg siRNA. Increased numbers of parasites can be subjected to RNAi by electroporation as long as the volume of buffer and the amount of siRNA is adjusted accordingly.
6. Routinely, knockdown at the RNA level is measured 2 days after RNAi treatment. Sustained gene suppression for up to 40 days has been observed in cultured parasites (14). Routinely, knockdown at the protein level is measured ≥ 7 days post treatment.
7. The experiment can be stopped at this step and parasites stored at –80°C.
8. To avoid any potential RNase contamination, a specific area in the laboratory should be designated for RNA work only. Before working with RNA, ensure that the lab bench, pipettes, and other lab equipment have been cleaned with an RNase decontamination solution (e.g., RNase*Zap* solution). Wear gloves and ensure that during this procedure nothing is touched with bare hands to prevent RNase contamination from the surface of the skin. It may be necessary to change gloves frequently.
9. It is advisable to make up a master mix of these reagents that is sufficient for one more than the total needed for the experiment. This ensures that there is sufficient mix for the final tube, which otherwise tends to come up short.
10. Sample containing ≥2.5 μg protein can be resolved by SDS-PAGE and subjected to Western blot analysis, if a suitable antibody is available that can detect the target protein. If the target is an enzyme, an appropriate enzymatic assay can be undertaken.
11. As an alternative to using trypan blue, parasite viability by vital dye staining for ~5 min with 4 μg/mL Hoechst 33258 dye in phosphate buffered saline (PBS) can also be used. Parasites are examined for viability by fluorescence microscopy with dead parasites exhibiting internal blue fluorescence using a 460-nm reading filter.
12. The same Percoll purification protocol (**Section 3.7**, Steps 2–5) can be used to separate live cultured schistosomula, which pellet, from dead ones, which do not.

13. As a control, an irrelevant long dsRNA whose sequence is not present in the *S. mansoni* genome must be selected and prepared following the same procedure used for parasite-specific long dsRNA. We have used sequence derived from a yeast expression vector that has no counterpart in the *S. mansoni* genome (13, 14).
14. It is likely that both Steps 1 and 2 can be conducted with Basch medium. However, we have not experimentally tested this.

Acknowledgments

Thanks to David Ndegwa for technical assistance. This work was supported by NIH R01 AI-056273.

References

1. Krautz-Peterson, G., Bhardwaj, R., Faghiri, Z., Tararam, C. A., and Skelly, P. J. (2010) RNA interference in schistosomes: machinery and methodology. *Parasitology* **137**, 485–495.
2. Yoshino, T. P., Dinguirard, N., and Mourão, Mde M. (2010) In vitro manipulation of gene expression in larval Schistosoma: a model for postgenomic approaches in Trematoda. *Parasitology* **137**, 463–483.
3. Hoffmann, K. F., and Dunne, D. W. (2003) Characterization of the Schistosoma transcriptome opens up the world of helminth genomics. *Genome Biol* **5**, 203.
4. Verjovski-Almeida, S., DeMarco, R., Martins, E. A., et al. (2003) Transcriptome analysis of the acoelomate human parasite Schistosoma mansoni. *Nat Genet* **35**, 148–157.
5. Oliveira, G. (2007) The Schistosoma mansoni transcriptome: an update. *Exp Parasitol* **117**, 229–235.
6. van Hellemond, J. J., van Balkom, B. W., and Tielens, A. G. (2007) Schistosome biology and proteomics: progress and challenges. *Exp Parasitol* **117**, 267–274.
7. Wilson, R. A., Ashton, P. D., Braschi, S., Dillon, G. P., Berriman, M., and Ivens, A. (2007) 'Oming in on schistosomes: prospects and limitations for post-genomics. *Trends Parasitol* **23**, 14–20.
8. Berriman, M., Haas, B. J., LoVerde, P. T., et al. (2009) The genome of the blood fluke Schistosoma mansoni. *Nature* **460**, 352–358.
9. Liu, F., Zhou, Y., Wang, Z. Q., et al. (2009) The Schistosoma japonicum genome reveals features of host-parasite interplay. *Nature* **460**, 345–351.
10. Hackett, F. (1993) The culture of *Schistosoma mansoni* and production of life cycle stages. In: Hyde, J. E. (ed.) *Protocols in Molecular Parasitology*. Humana, Totowa, New Jersey, pp. 89–99.
11. Basch, P. F. (1981) Cultivation of Schistosoma mansoni in vitro I. Establishment of cultures from cercariae and development until pairing. *J Parasitol* **67**, 179–185.
12. Livak, K. J., and Schmittgen, T. D. (2001) Analysis of relative gene expression data using real-time quantitative PCR and the 2(-Delta Delta C(T)) Method. *Methods* **25**, 402–408.
13. Ndegwa, D., Krautz-Peterson, G., and Skelly, P. J. (2007) Protocols for gene silencing in schistosomes. *Exp Parasitol* **117**, 284–291.
14. Krautz-Peterson, G., Radwanska, M., Ndegwa, D., Shoemaker, C. B., and Skelly, P. J. (2007) Optimizing gene suppression in schistosomes using RNA interference. *Mol Biochem Parasitol* **153**, 194–202.

Chapter 16

Performing the Labeled microRNA Pull-Down (LAMP) Assay System: An Experimental Approach for High-Throughput Identification of microRNA-Target mRNAs

Ren-Jun Hsu and Huai-Jen Tsai

Abstract

We developed a simple, direct, and cost-effective approach to search for the most likely target genes of a known miRNA in vitro. We term this method "Labeled microRNA (miRNA) pull-down assay system," or LAMP. Briefly, the pre-miRNA is labeled with digoxigenin (DIG), mixed with cell extracts, and immunoprecipitated by anti-DIG antiserum. We concluded that LAMP is an experimental approach for high-throughput identification of the target gene of known miRNAs from both *Caenorhabditis elegans* and zebrafish (*Danio rerio*), yielding fewer false-positive results than those produced by using only the bioinformatics approach.

Key words: microRNA, immunoprecipitation, target genes, *C. elegans*, zebrafish.

1. Introduction

A mature miRNA is a 19- to 30-nucleotide-long non-coding RNA excised from a 70-nucleotide pre-miRNA (dsRNA hairpin) by Dicer (1–4). It is clear that miRNAs play essential roles in gene expression, development, specification, and differentiation of cell fate in animals (5–8). The first miRNA found in *Caenorhabditis elegans*, *lin-4*, targets the 3′-untranslated regions (UTR) of *lin-28* and *lin-14* to impede translation (9, 10). Investigators have also reported miRNAs in mice, such as *miR-196*, which is involved in homeobox gene regulation (6), and *miR-208*, which is required for cardiomyocyte hypertrophy, fibrosis, and β-MHC expression (11). In zebrafish embryos, miRNAs are processed

J. Goodchild (ed.), *Therapeutic Oligonucleotides*, Methods in Molecular Biology 764,
DOI 10.1007/978-1-61779-188-8_16, © Springer Science+Business Media, LLC 2011

to silence genes during brain development and morphogenesis (7). Up to now, an estimated 5234 mature miRNAs have been found in primates, rodents, birds, fish, worms, flies, plants, and viruses (*see* ftp://ftp.sanger.ac.uk/pub/mirbase/sequences/CURRENT/README). These miRNAs may control up to 30% of genes in animals (12, 13, and 14). Thus far, however, only a few known miRNAs have been matched to their target genes (8).

The conventional methods that determine the target genes of an miRNA are completely dependent on bioinformatic and microarray analyses. Bypassing both the bioinformatics and microarray methods, we have developed an experimental approach to search for the target gene of a known miRNA. First, we label pre-miRNA with digoxigenin (DIG) and then mix it with cell extracts. The endogenous Dicer cuts this complex in vitro and generates mature miRNA. Next, this DIG-labeled miRNA is attached to its target gene(s) by the endogenous RNA-induced silencing complex (RISC). The mixtures of miRNA-target mRNA are then pulled down by anti-DIG antiserum. To finally identify the target genes of a given miRNA, we clone all cDNAs from total mRNAs after they are pulled down for further DNA sequencing or cloning out by reverse transcriptase-polymerase chain reaction (RT-PCR). In addition, to increase the degree of certainty in our method, we also employ microarray analysis to analyze all mRNAs after they are pulled down, further validating the target genes predicted by the (DIG)-*La*beled *mi*RNA *p*ull-down (LAMP) assay system we describe here.

2. Materials

2.1. Preparation of DIG-pre-miRNA

1. RNAse-free 70% ethanol.
2. Tris–HCl (pH 7.4) 10 mM, EDTA 1 mM
3. DIG RNA-labelling kit (Roche)
4. Biotin RNA-labelling kit (Roche) (If biotin labeling is preferred)
5. Transcription mix: 1 μg linearized DNA, 2 μL 10× transcription buffer, 2 μL NTP-DIG-RNA mix (Roche), 1 μL RNAse inhibitor (35 units/μL), 1 μL T3 or T7 RNA polymerase (20 units/μL), sterile water to give a final volume of 20 μL
6. RNAse-free DNAse
7. EDTA: 0.5 M, pH 8
8. LiCl: 4 M
9. Sterilized Diethylpyrocarbonate (DEPC) water

2.2. Cell Extracts

1. Extract buffer: 15 mM Tris–HCl, pH 7.4, 250 mM sucrose, 2 mM EDTA
2. Phenylmethylsulfonyl fluoride (PMSF): 0.2 M
3. RNasin: N251B (Promega)

2.3. LAMP Assay Protocol

1. RNasin: N251B (Promega)
2. Binding buffer: 25 mM Tris–HCl, pH7.4, 60 mM KCl, 2.5 mM EDTA, 0.2% Triton X-100
3. Anti-DIG agarose beads (Sigma-Aldrich)
4. Anti-biotin agarose beads (Sigma-Aldrich) (If biotin labeling is used)
5. Washing buffer: 20 mM Tris–HCl, pH 7.4, 350 mM KCl, 0.02% NP-40
6. DNase (M610A, Promega)
7. phenol:chloroform solution (1:1)
8. 70% alcohol
9. RNase-free distilled water
10. LiCl: 4 M

3. Methods

3.1. Preparation of DIG-pre-miRNA (see Note 1)

1. Linearize 5 μg of DNA (clone of the pre-miRNA) by digesting with the appropriate restriction enzyme for 6 h.
2. Precipitate the DNA with 1/10 volume of 3 M sodium acetate and 2× volume of 100% ethanol, centrifuge, and wash with RNAse-free 70% ethanol.
3. Re-suspend the DNA in 10 mM Tris–HCl (pH 8.0) and 1 mM EDTA.
4. Analyze an aliquot by electrophoresis on agarose gel.
5. Mix together 1 μg linearized DNA and 20 μL transcription mix buffer (*see* **Note 2**).
6. Incubate for 3 h at 37°C.
7. Digest the template DNA by adding 10 μL RNAse-free DNAse for 30 min at 37°C.
8. Stop the synthesis reaction and precipitate the RNA for 30 min with EDTA 1 μL, LiCl 2.5 μL, and 100% ethanol 75 μL at –70°C.
9. Centrifuge at 4°C for 30 min at 12,879 ×*g*.

10. Wash with 70% ethanol, dry, and re-suspend in 20 μL sterile DEPC water.
11. Analyze 1 μL on agarose gel (generally 1 μL will be used for the hybridization).

3.2. Cell Extracts (see Note 3)

1. Collect 1000 zebrafish embryos at the desired stage and place them in a centrifuge tube. Centrifuge the tube at low speed (805 ×*g*) for 5 min at room temperature to concentrate the embryos and remove water. The procedure of collecting the cell extract of *C. elegans* was the same as that used for zebrafish, except that approximately 5000 *C. elegans* at stage L1 or L2-L4 were collected, washed with M9 buffer, and spun down at 800×*g* for 10 min.
2. Add 1000 μL extract buffer to the tube and centrifuge at low speed (800×*g*) for 10 min. Wash the embryos several times with fresh extract buffer to remove impurities.
3. Remove extract buffer and add 495 μL of fresh extract buffer, 5 μL PMSF and 2 μL RNasin. Place the tube on ice, and sonicate at 50 duty cycles and a 5 micro-tip limit for 10 min to break embryos. If embryos are not completely homogenized, place the embryos on ice and let them cool down to 4°C and repeat the sonication process for another 10 min (*see* **Note 4**).
4. Centrifuge at high speed (10000×*g*) for 30 min at 4°C.
5. Save supernatant, transfer to a new microtube.

3.3. LAMP Assay Protocol Using Anti-DIG Agarose Beads (see Note 5)

1. To 400 μL cell extract add 100 μL DIG-labeled probe in RNase-free H_2O and 2 μL RNasin. Mix well and then place the microtube on ice for 30 min.
2. Make up the volume to 1000 μL with binding buffer. Then add in 1 μL RNasin and incubate the mixture at 30°C for 1 h.
3. Take 20 μL of anti-DIG agarose beads and immerse in 500 μL of binding buffer for at least 30 min.
4. Mix the incubated sample with the processed agarose beads and add 3 μL RNasin. Centrifuge the mixture at 4°C at low speed (1.6–41.9×*g*) overnight (*see* **Note 6**).
5. Precipitate the anti-DIG agarose beads at 12,879×*g* for 30 min at 4°C and remove supernatant.
6. Add in 1000 μL washing buffer and mix well. Centrifuge the mixture at 10,000×*g* for 10 min at 4°C. Save the pellet. Repeat the procedure three times (*see* **Note 7**).
7. Dissolve pellet in 200 μL binding buffer. Incubate the mixture at 95°C for 15 min.

8. Centrifuge the mixture at high speed (10,000×*g*) for 10 min. Transfer the supernatant to a new microtube. Then, add in 5 μL DNase and 2 μL RNasin. Incubate the mixture at 37°C for 30 min.
9. Add an equal volume of phenol/chloroform (1:1) solution and allow the mixture to sit at room temperature for 10 min.
10. Centrifuge the mixture at high speed (10,000×*g*) for 15 min at 4°C.
11. Take 200 μL of supernatant, add one tenth volume of LiCl solution and then add two volumes of 100% alcohol. Incubate the mixture at –20°C at least 4 h or overnight.
12. Centrifuge the mixture at high speed (10,000×*g*) for 10 min at 4°C.
13. Remove the supernatant, add in 500 μL 70% alcohol, and centrifuge the mixture at high speed (10,000×*g*) for 10 min at 4°C.
14. Remove the supernatant. Dissolve the pellet in 20 μL RNase-free distilled water and store the product at –80°C.

4. Notes

1. Instead of the 5′- or 3′-labeling method, we recommend using a random labeling method to label either DIG or biotin within pre-miRNA because a 5′- or 3′-labeled marker might be cut off by Dicer in the process from pre-miRNA to mature miRNA.
2. When random labeling is used, we do not recommend using DIG-UTP or Biotin-UTP for the reaction mixture. Instead, unlabeled UTP should be used. The ratio for labeled and unlabeled UTP could be between 1:3 and 1:1.25. In case the DIG/Biotin RNA labeling Mix (Roche) is used, it is not necessary to add the unlabeled UTP into the reaction mixture.
3. Cell extracts must be prepared freshly to use in experiments.
4. The intensity of sonication used to break cells or embryos can vary with equipment. Therefore, to avoid overheating, cells or embryos can be divided into several tubes and treated for a shorter period of time. Under these circumstances, proteins will not be denatured and mRNA will not be degraded.
5. If biotin-labeling is performed, the following steps should be modified:
   a. Step 3: Take 40 μL anti-biotin agarose beads, add into 300 μL binding buffer, and incubate the tube on ice for 15 min.

b. Step 10: Remove the supernatant, add 500 μL 70% alcohol, and centrifuge the mixture at high speed (10,000×*g*) for 10 min at 4°C.

c. Step 13 (additional step): Transfer the supernatant to a new microtube. Then add 5 μL DNaseI and 2 μL RNasin and incubate at 37°C for 20 min.

6. Add sample to anti-DIG/biotin agarose beads again and rotate the mixture at low speed (1×*g*) at 4°C overnight. A low speed of rotation is best.
7. If the sample size is small, such as a small number of cells or embryos, the washing steps can be reduced accordingly to once or twice.

Acknowledgment

This work was supported by the National Science Council of Taiwan, Republic of China, under the grant of NSC97-B001, ROC.

References

1. Grishok, A., Pasquinelli, A. E., Conte, D., Li, N., Parrish, S., Ha, I., Baillie, D. L., Fire, A., Ruvkun, G., and Mello, C. C. (2001) Genes and mechanisms related to RNA interference regulate expression of the small temporal RNAs that control *C. elegans* developmental timing. *Cell* **106**, 23–24.
2. Hutvágner, G., McLachlan, J., Pasquinelli, A. E., Bálint, E., Tuschl, T., and Zamore, P. D. (2001) A cellular function for the RNA-interference enzyme Dicer in the maturation of the *let-7* small temporal RNA. *Science* **293**, 834–838.
3. Lee, Y., Ahn, C., Han, J., Choi, H., Kim, J., Yim, J., Lee, J., Provost, P., Rådmark, O., Kim, S., et al. (2003) The nuclear RNase III Drosha initiates microRNA processing. *Nature* **425**, 415–419.
4. Bernstein, E., Caudy, A. A., Hammond, S. M., and Hannon, G. J. (2001) Role for a bidentate ribonuclease in the initiation step of RNA interference. *Nature* **409**, 363–366.
5. Chen, C. Z., Li, L., Lodish, H. F., and Bartel, D. P. (2004) MicroRNAs modulate hematopoietic lineage differentiation. *Science* **303**, 83–86.
6. Yekta, S., Shih, I. H., and Bartel, D. P. (2004) MicroRNA-directed cleavage of HOXB8 mRNA. *Science* **304**, 594–596.
7. Wienholds, E., Koudijs, M. J., Eeden, F. J., Cuppen, E., and Plasterk, R. H. (2003) The microRNA-producing enzyme Dicer1 is essential for zebrafish development. *Nat Genet* **35**, 217–218.
8. Giraldez, A. J., Cinalli, R. M., Glasner, M. E., Enright, A. J., Thomson, J. M., Baskerville, S., Hammond, S. M., Bartel, D. P., and Schier, A. F. (2005) MicroRNAs regulate brain morphogenesis in zebrafish. *Science* **308**, 833–838.
9. Lee, R. C., Feinbaum, R. L., and Ambros, V. (1993) The *C. elegans* heterochronic gene lin-4 encodes small RNAs with antisense complementarity to lin-14. *Cell* **75**, 843–854.
10. Moss, E. G., Lee, R. C., and Ambros, V. (1997) The cold shock domain protein LIN-28 controls developmental timing in *C. elegans* and is regulated by the *lin-4* RNA. *Cell* **88**, 637–641.
11. van Rooij, E., Sutherland, L. B., Qi, X., Richardson, J. A., Hill, J., and Olson, E. N. (2007) Control of stress-dependent cardiac growth and gene expression by a microRNA. *Science* **316**, 575–579.
12. Berezikov, E., Guryev, V., Belt, J., Wienholds, E., Plasterk, R. H., and Cuppen, E. (2005) Phylogenetic shadowing and

computational identification of human microRNA genes. *Cell* **120**, 21–24.

13. Lewis, B. P., Shih, I. H., Jones-Rhoades, M. W., Bartel, D. P., and Burge, C. B. (2003) Prediction of mammalian microRNA targets. *Cell* **115**, 787–798.

14. Wienholds, E., Kloosterman, W. P., Miska, E., Alvarez-Saavedra, E., Berezikov, E., Bruijn, E., Horvitz, H. R., Kauppinen, S., and Plasterk, R. H. (2005) MicroRNA expression in zebrafish embryonic development. *Science* **309**, 310–311.

Chapter 17

Synthesis, Purification, and Characterization of Oligoribonucleotides that Act as Agonists of TLR7 and/or TLR8

Tao Lan and Ekambar R. Kandimalla

Abstract

Viral single-stranded (ss) RNA is the natural ligand for TLR7 and TLR8. Synthetic ssRNA has been shown to act as a ligand for TLR7 and TLR8. We have previously reported a novel RNA structure, referred to as stabilized immune modulatory RNA (SIMRA), in which two short phosphorothioate oligoribonucleotides were linked through their 3′-ends via a linker. SIMRA compounds had greater stability in serum than unmodified ssRNA and induced immune responses via TLR7 and/or TLR8. SIMRA compounds were synthesized using phosphoramidite chemistry on controlled-pore glass solid support derivatized with a linker. After cleavage from the solid support and removal of protecting groups, SIMRA compounds were purified on an anion-exchange HPLC followed by desalting/dialysis, and lyophilization. SIMRA compounds were characterized for their purity and sequence integrity by anion-exchange HPLC, capillary gel electrophoresis, polyacrylamide gel electrophoresis, and MALDI-TOF mass spectrophotometric analysis. As SIMRA compounds induce TLR7- and/or TLR8-mediated Th1-type immune responses, they have potential utility as therapeutic agents for a broad range of diseases, including cancer, infectious diseases, asthma, and allergies, and as adjuvants with vaccines.

Key words: Analytical, characterization, immune-modulatory oligonucleotide, oligoribonucleotide, RNA, synthesis, toll-like receptor, TLR7, TLR8.

1. Introduction

The innate immune system of mammals consists of highly conserved receptors called pattern-recognition receptors (PRR). Toll-like receptors (TLRs) are one class of PRR which recognize pathogen-associated molecular patterns (PAMPs) present within the pathogens (1, 2). Among the ten TLRs identified in humans, TLR7 and TLR8 recognize viral single-stranded (ss) RNA (3–5).

J. Goodchild (ed.), *Therapeutic Oligonucleotides*, Methods in Molecular Biology 764,
DOI 10.1007/978-1-61779-188-8_17, © Springer Science+Business Media, LLC 2011

Both the receptors are expressed on endosomal membranes and their expression is cell-specific. TLR7 is expressed in human B cells and plasmacytoid dendritic cells (pDCs) and TLR8 is expressed in myeloid DCs (mDCs) and monocytes (6). Stimulation of TLR7/8 by their ligands leads to downstream activation of a number of transcription factors, including NF-κB. The immune response profiles induced by TLR7/8 agonists are dependent on the type of cells (expressing TLR7 vs. 8) and the signal transduction cascades they activate. In general, TLR7/8 activation have been shown to induce Th1-type immune responses (3–5, 7).

Viral and synthetic ssRNA as well as some purine nucleoside analogs have been shown to be ligands for TLR7 and 8 (3–5, 8). However, advancement of ssRNA as TLR7/8 agonists for therapeutic applications has been hampered by its susceptibility to nuclease degradation (3, 9, 10). To date all the studies with ssRNA have used formulation with lipids (3, 9, 10). We previously reported a novel class of RNA structure, referred to as stabilized immune modulatory RNA (SIMRA), in which two short phosphorothioate oligoribonucleotides were linked through their 3′-ends via a linker (**Fig. 17.1**) (11–15). We have also observed that the stability of ssRNA depends on their nucleotide composition: specific dinucleotide motifs such as UA, UC, and CA are more susceptible to degradation by endonucleases. Avoiding these dinucleotide motifs further enhances stability of ssRNA against endonucleases. These novel structures were more stable against 3′-*exo* and *endo* nuclease digestion in human serum than were unmodified ssRNAs (12–15). SIMRA compounds showed immune stimulatory activity in vitro in TLR8-transfected cell lines and human primary cells and in vivo in non-human primates (12). Incorporation of chemical modifications such as 7-deaza-guanosine or arabino-cytidine in place of guanosine or cytidine,

a **5'-N-N-N-N-N-N-N-N-N-N-N-X-N-N-N-N-N-N-N-N-N-N-N-5'**

b

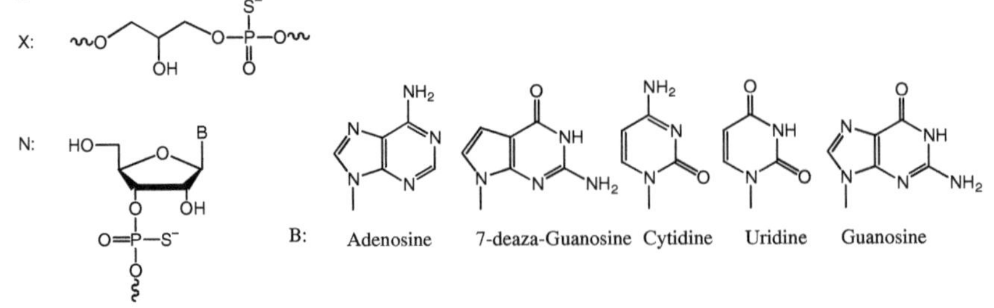

Fig. 17.1. (**a**) Structure of a typical SIMRA compound. Two 11-mer phosphorothioate oligoribonucleotides were linked through their 3′-ends via a linker (X). (**b**) Chemical structures of linker X and nucleotides N.

respectively, results in the activation of both TLR7 and TLR8 (12, 14). We have also identified specific RNA motifs that selectively activate TLR7 when 7-deazaguanosine is incorporated for guanosine (13).

SIMRA compounds were synthesized on automated DNA/RNA synthesizers using phosphoramidite chemistry (16). The instrument was programmed to couple phosphoramidite monomers to solid support bound linker one at a time using repeated cycles, according to SIMRA compound's sequence starting from the 3′-end. Each cycle consists of four steps: deblock, coupling, oxidation, and capping; plus washing with acetonitrile in between (**Fig. 17.2**). After completion of the synthesis of desired SIMRA compound, it was treated with 40% methylamine in water and concentrated ammonium hydroxide to cleave from solid support and deprotect exocyclic

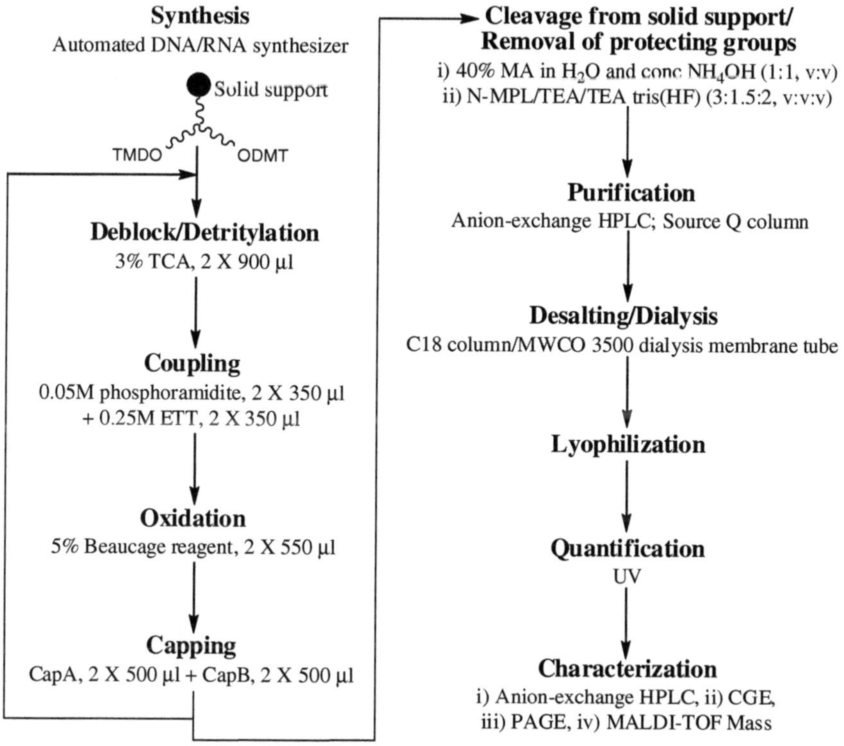

Fig. 17.2. Steps involved in the solid phase synthesis, purification, and characterization of SIMRA compound. TCA, trichloroacetic acid; ETT, ethylthiotetrazole; Cap A, capping reagent A (tetrahydrofuran/acetic anhydride/2,6-lutidine, 8:1:1, v:v:v); Cap B, Capping reagent B (methylimidazole/tetrahydrofuran, 16:84, v:v); MA, methylamine; *N*-MPL, *N*-methylpyrrolidinone; TEA, triethylamine; CGE, capillary gel electrophoresis; PAGE, polyacrylamide gel electrophoresis. There was an acetonitrile wash following each step of the synthesis cycle (detritylation/coupling/oxidation/capping). One nucleoside residue is added in each branch during each cycle. SIMRA compound is cleaved from solid support, deprotected, purified, and characterized following synthesis as shown in the *right column*.

amino groups. Then the 2′-protecting group was cleaved with *N*-methylpyrrolidinone/triethylamine/triethylamine tris(hydrogen fluoride) reagent. The crude SIMRA compound obtained was purified by anion-exchange HPLC. Subsequent to purification, SIMRA compound was desalted, filtered, and lyophilized then characterized by means of anion-exchange HPLC, capillary gel electrophoresis (CGE), polyacrylamide gel electrophoresis (PAGE), and MALDI-TOF mass spectrophotometry.

2. Materials

2.1. Automated Synthesis of SIMRA Compounds

1. Phosphoramidites of adenosine (A), cytidine (C), guanosine (G), 7-deaza-G, and uridine (U) (ChemGenes Corp.): Exocyclic amino groups were protected by acetate (A, C, and G) and isobutyrate (7-deaza-G). 2′-Hydroxy groups were protected by *tert*-butyldimethylsilyl (TBDMS) groups. Phosphoramidite solutions were prepared in low-water acetonitrile (<20 ppm) at a concentration of 0.05 M and were generally stable for up to 2 weeks when stored under argon (*see* **Notes 1** and **2**).
2. Solid support: Glycerol derivatized controlled-pore glass (CPG) (ChemGenes Corp.) 500 Å pore size and ~45 μmol loading were selected for optimized synthesis.
3. Ethylthiotetrazole (ETT): 0.25 M (Glen Research).
4. Capping reagent A: tetrahydrofuran (THF)/acetic anhydride/2,6-lutidine, 8:1:1, v:v:v (Glen Research).
5. Capping reagent B:methylimidazole/THF, 16:84, v:v (Glen Research).
6. Low water content acetonitrile (J.T. Baker Chemical Co.).
7. Beaucage reagent (R.I. Chemical Inc.): A 5% (w/v) solution in low water content acetonitrile was freshly prepared in-house prior to synthesis and stored in a silanized glass bottle (*see* **Note 3**).
8. Trichloroacetic acid: 3% w/v in dichloromethane.
9. Mermade 6 DNA/RNA synthesizer (BioAutomation) Synthesis was carried out on a 5-μmol scale using a 1-mL column (*see* **Note 4**).
10. Columns and frits for synthesizer (BioAutomation).
11. RNase-free water for irrigation (B. Braun Medical Inc.) was used throughout the synthesis, purification, and characterization of SIMRA compounds (*see* **Note 5**).

2.2. Cleavage and Deprotection

1. Cleavage reagent: 40% methylamine in water:concentrated ammonium hydroxide 1:1 v:v, freshly prepared
2. 2′-TBDMS deprotection reagent: *N*-methylpyrrolidinone: triethylamine:triethylamine tris(hydrogen fluoride) 3:1.5:2 v:v:v, freshly prepared
3. Tris–HCl: 0.1 M, pH 7.5
4. SpeedVac evaporating centrifuge

2.3. Purification

1. Waters 600 HPLC system: 600 controller, Delta 600 pumps, 2489 UV/vis detector (Waters Corp.) used for purification of crude RNA.
2. Anion-exchange column (10 × 250 mm) filled with Source 15Q resin (GE Healthcare Bio-sciences) was packed in house. The column was equipped with a column heater (ChromTech Inc.).
3. Buffer A: 20 mM Tris–HCl, pH 7.5, in 20% acetonitrile.
4. Buffer B: 2.5 M NaCl in Buffer A.

2.4. Desalting/ Dialysis

1. Sep-Pak Vac C18 6 cc/500 mg cartridge (Waters Corp.)
2. Sodium acetate: 0.3 M in RNase-free water
3. Acetonitrile, RNase-free water: 1:1, v:v
4. Microdisc 0.2-μm syringe filters (Pall Corp.)
5. Spectra/Por dialysis membrane with 3500 dalton molecular weight cutoff and universal closures (Spectrum Laboratories, Inc.)
6. Sterilized tubes: 15 mL and 50 mL (VWR International, LLC.)

2.5. Quantification

1. Hitachi U2000 UV/vis spectrophotometer (Hitachi High Technologies America, Inc.)
2. 1 cm path-length quartz cuvettes: 1 mL

2.6. Characterization by Anion-Exchange HPLC

1. Waters 600 HPLC system: 600E controller, 600 pumps with heater, 486 UV/vis detector
2. DNA-Pac PA-100 column: 4 × 50 mm (Dionex Corp.)
3. Buffer A: 100 mM Tris–HCl, pH 8.0, in 20% acetonitrile
4. Buffer B: 2.0 M LiCl in Buffer A

2.7. Capillary Gel Electrophoresis (CGE)

1. P/ACE Capillary Electrophoresis System (Beckman Coulter Inc.)
2. eCAP ssDNA R100 kit (Beckman Coulter Inc.) was used to fill the capillary with denaturing gel using the manufacturer's recommended loading protocol

2.8. Polyacrylamide Gel Electrophoresis

1. SequaGel-Urea gel concentrate (National Diagnostics)
2. Urea gel system diluent (National Diagnostics)
3. Urea gel system buffer (National Diagnostics)
4. Ammonium persulfate
5. *N*,*N*,*N*′,*N*′-tetramethylethlenediamine (TEMED)
6. 5× TBE buffer: 0.445 M Tris, 0.445 M boric acid, and 0.01 M EDTA

2.9. MALDI-TOF Mass Spectrometry

1. MALDI Micro MX instrument with a compatible 96-well stainless steel plate (Waters Corp.).
2. Matrix solution: 3-hydroxypicolinic acid (5%, w/v), ammonium citrate dibasic (1% w/v) in 50% acetonitrile/water (v:v).

3. Methods

3.1. Automated Synthesis of SIMRA Compound

For a 5-μmol scale synthesis on Mermade-6 DNA/RNA synthesizer, ~110 mg of solid support (at a loading of ~45 μmole/g) was packed into a 1-mL synthesis column (*see* **Note 6**). The synthesis cycle started with deblocking in order to remove 5′-dimethoxytrityl (DMT) groups from glycerol linker on the solid support. 0.9 mL of 3% TCA was delivered to column and allowed to react on the solid support for 30 s. The same process was repeated one more time immediately. After removal of 5′-DMT groups, the column was washed with acetonitrile (4 × 1 mL) followed by addition of 700 μL of a mixture (1:1, v:v) of 0.05 M phosphoramidite monomer solution and 0.25 M ETT. The coupling reaction was performed twice for 15 min each time. Then 550 μL of 5% Beaucage reagent was delivered to the column to oxidize the phosphorotriester to phosphorothioate (17). The sulfurization process was carried out twice and each cycle was 8 min long. The last step was to cap unreacted 5′-hydroxyl groups with 1 mL of a mixture of capping reagents A and B (1:1, v:v) twice. Each capping cycle lasted 60 s. After the entire sequence was synthesized, the 5′-DMT group of the last residue was also removed. The column was washed with acetonitrile (4 × 1 mL) and drained completely.

3.2. Cleavage from Solid Support and Removal of Protecting Groups

For cleavage and deprotection, solid support was transferred from the column to a 3.5-mL tube. Two mL of a freshly prepared mixture of 40% methylamine in water and concentrated ammonium hydroxide (1:1, v:v) was added to the tube. The mixture was left at ambient temperature for 2 h to cleave the compound from solid support and remove acetyl and cyanoethyl protecting groups from SIMRA compound (*see* **Note 7**). The solid support was removed by filtration and the solution was dried completely in a SpeedVac evaporating centrifuge (*see* **Note 8**). 500 μL of freshly prepared 2′-TBDMS deprotection reagent was added and the reaction sample was incubated at 65°C for 90 min (*see* **Notes 9** and **10**). After completion of the reaction, 15 mL 0.1 M Tris–HCl, pH 7.5, solution was added to quench the reaction (*see* **Note 11**).

3.3. Purification

SIMRA compound was purified on a Waters 600 HPLC system at 85°C with a flow rate of 10 mL/min (*see* **Note 5**). After quenching the 2′-deprotection reaction, the reaction mixture was loaded on to an anion-exchange column and washed with 100% buffer A for 2 min. Then buffer B was increased from 0 to 36% in 10 min and finally from 36 to 70% in 34 min (*see* **Note 12**). The main absorbance peak detected at 260 nm following elution was collected in fractions (*see* **Note 13**). Each fraction was evaluated for its purity by analytical HPLC as described below, and fractions with greater than 95% purity were combined and solvent was evaporated on a rotary evaporator.

3.4. Desalting/Dialysis and Lyophilization

Sodium chloride and other small molecule impurities can be removed by desalting. For synthesis larger than 5 μmole, dialysis was performed following desalting to further remove small molecules impurity whose molecular weights are less than the cut-off point (3500 Daltons for this application). Volatile component such as acetonitrile can be removed by lyophilization.

The SIMRA compound/NaCl mixture from chromatography was re-dissolved in RNase-free water and desalted using a 6-mL C18 Sep-Pac column as follows.

1. Wash Sep-Pak cartridge with 5 mL acetonitrile.
2. Wash Sep-Pak cartridge with 10 mL 0.3 M sodium acetate.
3. Load Sep-Pak cartridge with SIMRA compound solution.
4. Wash Sep-Pak cartridge with 20 mL RNase-free water.
5. Add 1.5 mL of 50% acetonitrile to cartridge and collect eluent in a 3.5-mL tube.

Acetonitrile in the eluent was removed by evaporation in a SpeedVac for 30 min and the remaining solution was filtered through a 0.2-μm syringe filter. The filtrate was collected in a 3.5-mL tube and lyophilized to dryness.

If synthesis scale was larger than 5 μmole, additional dialysis was carried out as follows.

1. Remove acetonitrile by evaporation in SpeedVac for 30 min.
2. Transfer remaining solution to 18-mm-wide dialysis membrane, one end of which was pre-sealed by a closure. Seal the other end with a closure.

3. Leave the membrane overnight in a 4-L autoclaved beaker filled with RNase-free water. Replace the water the next morning and continue dialysis for another 6 h.
4. Open one closure from membrane and transfer the solution to a 50-mL tube. Wash the membrane with 3 × 1 mL of RNase-free water and add washings to the tube.
5. Filter solution through a 0.2-μm syringe filter.
6. Collect the filtrate in a 15-mL tube and lyophilize to dryness (*see* **Note 14**).

3.5. Quantification of SIMRA Compound

After lyophilization, the white fluffy residue was re-suspended in 200 μL RNase-free water. Quantification of SIMRA compound was performed by measuring absorbance or optical density (O.D.) at 260 nm on a UV/vis spectrophotometer as follows:

1. Add 1 μL SIMRA compound solution to 999 μL of RNase-free water and vortex the mixture for 30 s.
2. Set Hitachi U2000 UV/vis spectrophotometer to single wavelength absorbance mode at 260 nm.
3. Fill two clean UV compatible quartz 1-mL cuvettes with 1.0 mL of RNase-free water and place them at sample and reference positions in the spectrophotometer. Close the lid and autozero the reading.
4. Empty the sample cuvette and refill it with 1 mL diluted SIMRA compound solution made in Step 1.
5. The absorbance reading equals the number of O.D. units of SIMRA compound in each μL of the concentrated solution. Multiply the absorbance by 1000 to get the absorbance of the concentrated solution of SIMRA compound.
6. The extinction coefficient of SIMRA compound is calculated using nearest neighbor method (18).
7. The concentration of SIMRA compound is calculated according to the following formula:

$$\text{Concentration (Mol/L)} = \text{Absorbance at 260 nm/Extinction coefficient}$$

After quantification SIMRA compound is ready for characterization and biological studies and should be stored at –20°C.

3.6. Characterization of SIMRA Compound by Anion-Exchange HPLC

Anion-exchange HPLC is able to separate oligonucleotides of different length. Purity of SIMRA compound was determined on a Waters 600 HPLC system at 85°C with a flow rate of 2 mL/min. Peaks were detected by UV absorbance at 260 nm. About 0.2 O.D. units of SIMRA compound in 200 μL water was injected onto the column, which was eluted with the following gradient.

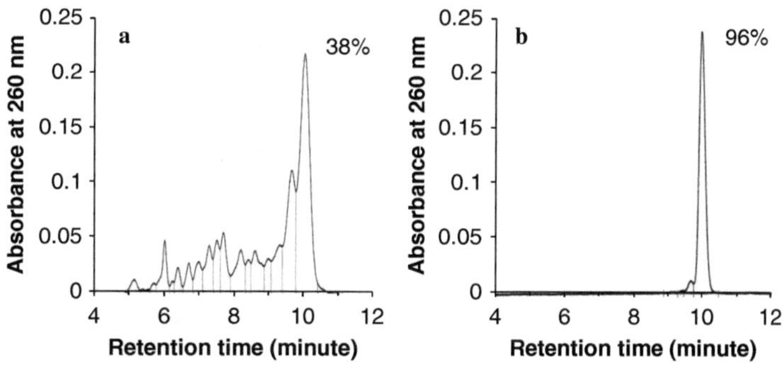

Fig. 17.3. Anion-exchange HPLC analytical profile of (**a**) crude (after cleavage and deprotection and before purification steps) and (**b**) purified SIMRA compound, 5′-ACUUG1AACUUG1-X-G1UUCAAG1UUCA-5′, where G1 is 7-deaza-guanosine, and X is 1,2,3-propanetriol (glycerol) linker.

0–2 min: 100% buffer A

2–10 min: 100% buffer A to 0% buffer A

10–15 min: 100% buffer B

Purity was calculated by integration using EmPower software. Analytical HPLC profiles of crude and purified SIMRA compound are shown in **Fig. 17.3a** and **b** that indicates purity of about 96% following HPLC purification. The remaining products contain one, two, or three nucleotides less than the full-length SIMRA compound.

3.7. Capillary Gel Electrophoresis

Capillary gel electrophoresis is another technique separating oligonucleotides of different length and is widely used to analyze oligonucleotide purity. CGE was performed on P/ACE Capillary Electrophoresis System. The capillary was filled with denaturing gel made from eCAP ssDNA R100 kit using the manufacturer's recommended loading protocol. About 0.2 O.D. units of SIMRA compound was dissolved in 200 μL of water and loaded onto the capillary by electrokinetic injection for 2–4 s at 6 kV. The separation was carried out at 14 kV for 30 min and the capillary temperature was set to 30°C. Purity was determined by integration using 32 Karate software. The CGE profile of a SIMRA compound is shown in **Fig. 17.4** that indicates purity of 96%.

3.8. Polyacrylamide Gel Electrophoresis

Polyacrylamide gel electrophoresis is a common method to separate oligonucleotides according to their length and to evaluate oligonucleotide quality. 20% polyacrylamide gel (20.5 × 20 × 0.75 cm) was cast with two glass plates using the following recipe: 48 mL SequaGel, 6 mL system diluent, 6 mL system buffer, 0.48 mL 10% APS, and 24 μL TEMED. After polymerization the gel was pre-run at 5 watts in TBE buffer for 30 min. About 0.25 O.D. units (purified) or 0.75 O.D. units (desalted, crude) SIMRA compound were loaded onto the gel and electrophoresis

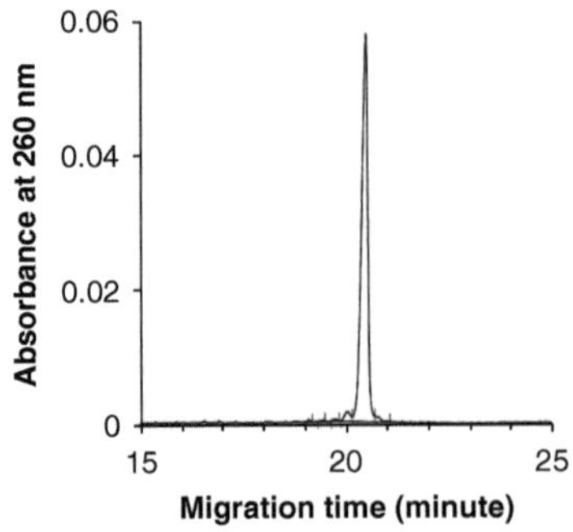

Fig. 17.4. CGE analytical profile of purified SIMRA compound described in **Fig. 17.3**.

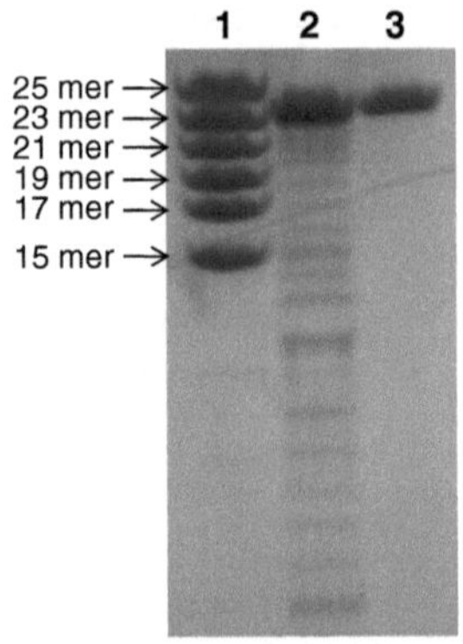

Fig. 17.5. Mobility of the SIMRA compound described in **Fig. 17.3** on 20% PAGE. Lane 1, molecular length markers as indicated; Lane 2, crude SIMRA compound before purification; and Lane 3, HPLC purified SIMRA compound.

was performed using constant power of 15 watts for 5–6 h. The bands were visualized by UV shadowing at 254 nm. The PAGE profile of a SIMRA compound along with a standard compound is shown in **Fig. 17.5**.

3.9. MALDI-TOF Mass Spectroscopy

MALDI-TOF mass spectroscopy measures the molecular weight of the compound and provides insight of the integrity of SIMRA compound synthesized. About 0.05 O.D. units SIMRA compound in 1 μL water was mixed with 10 μL matrix solution. 0.6 μL of the resulting mixture was transferred to a 96-well MALDI plate. After solvent had evaporated under ambient conditions, the plate was placed inside MALDI Micro MX spectrometer. This was calibrated and operated according to the manufacturer's instructions and the mass spectrum of the SIMRA compound was obtained. In general, a pattern of a molecular peak followed by a series of sodium adduct peaks was visualized in spectra and the molecular peak was compared with compound's calculated molecular weight to confirm the identity. The molecular peak is within 0.04 % of the calculated value (**Fig. 17.6**).

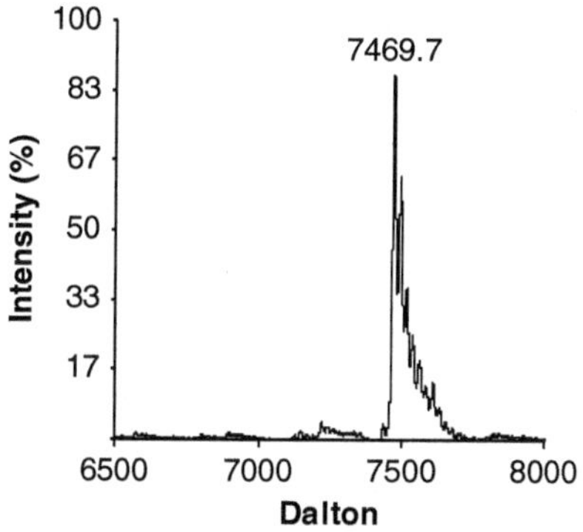

Fig. 17.6. MALDI-TOF mass spectrum of the SIMRA compound described in **Fig. 17.3**. The calculated molecular weight of the compound is 7467.

4. Notes

1. Always wear personal protections such as lab coat, safety glasses, and gloves when performing experiments. All chemical should be handled in a well-ventilated chemical hood.
2. Add molecular sieves to phosphoramidite solution, activator solution, and acetonitrile to remove trace amounts of moisture to improve synthesis quality and yield.
3. Beaucage reagent solution should be prepared in a silanized bottle to avoid precipitation.
4. Use argon rather than nitrogen or helium gas for operating Mermade DNA/RNA synthesizer.
5. SIMRA compounds are subject to nuclease degradation. Water used in experiments should be RNase free. All containers (bottles, tubes) and pipette tips should be autoclaved or sterilized. Wear clean gloves and use caution when handling RNA samples.
6. It is always better to use freshly prepared reagents for automated synthesis, deprotection, and purification of SIMRA compounds. Any solution older than 1 week should be discarded. Cleavage/deprotection solution and CE running buffer should be stored at 4°C. 2′-TBDMS deprotection cocktail, capping solution, Beaucage Reagent, deblock solution, sodium acetate solution, HPLC buffers, and TBE buffer should be stored at room temperature. Phosphoramidite and activator solutions should be stored at room temperature under argon atmosphere.
7. Frequent gentle shake or vortex during cleavage/deprotection will improve reaction yield.

8. Washing solid support with 2 × 0.5 mL of water after cleavage and deprotection reaction will improve final yield.
9. If 2′-TBDMS deprotection cocktail precipitated during storage, heat it at 65°C for 5 min to re-dissolve the precipitate.
10. Some RNA sequences are less soluble in 2′-TBDMS deprotection cocktail. In such case heating the reaction mixture at 65°C should resolve the issue. If the precipitate still persists after heating, sonicate the mixture for 5 min.
11. Some RNA sequences are less soluble in quenching solution after 2′-deprotection. Addition of acetonitrile improves the solubility.
12. 2′-deprotection cocktail has UV absorbance at 260 nm, thereby it will interfere HPLC's baseline during purification. Washing the column with 100% buffer A for 5 min after loading the compound on to column before starting the gradient will eliminate this problem.
13. Oligonucleotides containing 7-deaza-guanosine tend to have longer retention time than those containing natural guanosine on HPLC. This slow elution should be watched during purification/characterization.
14. SIMRA compounds should be routinely tested for endotoxin contamination by LAL-test. Only samples with less than 1 EU/mg of endotoxin should be used in biological assays.

Acknowledgments

The authors thank Dr. Sudhir Agrawal for support, encouragement, and suggestions.

References

1. Akira, S., Takeda, K., and Kaisho, T. (2001) Toll-like receptors: critical proteins linking innate and acquired immunity. *Nat Immunol* **2**, 675–680.
2. Pandey, S., and Agrawal, D. K. (2006) Immunobiology of toll-like receptors: emerging trends. *Immunol Cell Biol* **84**, 333–341.
3. Diebold, S. S., Kaisho, T., Hemmi, H., Akira, S., and Reis e Sousa, C. (2004) Innate viral responses by means of TLR7-mediated recognition of single-stranded RNA. *Science* **303**, 1529–1531.
4. Heil, F., Hemmi, H., Hochrein, H., Ampenberger, F., Kirschning, C., Akira, S., Lipford, G., Wagner, H., and Bauer, S. (2004) Species-specific recognition of single-stranded RNA via toll-like receptor 7 and 8. *Science* **303**, 1526–1529.
5. Lund, J. M., Alexopoulou, L., Sato, A., Karow, M., Adams, N. C., Gale, N. W., Iwasaki, A., and Flavell, R. A. (2004)

Recognition of single-stranded RNA viruses by toll-like receptor 7. *Proc Natl Acad Sci USA* **101**, 5598–5603.

6. Chuang, T. H., and Ulevitch, R. J. (2000) Cloning and characterization of a sub-family of human toll-like receptors: hTLR7, hTLR8, and hTLR9. *Eur Cytokine Netw* **11**, 372–378.
7. Iwasaki, A., and Medzhitov, R. (2004) Toll-like receptor control of the adaptive immune responses. *Nat Immunol* **10**, 987–995.
8. Lee, J., Chuang, T. H., Redecke, V., She, L., Pitha, P. M., Carson, D. A., Raz, E., and Cottam, H. B. (2003) Molecular basis for the immunostimulatory activity of guanine nucleoside analogs: activation of Toll-like receptor 7. *Proc Natl Acad Sci USA* **100**, 6646–6651.
9. Dow, S. (2008) Liposome–nucleic acid immunotherapeutics. *Expert Opin Drug Deliv* **5**, 11–24.
10. Forsbach, A., Nemorin, J. G., Montino, C., Müller, C., Samulowitz, U., Vicari, A. P., Jurk, M., Mutwiri, G. K., Krieg, A. M., Lipford, G. B., and Vollmer, J. (2008) Identification of RNA sequence motifs stimulating sequence-specific TLR8-dependent immune responses. *J Immunol* **180**, 3729–3738.
11. Agrawal, S., and Kandimalla, E. R. (2007) Synthetic agonists of toll-like receptors 7, 8, and 9. *Biochem Soc Trans* **35**, 1461–1467.
12. Lan, T., Kandimalla, E. R., Yu, D., Bhagat, L., Li, Y., Wang, D., Zhu, F., Tang, J. X., Putta, M. R., Cong, Y., Trombino, A. F., Sullivan, T., and Agrawal, S. (2007) Stabilized immune modulatory RNA compounds as agonists of toll-like receptors 7 and 8. *Proc Natl Acad Sci USA* **104**, 13750–13755.
13. Lan, T., Dai, M., Wang, D., Zhu, F., Kandimalla, E. R., and Agrawal, S. (2009) Toll-like receptor 7 selective synthetic oligoribonucleotide agonists: synthesis and structure-activity relationship studies. *J Med Chem* **52**, 6871–6879.
14. Lan, T., Bhagat, L., Wang, D., Dai, M., Kandimalla, E. R., and Agrawal, S. (2009) Synthetic oligoribonucleotides containing arabinonucleotides act as agonists of TLR7 and 8. *Bioorg Med Chem Lett* **19**, 2044–2047.
15. Lan, T., Putta, M. R., Wang, D., Dai, M., Yu, D., Kandimalla, E. R., and Agrawal, S. (2009) Synthetic oligoribonucleotides-containing secondary structures act as agonists of Toll-like receptors 7 and 8. *Biochem Biophys Res Commun* **386**, 443–448.
16. Beaucage, S. L., and Caruthers, M. H. (1981) Deoxynucleotide phosphoramidites – a new class of key intermediates for deoxypolynucleotide synthesis. *Tetrahedron Lett* **22**, 1859–1862.
17. Iyer, R. P., Egan, W., Regan, J. B., and Beaucage, S. L. (1990) 3H-1,2-Benzodithiole-3-one-1,1-dioxide as an improved sulfurizing reagent in the solid-phase synthesis of oligodeoxyribonucleoside phosphorothioates. *J Am Chem Soc* **112**, 1253–1254.
18. Puglisi, J. D., and Tinoco, I. Jr. (1989) Absorbance melting curves of RNA. *Methods Enzymol* **180**, 304–325.

Chapter 18

Synthesis, Purification, and Characterization of Immune-Modulatory Oligodeoxynucleotides that Act as Agonists of Toll-Like Receptor 9

Mallikarjuna Reddy Putta, Dong Yu, and Ekambar R. Kandimalla

Abstract

Methods and protocols for automated synthesis and purification of immune modulatory oligonucleotides (IMOs), a novel class of Toll-like receptor 9 (TLR9) agonists, are described. IMOs containing two short identical sequences of 11-mers with phosphorothioate linkages can be synthesized in parallel synthetic strategy. A C3-linker that mimics the natural inter-nucleotide distance was commonly used for joining the two segments of IMOs. NittoPhase solid support bearing a symmetrical C3-linker (glycerol) and nucleoside-β-cyanoethyl-*N,N*-diisopropylphosphoramidites were used for IMO synthesis. The parallel synthesis was carried out in a 3′→ 5′ direction with removal of the final dimethoxytrityl (DMT) protecting group. After synthesis, the IMO was cleaved and deprotected by treating with aqueous ammonia. The product was purified on anion-exchange HPLC, desalted, lyophilized, and characterized by anion-exchange HPLC, capillary gel electrophoresis, polyacrylamide gel electrophoresis, and MALDI-TOF mass spectral analysis.

Key words: Agonist, immune-modulatory oligonucleotide, NittoPhase, solid-phase synthesis, symmetrical linker, toll-like receptor 9.

1. Introduction

Toll-like receptor 9 (TLR9) is a member of the TLR group of type I transmembrane receptors that continuously scout for infections through pathogen-associated molecular patterns (PAMPs) (1). Bacterial and viral DNA containing unmethylated cytosine-phosphate-guanosine (CpG) motifs are the natural ligands for TLR9 (2). Synthetic oligodeoxynucleotides containing unmethylated CpG motifs (CpG oligos) have also been found to stimulate

J. Goodchild (ed.), *Therapeutic Oligonucleotides*, Methods in Molecular Biology 764,
DOI 10.1007/978-1-61779-188-8_18, © Springer Science+Business Media, LLC 2011

TLR9-mediated immune responses. Many CpG-oligos are currently being evaluated in clinical trials as therapies for various diseases, including cancer, asthma, allergies, and infectious diseases (3, 4).

Our extensive structure activity relationship studies have shown requirement of a number of structural and sequence motifs of DNA for TLR9-mediated immune responses (4–25). These studies have shown that certain modifications in the CpG motif also activate TLR9 and induce distinct immune response profiles (4, 7, 8). We showed that an accessible 5′-end is required for immune stimulatory activity and blocking the 5′-end of CpG-DNA either by 5′–5′-linkage or conjugation of certain ligands abrogates immunostimulatory activity (5, 6). Based on these results, we designed and synthesized a novel class of immune-modulatory oligonucleotides (IMOs) as TLR9 agonists (7). IMOs contain two identical sequence segments of CpG-DNA, bearing 7-deaza-dG modifications and joined 3′–3′ via a glycerol linker (**Fig. 18.1**). An IMO containing 11 nucleotides in each segment showed the highest activity (7, 8).

IMOs can be synthesized in either linear or parallel synthetic strategy (**Scheme 18.1**). Parallel synthesis has several advantages over linear synthesis: (i) it permits incorporation of two or more identical sequence segments; (ii) the segments are synthesized concurrently, thereby halving synthetic steps and reducing time required for synthesis; and (iii) with fewer synthetic cycles, the yield and purity of the final product improves. Thus, a parallel solid-phase synthetic technique for the IMO synthesis was developed (7). A C3-linker that mimics the inter-nucleotide distance was found to be optimal for joining the two segments of IMO through 3′–3′-linkage though other types of linkers can be used (25). This chapter describes the methods and protocols used for the automated solid-phase synthesis, purification, and characterization of IMO. A NittoPhase solid support derivatized with symmetrical linker (glycerol) **1** and nucleoside-β-cyanoethyl-*N*,*N*-diisopropylphosphoramidites **2a–e** (**Fig. 18.2**) were used for IMO synthesis (*see* **Notes 1–3**).

1.1. Synthesis of the NittoPhase Derivatized Symmetrical Linker

To execute the parallel solid-phase synthesis of IMO, solid support bearing and symmetrically protected C3-linker (glycerol) **1** is required (**Scheme 18.2**). Although long chain alkyl amine-controlled pore glass (LCAA-CPG) solid support is commonly used for the solid-phase synthesis of oligonucleotides, we found NittoPhase solid support gives superior crude and pure product yields. NittoPhase derivatized linker **1** was synthesized from glycerol (3) as shown in **Scheme 18.2**. Selective protection of the two primary hydroxyls of **3** (**Scheme 18.2**) with 4,4′-dimethoxytrityl chloride (DMTCl) in the presence of 4-(dimethylamino)pyridine (DMAP) afforded *bis*-DMT glycerol **4**. The third hydroxyl of the

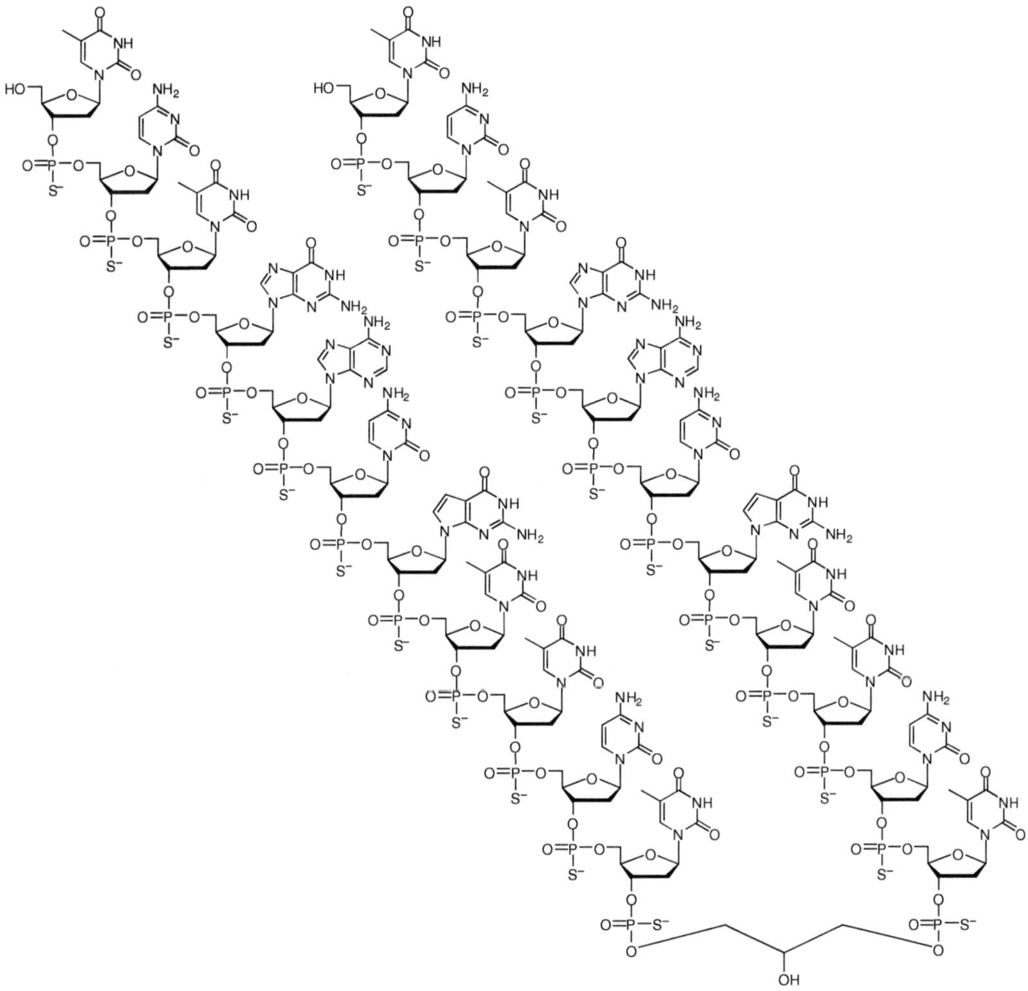

Fig. 18.1. Structure of a typical IMO (5′-TCTGAC**G̲**TTCT-X-TCTT**G̲**CAGTCT-5′, **G̲** = 7-deaza-dG and X = Glycerol).

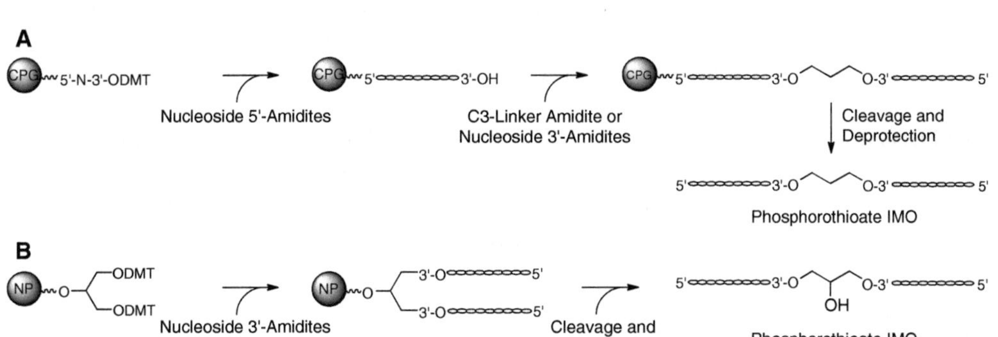

Scheme. 18.1. (**a**) Linear solid-phase synthesis and (**b**) parallel solid-phase synthesis.

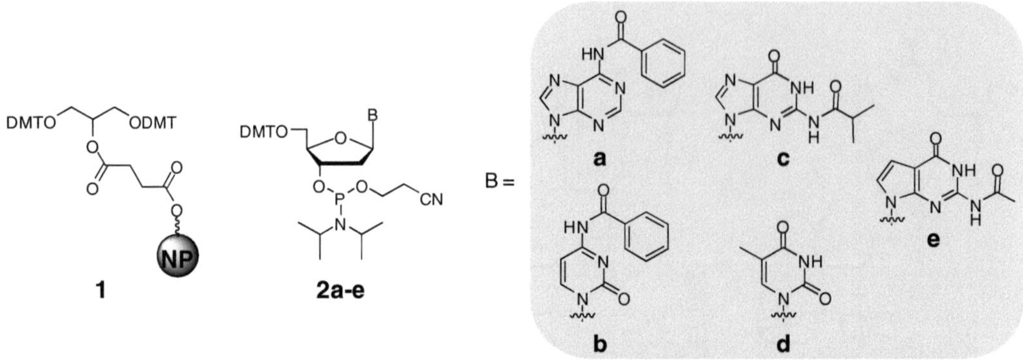

Fig. 18.2. Solid support derivatized symmetrical linker (Glycerol) **1**, and nucleoside-3′-phosphoramidites **2a–e**.

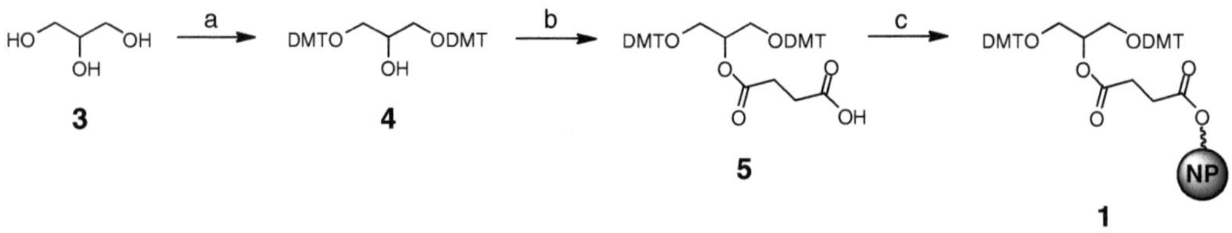

Scheme. 18.2. Synthesis of NittoPhase solid support. (**a**) DMTCl, DMAP, pyridine, 0°C – rt; (**b**) succinic anhydride, DMAP, pyridine; (**c**) NittoPhase (NP), DMAP, TBTU, pyridine, acetonitrile.

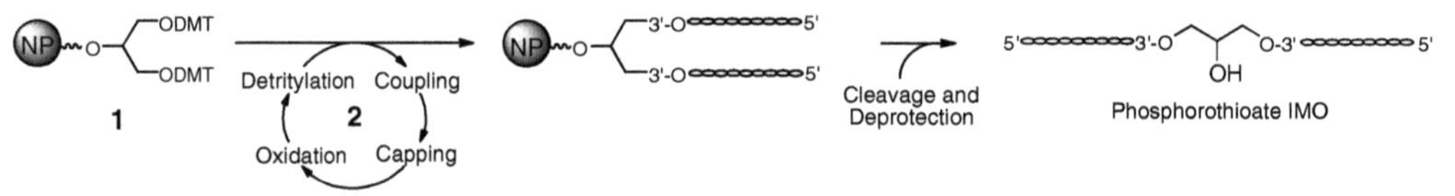

Scheme. 18.3. Automated synthesis of IMO. Reagents and conditions are given in **Section 3.2.1**.

linker was converted into its succinate **5** in quantitative yields by reacting with succinic anhydride in the presence of DMAP. Hydroxy-functionalized NittoPhase was derivatized with **5** in the presence of TBTU/DMAP in pyridine/acetonitrile mixture to afford the desired di-DMT-glycerol NittoPhase solid support **1**.

1.2. Synthesis of IMO

Synthesis of IMO was carried out on a 10-μmol scale using standard β-cyanoethyl-*N,N*-diisopropylaminophosphoramidite chemistry and NittoPhase solid support derivatized linker **1** as shown in **Scheme 18.3**.

2. Materials

1. NMR spectra were obtained on Varian 400 MHz Unity Inova instrument. Chemical shift (δ) values are expressed in ppm (parts per million) relative to TMS as an internal

standard and the coupling constants (*J*) are expressed in Hz.

2. High resolution mass spectra (electrospray ionization/time-of-flight) (HRMS ESI/TOF) were recorded on a Micromass LCT-TOF mass spectrometer.
3. TLC plates: silica gel 60 F_{254}
4. All reagents and chemical were purchased from Aldrich (St. Louis, MO) otherwise mentioned and used without further purification.

2.1. Synthesis of the NittoPhase Derivatized Symmetrical Linker

2.1.1. 1,3-bis-(4,4′-Dimethoxytrityloxy)-2-propanol, 4

1. 4-(Dimethylamino)pyridine (DMAP)
2. Pyridine: anhydrous
3. 4,4′-Dimethoxytrityl chloride (DMTCl)
4. Hexanes
5. Ethyl acetate
6. Triethylamine
7. NH_4Cl solution: saturated
8. Brine: saturated
9. Silica gel 60 for flash column chromatography: particle size 0.040–0.063 mm (230–400 mesh)
10. Hexanes/ethyl acetate eluent for flash column chromatography: 4:1 v/v containing 0.5% triethylamine

2.1.2. 1,3-bis-(4,4′-dimethoxytrityloxy)-2-propanol succinate, 5

1. Succinic anhydride
2. DMAP
3. Pyridine anhydrous
4. Citric acid solution: 10%
5. Dichloromethane
6. Methanol
7. Triethylamine

2.1.3. Preparation of 1,3-bis-DMT-glycerol Derivatized NittoPhase support, 1

1. NittoPhase solid support (NP): 90±15 μm particle size, 60±20 nm pore size, 450±50 μmol/g OH groups (Nitto Denko Corporation or Kinovate Life Sciences, Inc)
2. O-(Benzotriazol-1-yl)-*N,N,N′,N′*-tetramethyluronium tetrafluoroborate (TBTU)
3. Acetonitrile/pyridine 5:1 v/v and 95:1 v/v4. DMAP
5. Cap A: THF/acetic anhydride/2,6-lutidine 80/10/10 (v/v/v)
6. Cap B: *N*-methylimidazole/THF 16/84 (v/v)
7. Trichloroacetic acid: 3% in dichloromethane

2.2. Synthesis of IMO

2.2.1. Synthesis on Solid Support

1. MerMade 6 automated DNA synthesizer (BioAutomation, Plano, TX) equipped with a trityl monitor
2. Nucleoside-3′-phosphoramidites: dA^{Bz}, dC^{Bz}, dG^{ibu}, T, each 0.05 M in acetonitrile
3. Beaucage reagent: 3% solution in acetonitrile prepared freshly (*see* **Note 4**)
4. Tetrazole: 0.25 M in acetonitrile
5. Cap A: THF/acetic anhydride/2,6-lutidine 80/10/10 (v/v/v)
6. Cap B: *N*-methylimidazole/THF 16/84 (v/v)
7. Trichloroacetic acid: 3% in dichloromethane
8. Ammonium hydroxide: 28–30% w/w

2.2.2. Anion-Exchange HPLC Purification of IMO

1. Waters Delta 600 HPLC system (Milford, MA) equipped with UV detector and Empower software
2. Source 15Q 7.5 mm × 20 cm anion-exchange column (GE Healthcare)
3. United States Pharmacopea-quality sterile water
4. Buffer A: 25 mM Tris–HCl, pH 7.5, containing 20% acetonitrile
5. Buffer B: 2.5 M NaCl in buffer A
6. Reverse-phase C18 HPLC column: 1.5 cm × 28 cm, 37–55 μm Bondapak HC_{18} HA (Waters Corporation)
7. 50% acetonitrile in water.
8. Spectra Por MWCO 3500 regenerated cellulose dialysis tube
9. 0.2-μm syringe filter.

2.2.3. Analytical Anion-Exchange HPLC

1. Waters Delta 600 HPLC system (Milford, MA) equipped with UV detector.
2. DNA-Pac PA-100 column: 4 × 50 mm (Dionex).
3. Buffer A: 100 mM Tris–HCl, pH 8.0, containing 20% acetonitrile.
4. Buffer B: Buffer A containing 2 M LiCl.

2.2.4. Capillary Gel Electrophoresis

1. P/ACE MDQ system with 32 Karate software
2. eCAP ssDNA R100 denaturing gel kit (Beckman Coulter)
3. 7 M Urea solution (Beckman Coulter, Brea, CA)

2.2.5. Polyacrylamide Gel Electrophoresis (PAGE)

1. SequaGel Sequencing System Kit (EC-833) (National Diagnostics, Inc.)

2. Tris-borate-EDTA (TBE) buffer (10×) was prepared and diluted to 1× for running PAGE
3. Aqueous ammonium persulfate (APS): 10% solution w/v
4. *N,N,N′,N′*-Tetramethylethylenediamine (TEMED)
5. Bromophenol blue and xylene cyanol marker dyes
6. Formamide
7. Vertical gel electrophoresis apparatus
8. Molecular Imager VersaDoc MP 4000 System (Bio-Rad) equipped with CCD camera and QuantityOne software (version 4.6.3)

2.2.6. MALDI-TOF Mass Spectral Analysis

1. Matrix-assisted laser desorption/ionization-time of flight (MALDI-TOF) Micro MX mass spectrometer (Waters) with MassLynx software
2. Matrix: 5% 3-hydroxypicolinic acid, 0.5% ammonium citrate dibasic in acetonitrile/water 1:1 v/v
3. United States Pharmacopea-quality sterile water for irrigation/acetonitrile

3. Methods

3.1. Synthesis of Symmetrical Glycerol Linker and Derivatization of NittoPhase Solid Support

*3.1.1. 1,3-bis-(4,4′-Dimethoxytrityloxy)-2-propanol, **4***

1. Glycerol **3** (3.92 g, 42.6 mmol) and DMAP (5.2 g, 42.6 mmol) were dissolved in dry pyridine (100 mL), cooled to 0°C, and maintained under argon atmosphere.
2. DMTCl (30.3 g, 89.4 mmol) in pyridine (150 mL) was added drop-wise with constant stirring. Then, the reaction mixture was allowed to reach room temperature slowly (~4 h) and stirring continued overnight.
3. TLC in hexanes/ethyl acetate (3:1 v/v) containing 0.5% triethylamine indicated completion of reaction with R_f of the product = 0.63.
4. Pyridine was rotoevaporated to dryness; the residue was dissolved in ethyl acetate (500 mL) and washed successively with saturated NH_4Cl solution (500 mL) and brine (500 mL).
5. The organic layer was dried over anhydrous $MgSO_4$ and rotoevaporated to dryness.
6. The residue was purified on silica gel flash column chromatography using hexanes/ethyl acetate (4:1 v/v) containing 0.5% triethylamine to give *bis*-DMT glycerol **4** (26) as a white foam (91%).

7. ^{1}H-NMR ($CDCl_3$, 400 MHz): δ 2.50 (bs, 1H, -O*H*), 3.42 (m, 4H, 2 × $-CH_2O$-DMT), 3.83 (s, 12H, 4 × $-OCH_3$), 4.07 (m, 1H, –C*H*OH), 6.80–6.87 (m, 8H, Ar–*H*), 7.16–7.45 (m, 18H, Ar–*H*).
8. HRMS (ESI/TOF): Calculated for $C_{45}H_{44}O_7$ $[M+Na]^+$ 719.2985 and observed 719.2973.

*3.1.2. 1,3-bis-(4,4′-Dimethoxytrityloxy)-2-propanol succinate, **5***

1. Succinic anhydride (7.75 g, 77.5 mmol) was added in portions to a stirred solution of *bis*-DMT glycerol **4** (27.0 g, 38.75 mmol) and DMAP (4.74 g, 38.8 mmol) in dry pyridine (250 mL) at room temperature. The reaction mixture was stirred overnight and TLC with dichloromethane/methanol (9:1 v/v) containing 0.5% triethylamine indicated the completion of the reaction.
2. Pyridine was rotoevaporated to dryness; residue was dissolved in dichloromethane (500 mL) and successively washed with ice-cold 10% citric acid solution (2 × 250 mL) and water (250 mL).
3. The dichloromethane layer was dried over anhydrous $MgSO_4$, concentrated to ~50 mL using a rotoevaporator and purified by silica gel flash column chromatography using 0 → 2% methanol in dichloromethane containing 0.5% triethylamine.
4. The pure product was a white foam obtained as the triethylammonium salt of succinate **5** (26) (24.8 g, 83%).
5. TLC in dichloromethane/methanol 9:1 containing 0.5% triethylamine gave R_f 0.61.
6. ^{1}H-NMR ($CDCl_3$, 400 MHz): δ 1.16 (t, 9H, J = 7.4, $-N(CH_2CH_3)_3$), 2.53 (m, 2H, $-CH_2COOH$), 2.67 (m, 2H, $-OCOCH_2CH_2-$), 2.87 (m, 9H, $-N(CH_2CH_3)_3$), 2.25–3.34 (m, 4H, 2 × DMT-OCH_2-), 3.78 (s, 12H, 4 × $-OCH_3$), 5.23 (m, 1H, $-CH_2CHCH_2-$), 6.77 (dd, 8H, J = 9.2 & 2.6, Ar–*H*), 7.18–7.36 (m, 18H, Ar–*H*).
7. HRMS (ESI/TOF): Calculated for $C_{49}H_{48}O_{10}$ $[M+Na]^+$ 819.3145 and observed 819.3154.

*3.1.3. Preparation of 1,3-bis-DMT-glycerol Derivatized NittoPhase support, **1***

1. NittoPhase support was soaked for 30 min and thoroughly washed with acetonitrile. Succinate **5** (12 g, 15 mmol) in acetonitrile/pyridine 5:1 (360 mL) was added to NittoPhase (100 g) and then DMAP (9.2 g, 75 mmol) and TBTU (14.5 g, 45 mmol) were added in one portion.
2. The slurry was shaken overnight at room temperature and checked for loading by separating a small amount of support (*see* Step 6).

3. The NittoPhase support was thoroughly washed with acetonitrile containing 5% pyridine (2 × 400 mL) followed by acetonitrile (4 × 450 mL) to remove any trapped reagents in the swollen support.
4. Cap A (200 mL) and Cap B (200 mL) solutions were added simultaneously to the NittoPhase support and shaken for 4 h. Solutions filtered off and the capping reaction was repeated one more time.
5. The product was washed thoroughly with acetonitrile containing 5% pyridine followed by acetonitrile and dried under high vacuum for overnight to give dry NittoPhase support **1**.
6. Final loading was determined by treating small portion of dry support with 3% trichloroacetic acid in dichloromethane (note: this is the same reagent used for detritylation on automated DNA synthesizer) to release DMT that was quantitated from the absorbance at 498 nm. The final NittoPhase loading obtained was 301 μmol/g.
7. The linker loading was calculated by using the following equation.

$$\text{DMT loading mol/g of support} = A_{498} \times \nu/\varepsilon \times \text{g}$$

where A_{498} is absorbance of trityl cation at 498 nm, ν is the final volume of DMT solution used for A_{498} measurement, ε is 70,000 L/mol cm, the extinction coefficient of the DMT cation at 498 nm, and g is support weight in grams.

3.2. Synthesis of IMO

3.2.1. Synthesis on Solid Support

1. IMO was synthesized on a 10-μmol scale. The required amount of linker solid support **1** was packed in the appropriate column and placed on the automated DNA synthesizer.
2. Synthesis was carried out using the following program. ***Detritylation***: 3%TCA in DCM, 3 × 35 s.; ***Coupling:*** 0.05 M phosphoramidite **2a–d** in acetonitrile and 0.25 M tetrazole, 2 × 6 min, and for 0.05 M phosphoramidite **2e** in acetonitrile and 0.25 M tetrazole, 2 × 10 min; ***Oxidation***: 3% Beaucage reagent in acetonitrile, 2 × 4 min; ***Capping***; Cap A (acetic anhydride/pyridine/THF) and Cap B (*N*-methylimidazole in DCM), 2 × 2 min.
3. At the end of the synthesis, the product was detritylated and the support was washed with acetonitrile.
4. The solid support in the column was dried under a stream of dry nitrogen and transferred in to a bottle with a well-sealing screw cap to prevent loss of ammonia gas in the next step.
5. The support was treated with ~5 mL of 28–30% ammonium hydroxide solution and kept at 56°C for 16 h with occasional swirling (*see* **Note 5**).

6. The bottle was cooled to 0°C before opening and solid support was removed by filtration and rinsed with ~5 mL of water.
7. Excess ammonia was removed from the filtrate under vacuum and the solution was diluted with five volume equivalents of distilled water.
8. Crude IMO yield was checked by analytical anion-exchange HPLC (*see* **Section 3.2.3** and **Fig. 18.3a**) before purification on preparative anion-exchange HPLC as detailed below (*see* **Note 6**).

3.2.2. Anion-Exchange HPLC Purification of IMO

1. IMO was purified by HPLC equipped with a Source 15Q resin column (20 cm × 250 mm) at 85°C at a flow rate of 10 mL/min. The column was eluted with buffer A for 2 min and then with a gradient of buffer B from 0% → 30% in 10 min then 30% → 70% in 40 min.
2. The major absorbance peak detected at 260 nm was collected, solvent was removed by rotoevaporation, and the residue was re-dissolved in RNAse-free water.

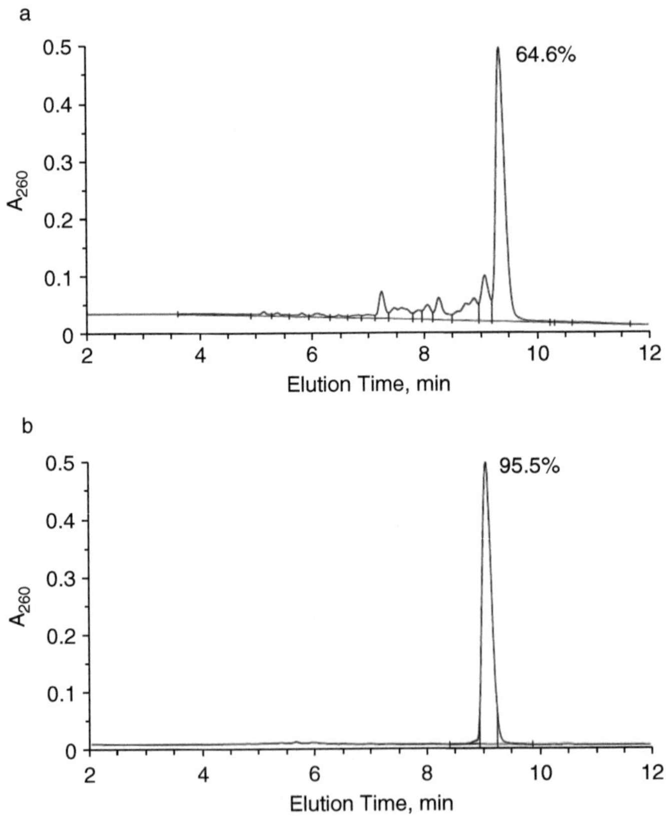

Fig. 18.3. Analytical anion-exchange HPLC profiles of (**a**) crude and (**b**) purified phosphorothioate IMO.

3. The purified IMO was desalted on a C18 reverse phase HPLC column (1.5 cm × 28 cm) by washing with RNAse-free water for 20 min at 25 mL/min flow rate.
4. IMO was eluted with 50% acetonitrile in water, rotoevaporated to dryness and dissolved in 5 mL of water.
5. Spectra Por regenerated cellulose dialysis membrane tube (~10 cm long) was rinsed thoroughly with RNAse free water and clamped at one end. The IMO solution was transferred into the dialysis tube with a Pasteur pipette and the end of the tube was tightly clamped.
6. The tube was stirred slowly in 2 L sterile water for 4 h. The water was changed after 4 h and dialysis was repeated two more times.
7. The solution was filtered through a 0.2-μm syringe filter and lyophilized to dryness giving pure IMO as a white solid (*see* **Note 7**).
8. The IMO was reconstituted in sterile water and the concentration was determined by measuring UV absorbance at 260 nm (*see* **Note 8**).

3.2.3. Analytical Anion-Exchange HPLC

1. The purity of the IMO was checked on a 4 × 50 mm DNA-Pac PA-100 column at 85°C using a flow rate of 2 mL/min and monitoring at 260 nm.
2. The column was eluted with 100% buffer A for 2 min then with a gradient of buffer B from 0 → 100% in 8 min followed by washing with buffer B for 5 min (**Fig. 18.3b**).

3.2.4. Capillary Gel Electrophoresis

1. The capillary was filled with denaturing gel made from an eCAP ssDNA R100 kit using the manufacturer's recommended loading protocol.
2. The sample was loaded by electrokinetic injection for 2–4 s at 6 kV and separation was carried out at 14 kV for 30 min in 7 M urea.
3. Purity was calculated by integration using 32 Karate software (**Fig. 18.4**).

3.2.5. Polyacrylamide Gel Electrophoresis (PAGE)

1. PAGE analysis of pure IMO along with crude product and marker oligonucleotides was performed on a 16% polyacrylamide gel prepared by mixing SequaGel concentrate, diluent and buffer according to the supplier's recommended protocol (*see* **Note 9**).
2. TEMED and freshly prepared ammonium persulfate solution were added; the mixture was swirled and quickly poured into the gel cast before polymerization occurs.

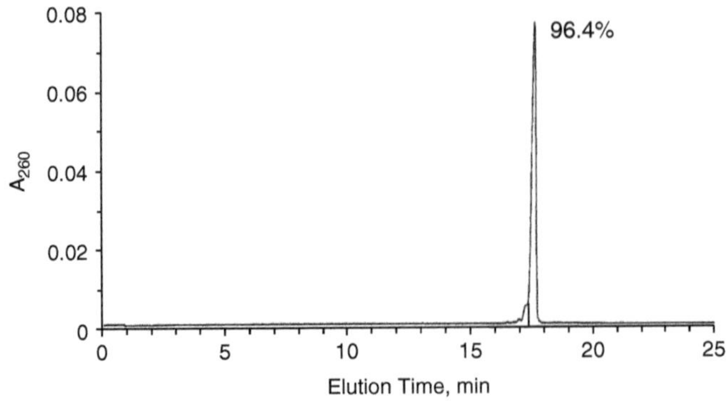

Fig. 18.4. Capillary gel electrophoresis profile of purified phosphorothioate IMO.

3. An appropriate comb was inserted immediately into the gel without trapping any air bubbles under the teeth and left 30–60 min at room temperature for polymerization.
4. The comb was removed and the wells were flushed with a syringe. The gel was pre-run for 15–20 min before loading the samples.
5. About 0.5 OD of sample in 10 μL of water and 5 μL of formamide was loaded into each well. A mixture of 45-mer, 25-mer, and 18-mer phosphorothioate linear DNAs were used as marker oligonucleotides and bromophenol blue and xylene cyanole were used as marker dyes.
6. Electrophoresis was carried out at constant 5 W, and the gel temperature during the run was maintained <40°C to avoid bowing of the bands.
7. After electrophoresis, the gel was carefully transferred onto a large TLC plate covered with plastic wrap. Bands ware visualized by UV shadowing at 254 nm and photographed (**Fig. 18.5**).

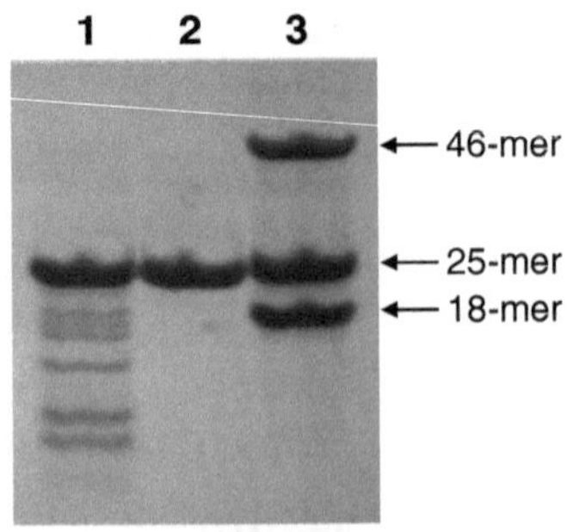

Fig. 18.5. Non-denaturing PAGE of crude and pure IMO along with oligonucleotide markers with different lengths. Lane 1 is crude IMO, lane 2 is purified IMO, and lane 3 is linear phosphorothioate oligonucleotide markers of indicated length.

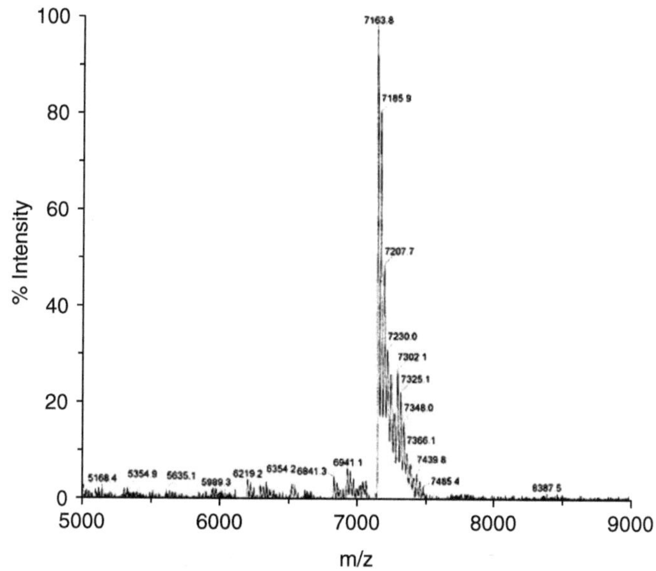

Fig. 18.6. MALDI/TOF mass spectrum of purified phosphorothioate IMO. Calculated mass 7163.8 matches with the observed value.

3.2.6. MALDI-TOF Mass Spectral Analysis

1. Matrix-assisted laser desorption/ionization time-of-flight (MALDI-TOF) mass spectral characterization of pure IMO was performed using 3-hydroxypicolinic acid (5%) and ammonium citrate dibasic (0.5%) in 1:1 mixture of acetonitrile/water as matrix.
2. About ~0.05 OD of IMO in 2 μL of water was mixed with 5 μL of matrix and 1 μL of the mixture spotted on the plate and dried.
3. The spectrum was acquired in a linear/positive mode and spectrum was processed with MassLynx software (**Fig. 18.6**).

4. Notes

1. Symmetrical linker derivatized NittoPhase solid support was suitable for IMO synthesis and the linker loading can be used up to 300 μmol/g. However, commonly used long chain alkyl amine-controlled pore glass (LCAA-CPG) solid support should be limited to much lower loading of ~45 μmol/g as it was found optimal for better coupling yields.
2. NittoPhase solid support was found to be superior over CPG solid support in IMO synthesis due to (i) higher coupling efficiency in each cycle, (ii) improved crude IMO yield and

pure IMO quality, and (iii) over 7.5-fold higher loading (compared with CPG support) translate to reduced cost of solid support and reagents.

3. Amidites of dA, dC, dG, T, and 7-deaza-dG are commercially available and they should be stored in the refrigerator preferably at –20°C.
4. Beaucage reagent is stable in acetonitrile for 1–2 weeks under dry conditions; however, it is to be discarded if any precipitation occurs. Silanized bottle should be used as container for Beaucage reagent.
5. For complete cleavage and deprotection, treatment with 28–30% ammonium hydroxide for 16 h at 56°C is enough. A longer treatment may result in 1–2% degradation of IMO.
6. Crude yields range from 60 to 75% for the IMO synthesized.
7. Lyophilized pure IMO should be handled under biosafety hood to avoid any endotoxins contamination.
8. The extinction coefficient, ε_{260}, was calculated for 22-mer DNA with internal C3 linker modification (27).
9. Acrylamide has been found to be neurotoxic. Protective gloves and eye ware should be used while handling these products.

Acknowledgments

The authors thank Dr. Sudhir Agrawal for support, encouragement, and suggestions.

References

1. Iwasaki, A., and Medzitov, R. (2010) Regulation of adaptive immunity by the innate immune system. *Science* **327**, 291–295.
2. Hemmi, H., Takeuchi, O., Kawai, T., Kaisho, T., Sato, S., Sanjo, H., Matsumoto, M., Hoshino, K., Wagner, H., Takeda, K., and Akira, S. (2000) A toll-like receptor recognizes bacterial DNA. *Nature* **408**, 740–745.
3. Krieg, A. M. (2006) Therapeutic potential of toll-like receptor 9 activation. *Nat Rev Drug Discov* **5**, 471–484.
4. Agrawal, S., and Kandimalla, E. R. (2007) Synthetic agonists of toll-like receptors 7, 8, and 9. *Biochem Soc Trans* **35**, 1461–1467.
5. Yu, D., Zhao, Q., Kandimalla, E. R., and Agrawal, S. (2000) Accessible 5′-end of CpG-containing phosphorothioate oligodeoxynucleotides is essential for immunostimulatory activity. *Bioorg Med Chem Lett* **10**, 2585–2588.
6. Kandimalla, E. R., Bhagat, L., Yu, D., Cong, Y., Tang, J., and Agrawal, S. (2002) Conjugation of ligands at the 5′-end of CpG DNA affects immunostimulatory activity. *Bioconjug Chem* **13**, 966–974.
7. Yu, D., Kandimalla, E. R., Bhagat, L., Tang, J. Y., Cong, Y., Tang, J., and Agrawal, S. (2002) "Immunomers" novel 3′–3′-linked CpG oligodeoxyribonucleotides as potent immunomodulatory agents. *Nucleic Acids Res* **30**, 4460–4469.
8. Kandimalla, E. R., and Agrawal, S. (2005) Synthetic agonists of Toll-like receptor 9. In Toll and Toll-like receptors: an immunologic

perspective (Rich, T. ed.). Kluwer/Plenum Publishers, New York, NY, pp. 181–212.
9. Agrawal, S., and Kandimalla, E. R. (2001) Antisense and/or immunostimulatory oligonucleotide therapeutics. *Curr Cancer Drug Target* **1**, 197–209.
10. Agrawal, S., and Kandimalla, E. R. (2003) Modulation of toll-like receptor 9 responses through synthetic immunostimulatory motifs of DNA. *Ann N Y Acad Sci* **1002**, 30–42.
11. Kandimalla, E. R., Bhagat, L., Li, Y., Yu, D., Wang, D., Cong, Y. P., Song, S. S., Tang, J. X., Sullivan, T., and Agrawal, S. (2005) Immunomodulatory oligonucleotides containing a cytosine-phosphate-2'-deoxy-7-deazaguanosine motif as potent Toll-like receptor 9 agonists. *Proc Natl Acad Sci USA* **102**, 6925–6930.
12. Kandimalla, E. R., Bhagat, L., Zhu, F. G., Yu, D., Cong, Y. P., Wang, D., Tang, J. X., Tang, J. Y., Knetter, C. F., Lien, E., and Agrawal, S. (2003) A dinucleotide motif in oligonucleotides shows potent immunomodulatory activity and overrides species specific recognition observed with CpG motif. *Proc Natl Acad Sci USA* **100**, 14303–14308.
13. Kandimalla, E. R., Yu, D., Zhao, Q., and Agrawal, S. (2001) Effect of chemical modifications of cytosine and guanine in a CpG-motif of oligonucleotides on immunostimulatory activity: structure-immunostimulatory activity relationships. *Bioorg Med Chem* **9**, 807–813.
14. Putta, M. R., Zhu, F., Li, Y., Bhagat, L., Cong, Y., Kandimalla, E. R., and Agrawal, S. (2006) Novel oligodeoxynucotide agonists of TLR9 containing N^3-Me-dC or N^1-Me-dG modifications. *Nucleic Acids Res* **34**, 3231–3238.
15. Putta, M. R., Zhu, F. G., Wang, D., Bhagat, L., Dai, M., Kandimalla, E. R., and Agrawal, S. (2010) Peptide conjugation at the 5'-end of oligodeoxynucleotides abrogates Toll-like receptor 9-mediated immune stimulatory activity. *Bioconjug Chem* **21**, 39–45.
16. Yu, D., Kandimalla, E. R., Bhagat, L., Tang, J. Y., Cong, Y., Tang, J., and Agrawal, S. (2002) 'Immunomers' – Novel 3'-3'-linked CpG oligodeoxynucleotides as potent immunomodulatory agents. *Nucleic Acids Res* **30**, 4460–4469.
17. Yu, D., Kandimalla, E. R., Cong, Y., Tang, J., Tang, J. Y., Zhao, Q., and Agrawal, S. (2002) Design, synthesis, and immunostimulatory properties of CpG DNAs containing alkyl-linker substitutions: role of nucleosides in the flanking sequences. *J Med Chem* **45**, 4540–4548.
18. Yu, D., Kandimalla, E. R., Roskey, A., Zhao, Q., Chen, L., Chen, J., and Agrawal, S. (2000) Stereo-enriched phosphorothioate oligodeoxynucleotides: synthesis, biophysical and biological properties. *Bioorg Med Chem* **8**, 275–284.
19. Yu, D., Kandimalla, E. R., Zhao, Q., Cong, Y., and Agrawal, S. (2001) Immunostimulatory activity of CpG oligonucleotides containing non-ionic methylphosphonate linkages. *Bioorg Med Chem* **9**, 2803–2808.
20. Yu, D., Kandimalla, E. R., Zhao, Q., Cong, Y., and Agrawal, S. (2001) Modulation of immunostimulatory activity of CpG oligonuclotides by site-specific deletion of nucleobases. *Bioorg Med Chem Lett* **11**, 2263–2267.
21. Yu, D., Kandimalla, E. R., Zhao, Q., Cong, Y., and Agrawal, S. (2002) Immunostimulatory properties of phosphorothioate CpG DNA containing both 3'-5'- and 2'-5'-internucleotide linkages. *Nucleic Acids Res* **30**, 1613–1619.
22. Yu, D., Kandimalla, E. R., Zhao, Q., Bhagat, L., Cong, Y., and Agrawal, S. (2003) Requirement of nucleobase proximal to CpG dinucleotide for immunostimulatory activity of synthetic CpG DNA. *Bioorg Med Chem* **11**, 459–464.
23. Yu, D., Putta, M. R., Bhagat, L., Li, Y., Zhu, F., Wang, D., Tang, J. X., Kandimalla, E. R., and Agrawal, S. (2007) Agonists of toll-like receptor 9 containing synthetic dinucleotide motifs. *J Med Chem* **50**, 6411–6418.
24. Yu, D., Putta, M. R., Bhagat, L., Dai, M., Wang, D., Trombino, A. F., Sullivan, T., Kandimalla, E. R., and Agrawal, S. (2008) Impact of secondary structure of toll-like receptor 9 agonists on interferon-α induction. *Antimicrob Agents Chemother* **52**, 4320–4325.
25. Putta, M. R., Yu, D., Bhagat, L., Wang, D., Zhu, F. G., and Kandimalla, E. R. (2010) Impact of nature and length of linker incorporated in agonists on Toll-like receptor 9-mediated immune responses. *J Med Chem* **53**, 3730–3738.
26. De Napoli, L., Di Fabio, G., Messere, A., Montesarchio, D., Musumeci, D., and Piccialli, G. (1999) Synthesis and characterization of new 3′-3′ linked oligodeoxyribonucleotides for alternate strand triple helix formation. *Tetrahedron* **55**, 9899–9914.
27. Puglisi, J. D., and Tinoco, I., Jr (1989) Absorbance melting curves of RNA. *Methods Enzymol* **180**, 304–325.

Chapter 19

Surface Plasmon Resonance Investigation of RNA Aptamer–RNA Ligand Interactions

Carmelo Di Primo, Eric Dausse, and Jean-Jacques Toulmé

Abstract

Instruments based on the surface plasmon resonance (SPR) principle allow label-free detection of interactions between targets immobilized at a solid–liquid interface and partners in solution. This method is well suited to determine the kinetic parameters, the equilibrium constant and the stoichiometry of a reaction. Aptamers are ligands identified from random libraries of RNA, DNA or chemically modified oligonucleotides by in vitro selection (SELEX). Aptamers can be raised against a great variety of targets (small molecules, proteins, nucleic acids, cells, viruses, bacteria). SPR is routinely used in our laboratory for the analysis of RNA aptamer–RNA target complexes. To illustrate SPR investigation of such complexes, we describe here methods that were successfully used to monitor the interaction between the trans-activating responsive element of HIV-1 and an RNA aptamer.

Key words: Aptamers, RNA hairpin, TAR RNA, HIV-1, loop–loop complex, surface plasmon resonance, sensor chip, interaction, kinetics.

1. Introduction

SELEX is a combinatorial strategy that was described about 20 years ago: alternative rounds of selection and amplification ultimately lead to the identification of oligonucleotides – so-called aptamers – displaying a predetermined property, contained in an initial pool of up to 10^{14} randomly synthesized oligomers (1, 2). Successful selections have been achieved against a wide range of targets encompassing small molecules (amino acids, nucleic acid bases, etc.) and live cells (3). Aptamers are very generally characterized by both high affinity (K_d in the low nanomolar range for proteins) and exquisite specificity of interaction: aptamers discriminate between enzymes from different species showing the

J. Goodchild (ed.), *Therapeutic Oligonucleotides*, Methods in Molecular Biology 764,
DOI 10.1007/978-1-61779-188-8_19, © Springer Science+Business Media, LLC 2011

same catalytic activity. For instance, the aptamer raised against the RNaseH domain of the HIV-1 reverse transcriptase does not bind to the human RNaseH1 and vice versa (4, 5). Therefore aptamers rival antibodies and offer several advantages: nucleic acids are easy to synthesize using standard phosphoramidite chemistry, ensuring high reproducibility from batch to batch. They can be conjugated to many different pending groups (e.g. fluorophores and polyethylene glycol) and can be chemically modified in order to improve their resistance to nucleases (6). They are easy to store and can be heat-denatured and renatured at will. In addition functional aptamers are small molecules compared to antibodies and are not immunogenic. Consequently aptamers attract an increasing interest for a number of different applications in the therapeutic as well as analytical fields (7, 8). Interestingly in contrast to other therapeutic oligonucleotides (antisense, siRNA) aptamers can be targeted to extracellular targets. A 2′fluoropyrimidine aptamer targeted to the vascular endothelial growth factor is being used for the treatment of wet-age related macular degeneration (9).

Aptamer-based biosensors have gained interest to detect biomolecules (8, 10). Numerous devices use electrochemistry (11), fluorescence (12, 13), atomic force microscopy (14) and quartz crystal microbalance (15) to monitor the interaction between the aptamer and the target. Twenty years ago, BIAcore AB was the first company to commercialize label-free detection instruments based on the surface plasmon resonance (SPR) principle (16, 17). The major technical advantage is that the measured signal is caused by refractive index changes due to the mass of a compound in solution that interacts with a target immobilized at a liquid–solid interface, in an evanescent wave field. Then, all kinds of complexes can be analysed from those formed with small molecules to those with high molecular weight compounds (proteins and nucleic acids), whole organisms (viruses, bacteria) and cells (for review see Reference (18)). The interest for this technology grew up significantly in the last decade. Other companies have now launched instruments that can rival BIAcore instruments for detection, sample handling, and last but not least, price. Of note SPR instruments were also used to carry out in vitro selection (5, 19). These applications will not be covered in the present chapter.

Instruments were historically developed to investigate protein–protein interactions. Although this biophysical tool is routinely used in research laboratories, aptamer–ligand interactions, between RNA molecules in particular, are seldom analysed by SPR. The main reason is likely due to the fear to manipulate molecules that can be degraded either because of the presence of magnesium ions or contaminating nucleases. Compared with proteins, nucleic acids actually offer several advantages. They can be easily end-functionalized with biotin allowing them to be captured on the sensor chips by streptavidin. Therefore the target will display a unique orientation at the solid–liquid interface reducing artefacts that could result from non-homogeneous functionalization. Nucleic acids offer another crucial advantage when interactions are analysed in heterogeneous medium. Being polyanions, in buffers containing ions, they do not interact non-specifically with most of the available surfaces that are negatively charged at physiological pH. This makes the experiments easier compared with molecules that can be really sticky, basic proteins for instance. SPR was successfully used in our laboratory to investigate RNA–RNA complexes formed between RNA aptamers and viral RNA targets from human pathogens (20–23), and nucleic acid complexes recognized by specific proteins (5, 23–26). SPR allowed the determination of the kinetic parameters and of the stoichiometry of the reaction (26).

This chapter presents protocols for investigating RNA aptamer–RNA ligand interactions with a BIAcore^{TM} 3000, one of the most popular SPR instrument (BIAcore, GE Healthcare, Uppsala, Sweden). We provide a description for preparing the instrument, the sensor chips, the RNA samples, for carrying out the regeneration of the surface and for running the binding kinetics. These protocols can be used for performing experiments with RNA molecules on other SPR instruments except for the automatic methods that will depend on the machine–computer interface.

To illustrate SPR investigation of aptamer–ligand complexes, we describe the interaction between the trans-activating responsive element TAR, an imperfect RNA hairpin located at the very 5′ end of the human immunodeficiency virus type 1 (HIV-1) that is crucial for the processive transcription of the viral genome, and a well-characterized RNA aptamer, R06, identified by SELEX (**Fig. 19.1**) (20, 27). The binding experiments were carried out on sensor chips of different geometries and matrix composition purchased from XanTec Bioanalytics (Duesseldorf, Germany). The results were compared to those obtained from a previous work (25) using a streptavidin-coated sensor chip from BIAcore.

SPR is a sensitive, reliable and fast label-free detection method for investigating RNA aptamer–RNA target complexes. SPR is also a technique of choice for investigating complexes formed between aptamers and proteins or small molecules. Instruments are sensitive enough to allow detection of low molecular weight analytes (>100 Da) that are injected over aptamer-functionalized surfaces. One limitation of this versatile method is the non-specific binding to the surface that can be observed sometimes when a positively charged protein is injected over an

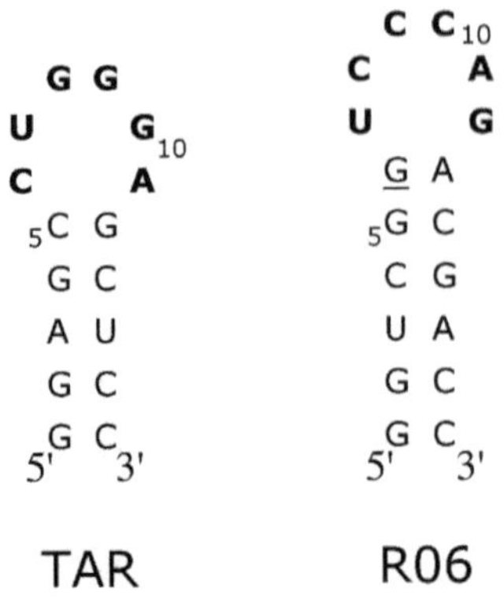

Fig. 19.1. Sequence and secondary structure of the RNA molecules. TAR is a shortened version of the full-length TAR imperfect hairpin that trans-activates the transcription of the HIV-1 genome. TAR was modified with a biotin-TEG at its 3′ end (not shown) to allow its capture on streptavidin-coated sensor chips. R06 is the RNA aptamer identified by in vitro selection against TAR. The loop bases engaged in Watson–Crick pairs are shown in bold. The G residue of R06 that was chemically modified with a 2′-*O*-methyl moiety to generate an aptamer that displayed lower affinity compared with that of the parent aptamer is underlined. Bases are numbered from the 5′ end.

aptamer-functionalized sensor chip. This can be overcome for instance by using surfaces with decreased charge density or by immobilizing the protein instead of the aptamer. Therefore SPR is of general interest for the characterization of aptamer–ligand complexes.

2. Materials

2.1. Preparation of RNA and Analysis by PAGE

1. Expedite 8908 nucleic acid synthesizer (Millipore).
2. Reagents for automated synthesis: *O-tert*-butyldimethylsilyl-ribonucleoside, phosphoramidites (iBu-G, Bz-A, Bz-C and U), 2′-*O*-methyl-ribonucleoside phosphoramidites (Dmf-G, Bz,-A, Bz-C and U), 0.2 mmol columns (iBu-G, Bz-A, Ac-C and U), activator, anhydrous acetonitrile, cap mix A, cap mix B, oxidizing solution and deblocking mix (Glen Research).
3. 10 mM Tris–HCl buffer, pH 7, 25 mM NaCl.
4. G-25 spin columns (GE Healthcare).
5. Spectra/Por MWCO: 3500 membrane tubing.
6. Denaturing loading buffer: 80% formamide, 10% glycerol, 0.025% bromophenol blue, 0.025% xylene cyanol, in milliQ water.
7. Stains-All solution: 30 mL stock Stains-All dye (Acros), 80 mL distilled water, 90 mL formamide; stock and diluted solutions should be stored at 4°C in a dark bottle.

2.2. Binding Studies

1. BIAcore™ 3000 apparatus (BIAcore) (*see* **Note 1**).
2. Streptavidin-coated sensor chips (BIAcore, XanTec Bioanalytics) (*see* **Note 2**).
3. Solution 1: 0.5% (w/v) SDS is milliQ water.
4. Solution 2: 50 mM glycine. Adjust pH to 9.5 with NaOH.
5. Sanitizer: 1% sodium hypochlorite.
6. HBS-EP running buffer for immobilizing the target (BIAcore).
7. 1.5-mL Eppendorf DNA LoBind tubes.
8. Running buffer for the binding experiments: 10 mM sodium phosphate, pH 7.2, at 20°C, 140 mM potassium chloride, 20 mM sodium chloride, 3 mM magnesium chloride, 0.005% of surfactant P20 (BIAcore).

3. Methods

3.1. Preparation of RNA Aptamer and RNA Target

The TAR HIV-1 RNA was conjugated at its 3′ end to a biotin residue through a linker, providing a 16-atom mixed polarity spacer based on a triethylene glycol containing four oxygen atoms.

3.1.1. Chemical Synthesis of RNA

1. The target and the RNA aptamer were synthesized on an expedite automated synthesizer following the manufacturers' instructions.
2. After removal of the 2′-protecting groups, purification was achieved by electrophoresis on a 7-M urea 20% polyacrylamide gel. This method led in our hands to better results than HPLC purification.
3. The product bands were visualized by UV-shadowing. In general one major band was observed corresponding to the oligonucleotide of the expected length, with minor bands corresponding to compounds lacking one or more nucleotides.
4. RNA from the excised band was recovered by passive elution at 4°C in 10 mM Tris–HCl buffer, pH 7, at 20°C, containing 1 mM EDTA and 25 mM NaCl. (High-yield recovery was achieved by electro-elution with a laboratory-made setup or a commercially available instrument.)
5. Samples were concentrated by ethanol precipitation by adding 0.1 volume of 3 M sodium acetate pH 5.3 and 2.5 volumes of cold ethanol. After storing overnight at −20°C followed by centrifugation at 8000×*g*, pellets were re-suspended in 50–100 μL of milliQ water.

6. Desalting was achieved either by G-25 spin columns or by dialysis using Spectra/Por MWCO: 3500 membrane tubing, in particular if buffer exchanges were required.
7. Samples were then analysed at 260 nm to determine their concentration using extinction coefficients calculated at http://www.ambion.com/techlib/misc/oligo_calculator.html.
8. Pure RNA samples of short length (<30 nt) are generally stored in water at –20°C with no significant alteration over time. Samples are also preferably aliquoted to avoid repeated cycles of freezing and thawing.

3.1.2. Synthesis of RNA by Transcription

1. Alternatively RNA molecules can be produced by in vitro transcription as described previously (28). Once produced the purification steps are the same as above.
2. RNAs produced by in vitro transcription will include a phosphate group at the 5′-end. If ^{32}P labelling is required, a phosphatase should be used first to remove the phosphate group.

3.1.3. RNA Obtained by Custom Synthesis

1. RNAs can be purchased to various suppliers. Cost will increase significantly if RNAs are ordered deprotected, purified, and desalted.
2. When nucleic acids are purchased from a supplier, always determine the concentration by measuring the absorbance at 260 nm. The quality of the sample can be rapidly checked by running a denaturing polyacrylamide gel.
3. Prepare a small size 7 M urea 20% polyacrylamide gel (20 cm × 20 cm × 1 mm thick).
4. Dissolve 1–3 μg of the RNA sample in 10 μL of a denaturing loading buffer.
5. Heat samples for 1 min at >95°C.
6. Pre-heat the gel for 30 min at 17 W, load the samples and run the electrophoresis.
7. Stain the gel in the dark for 30 min with a "Stains All" solution.
8. Wash the gel with distilled water. Pure samples should appear as a single band with a purple colouration. If samples are not at least 95% pure as quantitated by colour density, RNA should be purified again.

3.2. Binding Studies

3.2.1. Preparation of the BIAcore^{TM} 3000 Instrument

Compared with DNA oligonucleotides, RNA immobilized on surfaces can be very unstable if inappropriate experimental conditions are used. In particular, RNA can be rapidly degraded by RNases in a few minutes even at room temperature. The instrument requires an extensive cleaning procedure.

1. Set the temperature at least at 25°C.
2. Dock a maintenance sensor chip into the apparatus.
3. Disconnect the connector block from the internal microfluidic cartridge (IFC). Clean it with ethanol and distilled water. Dry it with compressed air.
4. Remove the injection needle from the delivery arm. Clean it with ethanol and distilled water. Dry it with compressed air.
5. Check the two syringes used for liquid handling on the apparatus. If necessary disconnect them from the pump and clean them. Check the O-ring and teflon seals. Exchange them if they are visibly damaged.
6. "Prime" three times with filtered autoclaved milliQ water. This procedure primes the liquid system by flushing pumps, IFC and autosampler with buffer.
7. Run twice "Desorb" as indicated by the manufacturer using as Solution 1, 0.5% (w/v) SDS in milliQ water and as Solution 2, 50 mM glycine-NaOH, pH 9.5. Desorb is a routine procedure to remove absorbed proteins from the autosampler and the IFC by flushing with SDS. The manufacturer advises to run it at least once a week. This procedure should be run systematically before any experiment with RNA.
8. "Prime" again the IFC with filtered autoclaved milliQ water.
9. Run "Sanitize" procedure using 1% sodium hypochlorite. Sanitize is a routine to disinfect the autosampler and the IFC that the manufacturer recommends to run at least once a month. From our own experience we advise to run it systematically prior to any experiment with RNA.
10. "Prime" again the IFC with filtered autoclaved milliQ water.
11. Change buffer: "Prime" once with the ready-to-use HBS-EP buffer (BIAcore) (*see* **Notes 3** and **4**).

3.2.2. Preparation of the Sensor Chips

1. Before inserting a new sensor chip into the instrument always inspect its surface by removing the chip from its protective cassette. Gently blow air on the surface to remove dust if necessary.
2. Dock a streptavidin-coated sensor chip into the apparatus.
3. Set the temperature of the instrument to the desired temperature. The TAR RNA–RNA aptamer complex was investigated at 23°C.
4. "Prime" twice with HBS-EP buffer.
5. Remove non-covalently bound streptavidin from the surface following the manufacturers' instructions. Start a sensorgram at 5 μL/min on one flow cell. For SA sensor chips (BIAcore) inject three 5-μL pulses ("INJECT" command) of a 50-mM

NaOH/1 M NaCl solution followed by a 10-μL pulse of HBS-EP buffer. For SAD, SAHC and SAP sensor chips (XanTec Bioanalytics), run 3–5 interaction/regeneration cycles before collecting data. Alternatively, inject one 5-μL pulse ("INJECT" command) of a freshly prepared 20 mM NaOH solution followed by a 10-μL pulse of HBS-EP buffer. Typically the signal will decrease reflecting removal of weakly bound proteins from the surface.

3.2.3. Preparation of the TAR HIV-1 Target and Immobilization on the Surface

1. 100 μL of 3′ biotinylated TAR were prepared in HBS-EP buffer at a final concentration of 50 nM in 1.5-mL Eppendorf DNA LoBind tubes (**Note 5**).
2. Heat the solution for 90 s at >95°C, put on ice for 5 min, leave at room temperature for 10 min and spin the tube 1 min at 5000 rpm in a microfuge.
3. Put the tube on the autosampler.
4. Run a sensorgram on one flowcell at 5 μL/min.
5. Wait at least 1 min to check for stability of the baseline.
6. Inject 5 μL of the biotinylated RNA ("QUICKINJECT" command).
7. Stop the injection when the desired amount of target immobilized on the surface is reached (inject again if necessary) (*see* **Note 6**).
8. Wash the autosampler and the IFC by injecting twice as much HBS-EP buffer as biotinylated target.
9. Stop the sensorgram and change buffer.
10. "Prime" three times with running buffer (Step 8 of **Section 2.2**).
11. Leave the instrument on standby.

3.2.4. Preparation of the RNA Aptamer Sample

1. Prepare the RNA aptamer sample by diluting a volume of the stock solution (100–200 μM) directly in the running buffer. Typically, for preliminary experiments, 100 μL solutions at 400–600 nM were prepared. For kinetic analysis at least three different concentrations including a saturating value should be prepared and injected (*see* following sections).
2. Heat the solution at >95°C for 90 s, put on ice for 5 min, leave at room temperature for 10 min and spin the tube 1 min at 5000 rpm in a microfuge (*see* **Note 7**).
3. Put the tube on the autosampler.

3.2.5. Preliminary Experiments

The goal is to confirm that the selected aptamers do interact with the immobilized target and to check how fast they bind and dissociate (*see* **Note 8**).

1. Start a sensorgram flowing buffer at 20 μL/min in the reference and TAR-functionalized cells.
2. Inject 20–50 μL of RNA aptamer at 400 nM prepared as indicated in the previous section. Use the "QUICKINJECT" command to reduce consumption of the sample.

3.2.6. Setting Regeneration Conditions

Several binding experiments can be performed with the same sensor chip if a regeneration step is run to dissociate the RNA–RNA complex instantaneously. If dissociation is rapid and occurs within a few minutes, there is no need to use chemical compounds to regenerate the surface. In this case the sensor chip will last longer compared to functionalized surfaces that need harsh regeneration conditions. RNA–RNA complexes may prove to be highly stable and often display slow dissociation phases with rate constants of 10^{-4} s^{-1} or less. Regeneration is required. A good control of this step is crucial to collect high-quality data. Ideally the regenerating compound should completely dissociate the complex without altering the immobilized target. This is sometimes difficult to achieve with aptamers targeting RNA molecules. On the one hand such complexes can be very stable; on the other hand RNA molecules are naturally unstable and prone to hydrolysis. Some loss of the binding activity of the target either due to incomplete regeneration or degradation is acceptable under particular conditions (*see* **Notes 9** and **10**).

To validate a regenerating compound, several (at least three) binding and regeneration cycles must be run.

1. Set the instrument to the temperature and to the flow rate that will be used for the binding experiments (23°C and 20 μL/min, respectively, were used for the TAR–aptamer complex).
2. Immobilize the RNA target on a sensorchip as described in **Section 3.2.3**.
3. Prepare the RNA binding partner as described in **Section 3.2.4**. Prepare a volume large enough (100–200 μL) so at least three injections can be performed. Place the tube on one rack on the sample holder.
4. Prepare 1 mL of 10 mM NaOH and 1 mL of 10 mM HCl in milliQ water. Most of the regenerating compounds are chemicals that can be prepared in milliQ water. Place these tubes on the sample holder.
5. Write an automatic method if available for the instrument. For the BIAcore™ 3000 the BIAcore Method Definition Language can be used to run an automatic procedure (*see* **Section 3.2.7** for more details and the BIAcore™ 3000 Instrument Handbook). The method should contain a loop for several injections of the binding partner followed by the

regeneration solution. An injection of running buffer of same duration is systematically performed afterwards to clean the IFC and the tubing of the instrument. For instance, on a BIAcore™ 3000 apparatus, the loop will contain the following main commands:

FLOW 20 (the flow rate of the sensorgram is set to 20 μL/min).

WAIT 04:00 (wait for 4 min before the aptamer is injected to control that the baseline is stable).

KINJECT r1d1 20 200 (20 μL of a 400-nM aptamer solution is injected from Rack 1 position d1 with a dissociation phase of 200 s, the same solution is used for every cycle).

QUICKINJECT r2f3 20 (20 μL of regenerant is injected from Rack 2 position f3).

QUICKINJECT r2f4 20 (20 μL of buffer is injected from Rack 2 position f4).

Figure 19.2 shows an example of sensorgrams obtained using 10 mM NaOH or 10 mM HCl to dissociate the TAR RNA–RNA aptamer complex on a SA sensor chip (BIAcore). Six interaction/regeneration cycles of 1150 s duration each are reported. Using 10 mM NaOH (**Fig. 19.2a**) the binding capacity of the target slightly increases then decreases. If 10 mM HCl is injected (**Fig. 19.2b**). The SPR signal decreases by 16 RU but remains constant for the next cycles.

6. The choice of NaOH as regenerant might be surprising because of the known sensitivity of RNA to basic pH conditions. The results demonstrate that at the time scale of the experiment (about 2 h), the immobilized RNA tolerates well the mild basic conditions (1 min pulse, low concentration, room temperature). The results obtained with HCl indicate that somehow part of the target was either degraded or that part of the complex was not accessible to the regeneration solution.
7. XanTec Bioanalytics advises running several cycles of binding and regeneration to stabilize the surface before collecting the data. We did observe using either BIAcore or XanTec Bioanalytics sensor chips that for some aptamer–ligand complexes data quality increased after the first cycles. RNA–RNA complexes that require magnesium ions for stability can be also regenerated with 20–50 mM solutions of EDTA.
8. Alternatively, nucleic acid complexes can be dissociated with a mixture of 40% formamide, 3.6 M urea and 30 mM EDTA prepared in milliQ water.

3.2.7. BIAcore Operation

The instrument can be controlled from an interfaced computer. Automatic operation can be performed through methods written

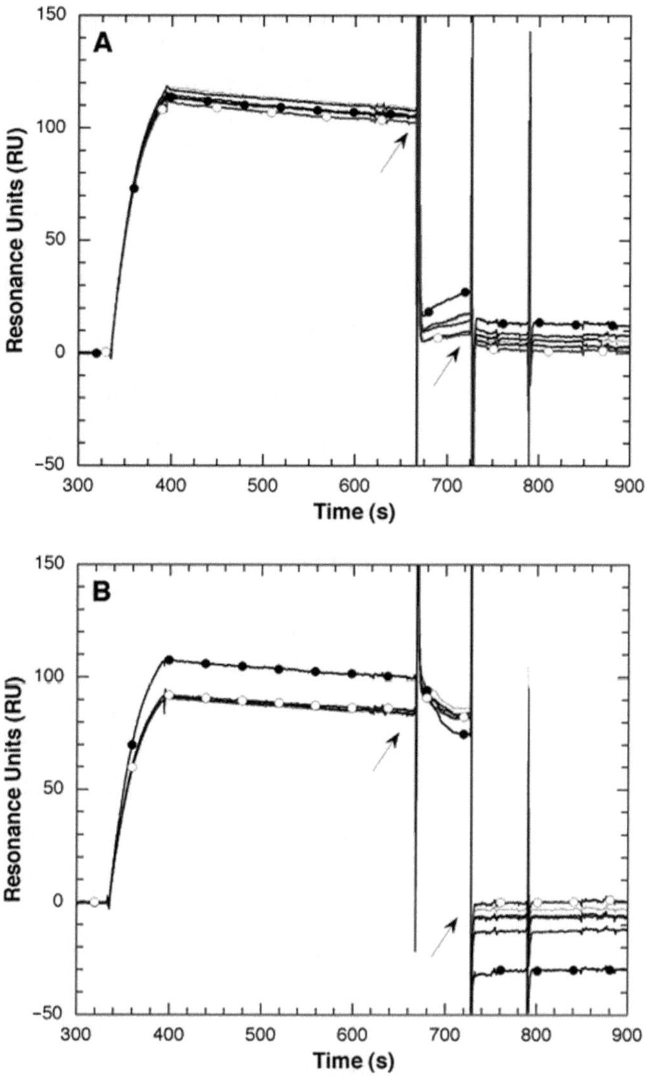

Fig. 19.2. Cycles of binding and regeneration analysed with the TAR-R06 complex. Effect of 10 mM sodium hydroxide (**a**) and hydrochloric acid (**b**) solutions on the dissociation of the target–aptamer complex. Six cycles of binding and regeneration were performed on a SA sensor chip (BIAcore) functionalized with biotinylated TAR. The R06 aptamer was prepared in the running buffer at 400 nM and was injected over the TAR surface at 20 μL/min, at 23°C. After the slow dissociation phase, the regeneration solutions were injected for 1 min (*first arrow*). A 1-min pulse of running buffer was then performed to rinse the micro-fluidic cartridge and the surface (*second arrow*). The SPR response from the blank flow cell was subtracted from that obtained with TAR-R06. The reported sensorgrams were normalized in *y*-axis only and were not double-referenced. Sensorgrams with filled and open circles correspond to the first and last cycles, respectively.

in a text file format. The simplest method contains two blocks, the DEFINE APROG and the MAIN blocks. The DEFINE APROG describes the sequence of operations for a run. The MAIN block defines the sequence of operations in the whole method. The method written below was used to operate the apparatus for the single-cycle kinetics experiments described in **Section 3.3**. This method provides a way to determine the kinetic parameters within one single binding cycle by injecting sequentially increasing amounts of the analyte (26, 29).

1. A sequence of operations named “Buffer” is entered to inject buffer for referencing the data as follows.

   DEFINE APROG Buffer

   MODE –d0.1–15000 (sensorgrams are recorded at 5 Hz data collection rate)

   FLOW 20 (the flow is set to 20 μL/min)

   FLOWPATH 3,4 (set the buffer flow path to channels 3 and 4. The biotinylated target was immobilized on channel 4)

   WAIT 02:00 (wait for 2 min before the aptamer is injected)

   KINJECT r1d6 30 350 (inject 30 μL of running buffer from position r1d6, waits 350 s for the dissociation phase)

   KINJECT r1d6 30 350 (as above)

   KINJECT r1d6 30 350 (as above)

   INJECT r1d5 20 (inject 20 μL of regeneration solution from position r1d5)

   QUICKINJECT r1d6 20 (injects 20 μL of running buffer from position r1d6)

   WASH l (wash the loop of the IFC)

   WASH s (wash the sample line of the IFC)

   WAIT 02:00 (wait for 2 min before the end of the program)

   END

2. Enter a second sequence of operations named “Interaction” as follows to inject three aptamer samples sequentially in order of increasing concentration.
   DEFINE APROG Interaction. The operations are similar except that the RNA samples were injected (KINJECT command) from three different rack positions.

3. Define the MAIN block to set the temperature of the instrument, to detect channels and to run the sequence of operations to inject buffer and the RNA samples.

   MAIN

   TEMPERATURE 23 (set the temperature of the instrument at 23°C)

DETECTION 4-3 (subtract the reference channel)

APROG Buffer (run the program to inject buffer)

APROG Interaction (run the program to inject the aptamer samples)

APPEND STANDBY (leave the instrument on standby flowing buffer)

END

4. For repetitive operations and handling more samples, sophisticated and flexible methods can be written using variables (*see* BIAcore^{TM} 3000 Instrument Handbook for more details).

3.2.8. Analysing the Data

The evaluation program, BiaEval 4.1, provided by the manufacturer is used to prepare the recorded sensorgrams for the kinetic analysis (*see* BiaEvaluation Software Handbook for more details).

1. The sensorgrams are normalized in *x*- and *y*-axis to account for small variations of the baseline and of the start injection points.
2. The sensorgrams are double-referenced to remove instrument noise and the buffer contribution to the signal: the referenced sensorgram (channel 4 minus channel 3) obtained when buffer is injected over channel 3 (reference) and channel 4 (target) is subtracted from that obtained when the aptamer samples are injected over the same channels (30). Despite the normalization and the double referencing of the data, small spikes at the beginning and at the end of the injections may still be present but will not affect the results (26) (*see* **Note 11**).
3. The sensorgrams are fitted with a Langmuir binding model. This corresponds to the simplest model for 1:1 interaction between the injected RNA aptamer and the immobilized TAR RNA target.
4. The association and dissociation rate constants, k_a and k_d, respectively, are determined from direct curve fitting of the sensorgrams with BiaEval 4.1 (*see* BiaEvaluation software handbook for more details) according to equations [1] and [2], for the association and the dissociation phases, respectively:

$$\frac{d\mathrm{RU}}{dt} = k_a[\mathrm{Aptamer}]\mathrm{RU}_{\max} - (k_a[\mathrm{Apamer}] + k_d)\mathrm{RU} \quad [1]$$

$$\frac{d\mathrm{RU}}{dt} = -k_d\mathrm{RU}_{t0} \cdot e^{-k_d(t-t_0)} \quad [2]$$

where RU is the signal response; RU_{max}, the maximum response level; RU_{t0}, the response at the beginning of the dissociation

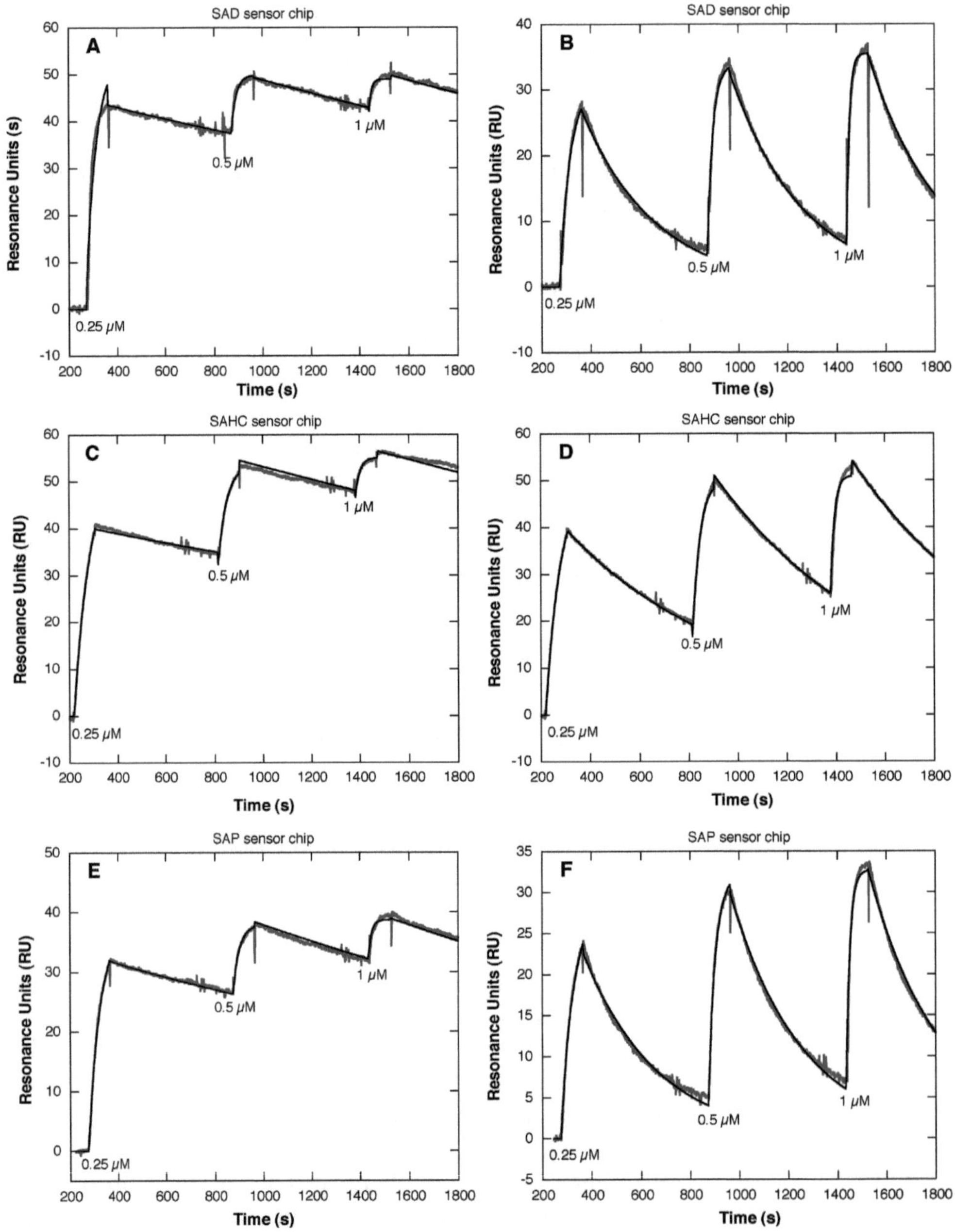

Fig. 19.3. Kinetic analysis of RNA aptamers binding to TAR immobilized on three different sensor chips. Eighty-six to hundred RU of biotinylated TAR were immobilized onto streptavidin-coated sensor chips from XanTec Bioanalytics (*see* **Table 19.1**). The binding experiments were performed at 23°C. Hairpin aptamers, prepared in the running buffer, were injected over the surface at 20 μL/min using the single-cycle kinetics method (*see* main text). The sensorgrams were fitted assuming a pseudo-first order kinetic model of the aptamer binding to TAR. The *gray curves* represent the recorded data and the black one the fit to a kinetic titration dataset of three analyte injections (*see* main text for details). Injection of the unmodified (**a**) and chemically modified (**b**) aptamers (0.25, 0.5 and 1 μM) across the TAR-coated surface on a SAD50m sensor chip. Injection of the unmodified (**c**) and chemically modified (**d**) aptamers (0.25, 0.5 and 1 μM) across the TAR-coated surface on a SAHC30m sensor chip. Injection of the unmodified (**e**) and chemically modified (**f**) aptamers (0.25, 0.5 and 1 μM) across the TAR-coated surface on a SAP sensor chip. The TAR surface was regenerated with 10 mM NaOH solutions.

phase; and [Aptamer], the molar concentration of the injected RNA aptamer. The dissociation equilibrium constant, K_D, is calculated as k_d/k_a.

3.3. Kinetic Analysis of the RNA aptamer–TAR Complex

1. The biotinylated TAR RNA was prepared in the running buffer and was immobilized as described in **Section 3.2.3** on streptavidin-functionalized sensor chips from XanTec Bioanalytics. SAHC30m, SAD50m and SAP sensor chips were functionalized with 100 RU, 86 RU and 97 RU of this target, respectively.
2. The RNA aptamer selected against TAR and the aptamer chemically modified with a 2′-*O*-methyl residue at position G6 (**Fig. 19.1**) were prepared in the running buffer (Item 8 of **Section 2.2**) These aptamers were injected sequentially over the TAR surface in order of increasing concentration (250, 500, and 1000 nM) using the single-cycle kinetics method described in **Section 3.2.7**. The results are reported in **Fig. 19.3**.
3. The obtained sensorgrams were fitted by direct curve fitting assuming a 1:1 interaction model, as described in **Section 3.2.8**. The kinetic parameters, k_a and k_d, and the binding equilibrium constant, K_D, are reported in **Table 19.1**.
4. Binding kinetics obtained using streptavidin-functionalized sensor chips from XanTec Bioanalytics were compared to those obtained with the SA sensor chip from BIAcore from a

Table 19.1
Equilibrium and rate constants of TAR-R06 complexes

Sensor chip	RU of TAR immobilized[a]	Complexes	k_a (10^4 M^{-1} s^{-1})	k_d (10^{-3} s^{-1})	K_D(nM)
SA (3D)[b]	80	TAR-R06	9.2 ± 0.6	0.3 ± 0.1	3.1 ± 0.1
		TAR-R06*[c]	7.9 ± 1.5	3.3 ± 0.1	44.9 ± 7.4
SAD50m (3D)	86	TAR-R06	7.7 ± 1.2	0.3 ± 0.1	4.4 ± 0.9
		TAR-R06*	7.3 ± 0.3	3.4 ± 0.1	49.3 ± 3.2
SAHC30m (3D)	100	TAR-R06	6.3 ± 0.4	0.4 ± 0.1	6.6 ± 1.6
		TAR-R06*	7.3 ± 0.3	3.4 ± 0.1	49.3 ± 3.2
SAP (2D)	97	TAR-R06	6.7 ± 1.0	0.3 ± 0.1	4.6 ± 0.2
		TAR-R06*	4.8 ± 1.1	3.1 ± 0.3	66.1 ± 7.5

[a]The value represents the amount of biotinylated TAR, in resonance units (RU), immobilized on the sensor chip. [b]Results are from reference *25*. [c]R06* is a chemically modified version of the R06 aptamer with a 2′-*O*-methyl residue at position G6 (*see* **Fig. 19.1**). k_a, k_d and K_D are the association and dissociation rate constants and the dissociation equilibrium constant, respectively. Values are the average and standard deviation of at least two independent experiments

previous work (25). The results (**Table 19.1**) show that neither the geometry (2D versus 3D) nor the chemical nature of the matrix has a significant influence on the binding parameters for the aptamer–target complex formed between TAR and the parent or the modified RNA aptamer (*see* **Note 12**).

4. Notes

1. The BIAcore^{TM} 3000 is a 4-channel instrument with adjustable temperature between 5 and 40°C, at ±0.1°C. For years BIAcore had the monopoly of SPR instruments. Two- or three-channel instruments are now available from other companies (Reichert Life Sciences, SensiQ). Multiplex apparatus from ForteBio, Bio-Rad, Ibis Technologies, Genoptics and BIAcore allow direct monitoring of at least eight to more than one hundred interactions simultaneously.
2. Biotinylated RNA targets can be easily immobilized on commercially available streptavidin-functionalized gold surfaces. We used sensor chips either from BIAcore or XanTec Bioanalytics companies. The following sensor chips were used to investigate RNA–RNA interactions:
   i. Standard SA (BIAcore). The attached matrix on the gold surface is a three-dimensional (3D) carboxymethylated dextran pre-functionalized with streptavidin.
   ii. SAD50m (XanTec Bioanalytics) has a 3D matrix made of dextran-based hydrogel 50 nm thick and medium chain density, pre-immobilized with streptavidin.
   iii. SAHC30m (XanTec Bioanalytics) has a 3D matrix made of a linear polycarboxylate hydrogel 30 nm thick and medium density; pre-immobilized with streptavidin. Compared to dextrans with low level of branching, these sensor chips are better defined and are expected to show higher signal-to-noise ratio, improved diffusion characteristics and higher chemical stability.
   iv. SAP (XanTec Bioanalytics) is a sensor chip with covalently immobilized streptavidin on a two-dimensional (2D) carboxymethyldextran surface.
   More information on streptavidin-coated surfaces is available at http://www.biacore.com and http://www.xantec.com.
3. Despite intensive cleaning of the instrument, decrease of the binding capacity of the surface functionalized with biotinylated RNA targets is systematically observed. How fast this

occurs depends on how clean were the instrument, the solutions and the materials (plastic vials and tips). It may also depend on the RNA structure and length. Magnesium ions will also negatively influence the integrity of the RNA. Stabilization can be improved by performing the experiments at low temperature (5–10°C) and by decreasing the magnesium ion concentration below 0.5 mM. Immobilization of the RNA target by capturing rather than by direct binding to the surface is also an alternative (*see* **Note 4** for details).

4. Capture of the RNA target through a biotinylated DNA anchor (i.e. an oligonucleotide complementary to part of the target sequence) is an alternative to direct binding. If the target does not display a single-stranded sequence at either end, at least 20-nt long, the RNA molecule can be synthesized with an extension at its 5′ or 3′ end, for instance a poly$(rA)_{25}$, containing a single G nucleotide to lock the anchor at a fixed position. The added sequence should not interfere with the proper folding of the target. Immobilization by capturing will help regenerating the surface when the interaction between the captured target and the injected partner is very stable. A 1-min pulse of 10 mM NaOH will easily dissociate the captured target from the DNA anchor. Capturing is more sample consuming than direct immobilization.

5. RNA aptamers identified against RNA targets are expected to bind with affinity in the low nanomolar range. This implies that aptamer will be injected at low concentration over the immobilized RNA target. Depending on the buffer and the plastic tubes used for preparing the samples, at concentration below 100 nM, part of the aptamer may stick on the plastic walls in particular if the tubes are centrifugated at high speed (14,000 rpm in a microfuge) leading to overestimation of the aptamer concentration really injected. The use of low binding tubes as DNA LoBind 1.5-mL tube from Eppendorf can prevent this artefact.

6. If the target is of the highest purity (100% biotin-TEG labelled), 100–200 RU are reached in less than 2 min at 5 μL/min with a 50-nM solution of biotinylated RNA. The sample concentration should not exceed 50 nM to control low levels of immobilization. In general manufacturers advise to block free streptavidin by injecting free biotin. Binding experiments performed at low or medium level of immobilization, performed with or without this blocking step, do not show any difference. If free biotin is injected, do not use concentration above 200 nM, otherwise the IFC will be contaminated requiring an intensive cleaning

procedure (injection of free streptavidin will help removing free biotin from the IFC). Do not perform blocking steps for preliminary experiments. This will allow immobilization of the biotinylated target on the same flow cell to increase density or, for instance, immobilization of a new biotinylated RNA molecule in the flow cell that was used as a reference.

7. Preparation of RNA aptamers and RNA in general in buffers containing magnesium ions is always subject to discussion. At high temperature RNA is prone to hydrolysis. But integrity actually strongly depends on the RNA length, on its sequence and its structure. RNA solutions can be prepared by two different ways. Either the RNA is diluted first in 4/5 volumes of milliQ water and heated and then 1/5 volume of 5× running buffer is added or the RNA is directly prepared and heated in the running buffer. Irrespective of the choice, keep the conditions used during the in vitro selection, both for the RNA target and the RNA aptamer. Changes in the heating procedure may affect the structure of the RNAs and their binding properties. Direct heating in the running buffer of a sample diluted at least 100-fold from the RNA stock solution will decrease artefacts due to refractive index differences between the running buffer and the injected sample. One good alternative is to prepare and to heat the RNA libraries used for SELEX directly in the buffer even if it contains magnesium but setting the temperature below 65°C, for no more than 3 min. Long RNAs prepared in this manner displayed higher SPR responses suggesting that heating in the presence of magnesium ions was thermodynamically more favourable to fold the RNA in the structure displaying the highest affinity for the immobilized target. Fear to cleave RNA is justified. Test different conditions to analyse how the samples behave in the presence of magnesium ions.
8. Diffusion of the analyte from the bulk solution to the sensor chip surface can be a limiting factor for determining accurately the rate constants. Mass transport limitations are generally observed when the binding of the analyte to the immobilized target is very fast (>10^5 to 10^6 M^{-1} s^{-1}) and when the surface density is high. Mass transport-controlled binding is used for measuring analyte concentration (*see* BIAtechnology and BIAapplication handbooks for details). In preliminary experiments, always run sensorgrams at different flow rates and surface densities. If diffusion is limiting, the binding kinetics will depend on these factors. Choose appropriate flow rate and surface density to overcome mass transport effects. If this cannot be eliminated, the BIAevaluation

software includes kinetic models to fit the data with mass transport terms.

9. A regeneration solution that keeps constant the binding capacity of the surface (RU_{max}) should be used if sensorgrams collected from several binding/regeneration cycles of the aptamer–ligand interactions are analysed by global fitting. This will avoid poor fitting of the data to the kinetic model due to variations of the RU_{max}. If the kinetics of aptamer–ligand complexes are investigated by the single-cycle kinetics method, slight loss of activity between experiments due to incomplete regeneration or degradation of the immobilized target, as it can be observed with RNA targets flowed with running buffers containing magnesium ions, has no influence on the results. This is obviously true if degradation is also negligible during the time window of the experiment. In addition the binding kinetics should not depend on the immobilization level. This point can be easily addressed by using two levels of ligand density.
10. Old buffers left at room temperature for days and an instrument that has not been cleaned properly will be responsible for a rapid degradation (within minutes) of the immobilized RNA targets due to the presence of RNases. If interaction between the aptamer and the ligand is expected because it was previously demonstrated by another technique and no signal is observed with SPR, the preparation of the instrument and the buffers are the first things to address. Clean the instrument as described in **Section 3.1** and prepare fresh solutions. If with a newly immobilized RNA target the sensorgram remains flat after these procedures, increase 10-fold the concentration of the injected partner to ensure that the concentration used for the preliminary experiment was not too low. Lastly, revert the experiment by immobilizing the RNA aptamer instead of the RNA target and inject the corresponding partner.
11. Spikes at the beginning and at the end of the association phase and air bubbles can really affect data quality. Running buffers should always be filtered and degassed. Some of the last generation instruments include an integrated buffer degasser. Spikes observed in referenced sensorgrams result mainly from refractive index differences between the injected solutions and the running buffer. This typically occurs when the injected sample is prepared from a stock solution the composition of which differs from that of the running buffer. Even if a high dilution factor is applied to the stock for preparing the sample in the running buffer,

stock solutions containing polyols such as glycerol will generate large spikes. Dialysis of the stock solutions against the running buffer is the best way to avoid refractive index changes when the sample is injected. Spikes can be also reduced by increasing the number of points collected per unit time during the sensorgram and by increasing the flow rate.

12. The choice of the sensor chip can be critical depending on the size and the nature of the injected analytes. Generally 2D coating will allow immobilization of ligand at low densities. Primary applications include binding studies with whole organisms (viruses and bacteria), cells and kinetic analysis with high molecular weight analytes. 3D coating will allow multilayer immobilization which leads to signal amplification. 3D hydrogel matrices exist in different thicknesses and porosities. The matrix should not prevent diffusion of the injected analyte. The smaller the analyte, the thicker and denser the hydrogel structure can be. Most of the available surfaces are negatively charged at physiological pH. This negative field can be responsible for non-specific interactions of positively charged compounds to the surface. We never observed non-specific binding with DNA, RNA or 2′-*O*-methyl oligonucleotides. A biotinylated nucleic acid desalted on a G-25 spin column and prepared in milliQ water at concentration lower than 50 nM cannot be immobilized on a streptavidin sensor chip! It must be prepared in solution of minimum ionic strength to shield the negative charges of the backbone. In contrast, non-specific binding to the surface is often observed with basic proteins prepared in buffers at physiological pH. Coatings of decreased thickness and density are recommended to reduce the negative charge of the sensor chip surface. In some case increasing the ionic strength is sufficient to avoid non-specific binding. Consult http://www.biacore.com and http://www.xantec.com prior ordering sensor chips.

Acknowledgements

We are grateful to the structural biology facility (UMS 3033/US001 CNRS/Inserm) of the Institut Européen de Chimie et Biologie (Pessac, France) for access to the BIAcore instrument that was acquired with the support of the “Conseil Régional d’Aquitaine.” We thank Nathalie Pierre for oligonucleotides synthesis.

References

1. Ellington, A. D., and Szostak, J. W. (1990) In vitro selection of RNA molecules that bind specific ligands. *Nature*, **346**, 818–822
2. Tuerk, C., and Gold, L. (1990) Systematic evolution of ligands by exponential enrichment: RNA ligands to bacteriophage T4 DNA polymerase. *Science*, **249**, 505–510
3. Gopinath, S. C. (2007) Methods developed for SELEX. *Anal Bioanal Chem*, **387**, 171–182
4. Andréola, M. L., Pileur, F., Calmels, C., Ventura, M., Tarrago-Litvak, L., Toulmé, J. J., and Litvak, S. (2001) DNA aptamers selected against the HIV-1 RNase H display in vitro antiviral activity. *Biochemistry*, **40**, 10087–10094
5. Pileur, F., Andréola, M. L., Dausse, E., Michel, J., Moreau, S., Yamada, H., Gaidamakov, S. A., Crouch, R. J., Toulmé, J. J., and Cazenave, C. (2003) Selective inhibitory DNA aptamers of the human RNase H1. *Nucleic Acids Res*, **31**, 5776–5788
6. Mayer, G. (2009) The chemical biology of aptamers. *Angew Chem Int Ed Engl*, **48**, 2672–2689
7. Dausse, E., Da Rocha Gomes, S., and Toulmé, J. J. (2009) Aptamers: a new class of oligonucleotides in the drug discovery pipeline? *Curr Opin Pharmacol*, **9**, 602–607
8. Tombelli, S., and Mascini, M. (2009) Aptamers as molecular tools for bioanalytical methods. *Curr Opin Mol Ther*, **11**, 179–188
9. Ruckman, J., Green, L. S., Beeson, J., Waugh, S., Gillette, W. L., Henninger, D. D., Claesson-Welsh, L., and Janjic, N. (1998) 2′-Fluoropyrimidine RNA-based aptamers to the 165-amino acid form of vascular endothelial growth factor (VEGF165). Inhibition of receptor binding and VEGF-induced vascular permeability through interactions requiring the exon 7-encoded domain. *J Biol Chem*, **273**, 20556–20567
10. Sefah, K., Phillips, J. A., Xiong, X., Meng, L., Van Simaeys, D., Chen, H., Martin, J., and Tan, W. (2009) Nucleic acid aptamers for biosensors and bio-analytical applications. *Analyst*, **134**, 1765–1775
11. Cheng, A. K., Sen, D., and Yu, H. Z. (2009) Design and testing of aptamer-based electrochemical biosensors for proteins and small molecules. *Bioelectrochemistry*, 77, 1–12
12. Nutiu, R., and Li, Y. (2005) Aptamers with fluorescence-signaling properties. *Methods*, **37**, 16–25
13. Wu, Z. S., Zhang, S., Zhou, H., Shen, G. L., and Yu, R. (2010) Universal aptameric system for highly sensitive detection of protein based on structure-switching-triggered rolling circle amplification. *Anal Chem*, **82**, 2221–2227
14. Basnar, B., Elnathan, R., and Willner, I. (2006) Following aptamer-thrombin binding by force measurements. *Anal Chem*, **78**, 3638–3642
15. Yao, C., Qi, Y., Zhao, Y., Xiang, Y., Chen, Q., and Fu, W. (2009) Aptamer-based piezoelectric quartz crystal microbalance biosensor array for the quantification of IgE. *Biosens Bioelectron*, **24**, 2499–2503
16. Löfås, S., and Johnsson, B. (1990) A novel hydrogel matrix on gold surfaces in surface plasmon resonance sensors for fast and efficient covalent immobilization of ligands. *J Chem Soc Chem Commun*, **21**, 1526–1528
17. Fägerstam, L. G., Frostell, A., Karlsson, R., Kullman, M., Larsson, A., Malmqvist, M., and Butt, H. (1990) Detection of antigen-antibody interactions by surface plasmon resonance. Application to epitope mapping. *J Mol Recognit*, **3**, 208–214
18. Rich, R. L., and Myszka, D. G. (2010) Grading the commercial optical biosensor literature-class of 2008: 'the mighty binders'. *J Mol Recognit*, **23**, 1–64
19. Misono, T. S., and Kumar, P. K. (2005) Selection of RNA aptamers against human influenza virus hemagglutinin using surface plasmon resonance. *Anal Biochem*, **342**, 312–317
20. Ducongé, F., Di Primo, C., and Toulmé, J. J. (2000) Is a closing "GA pair" a rule for stable loop-loop RNA complexes? *J Biol Chem*, **275**, 21287–21294
21. Darfeuille, F., Arzumanov, A., Gait, M. J., Di Primo, C., and Toulmé, J. J. (2002) 2′-O-methyl-RNA hairpins generate loop-loop complexes and selectively inhibit HIV-1 Tat-mediated transcription. *Biochemistry*, **41**, 12186–12192
22. Darfeuille, F., Hansen, J. B., Orum, H., Di Primo, C., and Toulmé, J. J. (2004) LNA/DNA chimeric oligomers mimic RNA aptamers targeted to the TAR RNA element of HIV-1. *Nucleic Acids Res*, **32**, 3101–3107
23. Di Primo, C., and Lebars, I. (2007) Determination of refractive index increment ratios for protein-nucleic acid complexes by surface plasmon resonance. *Anal Biochem*, **368**, 148–155
24. Aldaz-Carroll, L., Tallet, B., Dausse, E., Yurchenko, L., and Toulmé, J. J. (2002) Apical loop-internal loop interactions: a

new RNA-RNA recognition motif identified through in vitro selection against RNA hairpins of the hepatitis C virus mRNA. *Biochemistry*, **41**, 5883–5893
25. Lebars, I., Legrand, P., Aimé, A., Pinaud, N., Fribourg, S., and Di Primo, C. (2008) Exploring TAR-RNA aptamer loop-loop interaction by X-ray crystallography, UV spectroscopy and surface plasmon resonance. *Nucleic Acids Res*, **36**, 7146–7156
26. Di Primo, C. (2008) Real time analysis of the RNAI-RNAII-Rop complex by surface plasmon resonance: from a decaying surface to a standard kinetic analysis. *J Mol Recognit*, **21**, 37–45
27. Ducongé, F., and Toulmé, J. J. (1999) In vitro selection identifies key determinants for loop-loop interactions: RNA aptamers selective for the TAR RNA element of HIV-1. *RNA* **5**, 1605–1614
28. Watrin, M., Dausse, E., Lebars, I., Rayner, B., Bugaut, A., and Toulmé, J. J. (2009) Aptamers targeting RNA molecules. *Methods Mol Biol*, **535**, 79–105
29. Karlsson, R., Katsamba, P. S., Nordin, H., Pol, E., and Myszka, D. G. (2006) Analyzing a kinetic titration series using affinity biosensors. *Anal Biochem*, **349**, 136–147
30. Myszka, D. G. (1999) Improving biosensor analysis. *J Mol Recognit*, **12**, 279–284

Chapter 20

Practical Considerations for Analyzing Antigene RNAs (agRNAs): RNA Immunoprecipitation of Argonaute Protein

Jacob C. Schwartz and David R. Corey

Abstract

Target validation for small RNAs in cells can be a confusing task wrought with pitfalls and false leads. One technique for validating in vivo targets of small RNAs is immunoprecipitation of target RNAs using antibodies again the RNAi machinery. Antigene RNAs (agRNAs) regulate transcription in human cells using machinery from the RNAi regulatory pathway – namely argonaute proteins. Here we describe a technique for validating targets of agRNAs using RNA immunoprecipitation with antibodies against human argonaute proteins. This technique can be used to detect interactions of argonaute proteins in the cell nucleus with their targets, lowly expressed noncoding RNA transcripts.

Key words: Noncoding RNA, RNA immunoprecipitation, argonaute, antigene RNAs, transcriptional gene silencing.

1. Introduction

Antigene RNAs (agRNAs) are duplex RNAs that modulate gene expression (1–5). Depending on the identity and basal expression level of their target gene, agRNAs can cause either gene activation (4, 5) or gene silencing (1, 2, 5). They are chemically and structurally identical to standard siRNAs used for gene silencing experiments, but are not complementary to their target gene's mRNA. Instead, we have observed that they can bind to noncoding transcripts that overlap gene promoters (1, 2, 4, 5) or regions beyond the 3′ termini of genes (6).

Double-stranded RNAs (agRNAs or siRNAs) can modulate gene expression by recognition of their intended targets, but it is also important to recognize that observed phenotypes

J. Goodchild (ed.), *Therapeutic Oligonucleotides*, Methods in Molecular Biology 764,
DOI 10.1007/978-1-61779-188-8_20, © Springer Science+Business Media, LLC 2011

can be caused by "off-target" interactions. These unintended phenotypes can confound experiments and lead to mistaken conclusions (7–9). Guarding against off-target effects is especially important when agRNAs are involved because their mechanism and use have received far less study than have siRNAs.

Argonaute (AGO) proteins are involved in RNA interference (10, 11). There are four variant AGO proteins in mammalian cells (AGO1-4) and we have observed that AGO2 is the best candidate for involvement in agRNA-mediated gene silencing (3) or activation (12). We have used RNA immunoprecipitation (RIP) (13) (*see* **Note 1**) to demonstrate that addition of agRNAs to cells causes recruitment of AGO protein to noncoding transcripts that overlap the gene target for the agRNA (3, 5). Thus, RIP provides important support for believing that agRNA-mediated expression is not an off-target effect and is a useful experimental tool for probing mechanism.

Here we describe using RIP as a tool to validate agRNAs and investigate their mechanism of action. This protocol can be simply modified to (1) immunoprecipitate with different antibodies to study other proteins of interest or (2) use whole-cell lysates to study cytoplasmic interactions. Demonstrating that the addition of a duplex RNA to cells mediates the interaction with an RNA transcript at the target gene provides an important piece of evidence supporting belief that gene modulation by agRNAs is likely not an off-target effect.

1.1. Design of agRNAs

agRNAs, by definition, are not complementary to the mRNA for their target gene. The first step in their design, therefore, is to collect all published information on RNA transcription start and termination sites. This information is then supplemented by examination of sequence databases including NCBI, USC genome browser, DBTSS, and Fantom4 (*see* **Note 2**). To supplement this information, 5′ and 3′ rapid amplification of complementary DNA ends (RACE) can be used to confirm the transcription start and termination sites in the experimental cell line of interest. Especially for genes with TATA-less promoters, investigators should expect to observe multiple transcription start sites. Quantitative reverse transcriptase PCR using multiple primer sets targeting upstream and downstream of a presumed transcription start site or terminus provides another simple check for the boundaries of mRNA. Greatly decreased RNA levels will be observed as primers are designed beyond the boundaries of mRNA.

Once the boundaries of the target gene mRNA are clearly established, agRNAs can be designed. We typically design six to ten agRNAs to maximize the likelihood that one or two will successfully modulate gene expression. Typically we design some agRNAs to overlap the most upstream transcription start site. Others are designed to sequences further upstream. Target sequences are chosen to, as much as possible, possess a balanced AT/CG content. We have, however, observed some active agRNAs that had higher CG content than would have been predicted to be optimal for standard siRNAs. Sequences should also be examined using BLAST to minimize obvious complementarity to other genes.

It is essential that experimenters possess a positive control duplex RNA to ensure that their experimental system and protocols are adequate. We examine the literature or experimentally screen duplex RNAs to identify traditional mRNA-directed siRNAs that are effective inhibitors of target gene expression. Inhibition of expression by the siRNA indicates that our transfection procedure has been successful. Having a positive control is especially important if addition of agRNAs yields only negative results.

Negative controls are also critical components of an agRNA experiment (7, 8). One can never have enough of these. We use duplexes containing three or four mismatched bases. Mismatches can be spread throughout the duplex or clustered at either end to break up or preserve the potential for seed-sequence identity. We also use scrambled control duplexes where the sequence is mixed to maintain nucleotide base composition and noncomplementary controls that target other genes or nonhuman genes.

For any agRNA experiment, our goal is to identify multiple duplexes that modulate gene expression. We aim for multiple active duplexes because we recognize that the activity of any one active duplex might be due to an off-target effect. Off-target effects become a less likely explanation when two different agRNAs produce the same phenotype. Our criteria for believing that a phenotype is not an off-target effect also requires that multiple scrambled and mismatch containing duplexes be inactive.

1.2. Detection of Noncoding Transcripts

Once active agRNAs and relevant negative control oligomers are identified, the next step in analysis is to identify a potential nucleic acid target. To do this we design multiple RT-PCR primer sets upstream of the transcription start site to detect RNA transcripts in the promoter that may serve as a substrate for agRNAs. For this experiment, RNA should be treated with DNase to remove contaminating DNA. To further reduce the likelihood that signal might be due to contaminating DNA, a control sample to which no RT has been added should be tested in parallel.

Cloning and sequencing noncoding RNAs that potentially serve as a substrate for agRNAs helps confirm data from RIP experiments. Knowing the full sequence of an RNA transcript also provides useful information for additional experiments. For example, knowledge of the RNA sequences allows the design of primers complementary to upstream or downstream regions of the transcript that may work better with PCR. Many regions of the genome have multiple sense and antisense transcripts and these often overlap. By identifying the exact sequence of the transcripts, it is possible to design primers that will discriminate between them.

To clone noncoding RNAs from a region of interest, we use 5′ and 3′ RACE. We use the GeneRACER® kit from Invitrogen for 5′ and 3′ RACE. This kit selects to clone only capped and polyadenylated RNA transcripts, to ensure that only full-length, undegraded RNAs are sequenced. We design primers to detect sense and antisense transcripts overlapping the region targeted by agRNAs (*see* **Note 3**). After identification of the 5′ and 3′ ends of RNA transcripts, we design primers to target each 5′ and 3′ end identified to test whether these ends unite to form a continuous transcript.

1.3. Choice of Anti-argonaute Antibodies

Successful RIP requires use of an antibody capable of potent and selective recognition of a target protein (*see* **Note 4**). For many of our experiments we used an antibody developed and characterized by Z. Mourelatos to immunoprecipitate all four argonaute proteins (14). The Corey lab has also performed successful RIP experiments using polyclonal antibodies from Millipore against Agol (07-599) and Ago2 (07-590). Also, monoclonal antibodies against Ago2 that we have acquired from the labs of M. Siomi and G. Meister have worked well (Y. Chu unpublished results). In our hands, other anti-AGO antibodies have been less useful.

1.4. Interpretation of RIP Results

Figure 20.1a shows an outline of a typical RIP experiment. RIP experiments test for an interaction between a protein and a RNA transcript (15). RIP results do not suggest that an interaction is necessary for the regulation of transcription. Therefore, RIP provides only one piece of evidence, albeit an informative one, for the action of a particular agRNA. Instead, RIP argues for specificity of agRNAs and helps to suggest that gene modulation by an agRNA is not an off-target effect.

In a typical experiment RIP will show that an active agRNA associates with a noncoding transcript at a gene promoter and that a mismatch-containing duplex RNA does not associate (**Fig. 20.1b**). It is possible that RIP will appear to show association between mismatch containing duplex and noncoding RNA. Introduction of up to four mismatches should alleviate this problem. A nonspecific antibody is also used to ensure that detection of the noncoding transcript is specific to use of the anti-AGO antibody. We examine the input sample prior to immunoprecipitation as a positive control to ensure that the noncoding RNA can be detected.

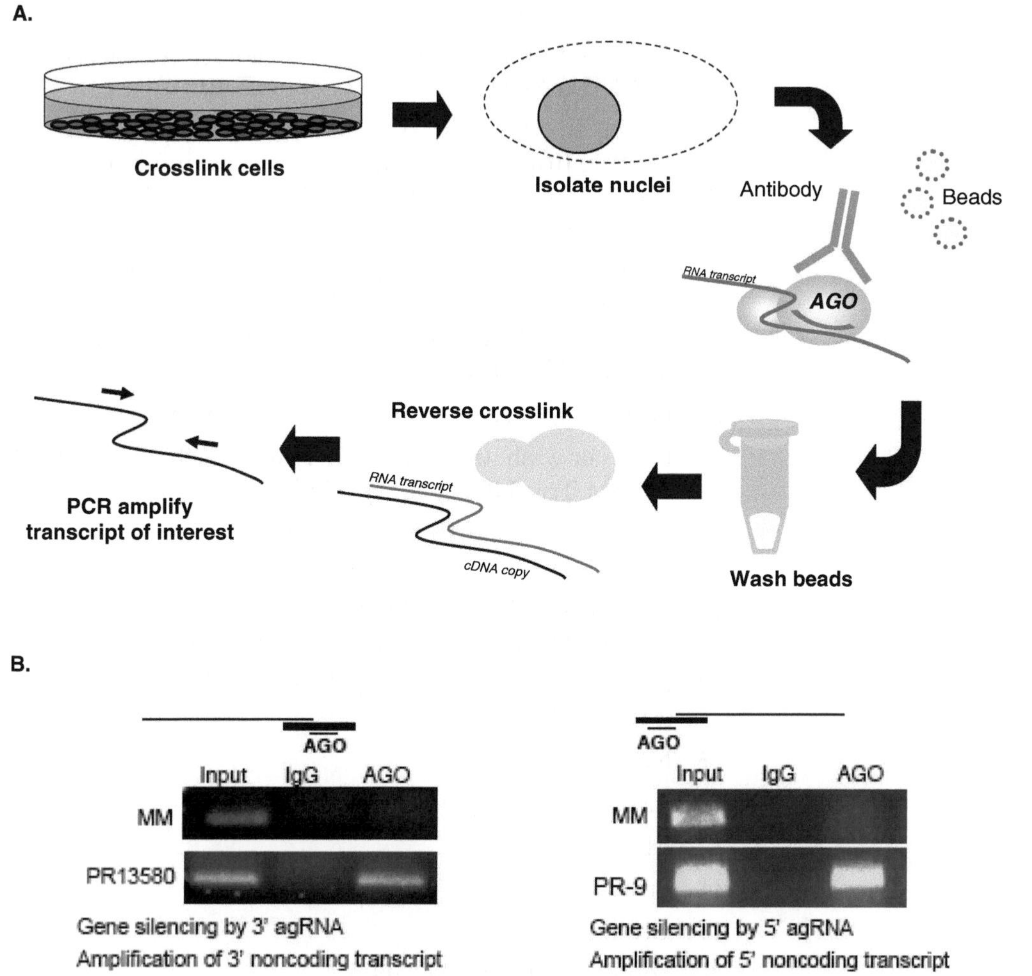

Fig. 20.1. Schematic for a typical RIP experiment and data. (**a**) The steps for a RIP experiment begin with formaldehyde crosslink of cells, isolation of nuclei, incubation of lysate with antibody followed by protein A/protein G conjugated beads. The beads are washed and complexes eluted and proteins dissociated by heat, high salt, and proteinase K digest. Detection is by RT-PCR. (**b**) Data from this experiment include an input control, a negative IgG control, and a specific pulldown with anti-AGO. Specificity is shown by comparing pulldown for a specific versus a mismatched RNA sequence. agRNAs can target noncoding RNAs at either the 5′ or 3′ ends of the mRNA coding region.

2. Materials

1. Anti-Argonaute antibodies: see **Section 1.3**
2. Cells: MCF7, T47D or cells of choice
3. Lipofection reagent: RNAi-Max® (Invitrogen) or reagent of choice
4. Phosphate buffered saline (PBS): modified Dulbecco's PBS lacking in magnesium or calcium (Sigma-Aldrich)
5. 1% Formaldehyde in PBS

6. 1.25 M Glycine
7. Hypotonic lysis buffer: 10 mM Tris (pH 7.5), 10 mM NaCl, 3 mM $MgCl_2$, 0.5% (v/v) NP-40
8. Buffer B: 1% SDS, 10 mM EDTA, 50 mM Tris–HCl, pH 8.1. Just before use, add 1× protease inhibitors cocktail (Roche) and RNaseIn^{TM} (50 U/mL) or RNase inhibitor of choice
9. IP Buffer: 0.01% SDS, 1.1% Triton X-100, 1.2 mM EDTA, 16.7 mM Tris (pH 8.1), 167 mM NaCl. Just before use add 1× protease inhibitors cocktail (Roche) and RNaseIn^{TM} (50 U/mL)
10. Protein A/G Agarose beads (Cal-Biochem)
11. Low-salt wash: 0.1% SDS, 1% Triton X-100, 2 mM EDTA, 20 mM Tris–HCl (pH 8.1), 150 mM NaCl
12. High-salt wash: 0.1% SDS, 1% Triton X-100, 2 mM EDTA, 20 mM Tris–HCl (pH 8.1), 500 mM NaCl
13. LiCl wash: 0.25 M LiCl, 1% NP40, 1% deoxycholate, 1 mM EDTA, 10 mM Tris–HCl (pH 8.1)
14. Elution buffer: 1% SDS, 1 M $NaHCO_3$, RNaseIn^{TM} (50 U/mL)
15. NaCl: 5 M
16. Tris–Cl: 1 M, pH 6.5
17. EDTA: 0.5 M
18. Proteinase K
19. Glycogen
20. Sodium acetate: 3 M, pH 5.2
21. Ethanol: 75%
22. Water: DEPC treated
23. TE Buffer: 10 mM Tris–HCl, pH 8.0, 1 mM EDTA
24. DNase I (Worthington)
25. DNase I reaction buffer: 500 mM Tris–HCl, 10 mM $MgSO_4$, 1 mM $CaCl_2$, pH 7.8
26. Total cDNA library preparation kit (ABI)
27. Agarose: 3 and 4%

3. Methods

3.1. Identify agRNAs that Regulate Transcription of a Gene of Interest

1. Analyze literature, NCBI databases, Fantom4 database, DBTSS, and UCSC genome browser to approximate the transcription start site of the gene of interest (**Note 2**).

It is preferable to design duplex RNAs to test for agRNA activity after the transcription start sites or termination sites have been determined for your cell line of interest (*see* **Section 3.2**). However, during initial screens for activity, this may not be possible if a large number of genes are being screened.

2. Design and order multiple RNA duplexes to target the regions of chromosomal DNA outside that coding for mRNA (outside of the region bracketed by the transcription start and termination sites). We typically test 10–20 RNA duplexes in order to find 3 or 4 active sequences. We often order multiple RNA duplexes that overlap a target region of interest but shifted slightly. For example, in our first agRNA paper we targeted multiple RNA duplexes shifted over by only 1–4 nucleotides targeting from 2 to 40 nucleotides upstream of the transcription start site for human progesterone receptor.
3. Design and order RNA positive and negative controls. For positive controls we typically order three siRNA sequences targeting the gene of interest mRNA near the AUG translation start site. After the initial screen, we choose the most potent siRNA to use as a positive control in subsequent experiments. For negative controls, a good starting point is to design a couple scrambles and mismatches of any of the sequences targeted by agRNAs. Ultimately, scrambled and mismatches will need to be tested using the chosen agRNA sequence of interest.
4. Test for activity of duplex RNAs by transfecting into the cell line of interest in 6-well dishes. Test should include multiple mismatched and scrambled negative controls (*see* **Note 5**). We typically do not test at concentrations higher than 50 nM due to the high chance for off-target effects at high concentrations. Gene activation or silencing can be detected by western or real time PCR. If any question remains of whether regulation occurs at the transcription or post-transcription levels, this can be resolved by real time PCR of the pre-mRNA, RNA polymerase II ChIP, or by nuclear run-on assay.
5. After identifying sequences that potentially activate or silence gene expression (*see* **Note 6**), experiments should be performed to test that the effect is dose-dependent and sequence specific. To test dose dependence, serial dilutions of duplex RNAs are tested with concentrations ranging from 1 to 50 nM. Sequence specificity is tested by designing mismatch controls based on the actual active sequence. This is a very important control to help establish that the effect

seen is not an off-target effect. Also, cells can be tested for interferon response using commercially available kits that include poly (I:C) RNA and real time PCR primers for interferon response genes such as OAS1, OAS2, MX1, IFITM1, and ISGF3r.

3.2. Identify Noncoding RNA Transcript of Interest

1. Analyze literature, NCBI databases, Fantom4 database, DBTSS, and UCSC genome browser to approximate the transcription start site of the gene of interest and evidence of any relevant transcripts. The transcription start sites for many genes are annotated in a misleading way by the NCBI reference sequence database. Furthermore, a well-characterized transcription start site for a gene might not be the site used in your cell line of interest. It is helpful for designing primers for RACE experiments to know the region where the transcription start site is expected to be found but it will often differ from annotated transcription start sites by dozens to hundreds or more nucleotides (16). In the NCBI, UCSC genome browser, and Fantom4 databases, EST data or transcription start sites for transcripts that appear unrelated to the gene of interest can be taken as evidence of potentially noncoding RNA transcripts.
2. Perform 5′ RACE using primers within your gene of interest to identify precisely the most upstream transcription start site for your gene of interest. For duplex RNAs that target regions overlapping mRNA sequences, it would be almost impossible to differentiate between siRNA and agRNA interactions.
3. Perform 5′ and 3′ RACE using primers within the region outside the mRNA coding region where agRNAs have been shown to regulate transcription. Clone and sequence any RNA transcripts found.
4. Perform long PCR on 5′ and 3′ ends of transcripts identified to discover whether the ends correspond to a continuous transcript. If multiple 5′ and 3′ ends are discovered, design primers that can distinguish between the ends to discover which end corresponds to which end.
5. Design and test the efficiency of multiple primer sets designed to detect your noncoding RNA transcript that you have identified. Identify the most efficient primer set for use in RIP experiments.

3.3. Transfection

1. Before actually planning a transfection it is useful to optimize seeding. Cell lines grow at different rates and have different sizes. Seeding numbers should be optimized so that cell numbers are maximized at harvest time, 5–7 days later,

but not over confluent, which can lead to contact induced inhibition of cell growth and changes in gene expression.

2. Seed cells in 15-cm-diameter dishes. The number of dishes can be modified in order to have enough material at harvest for immunoprecipitation. For our cells lines, MCF7 and T47D, we seed 5 million cells for MCF7 and 3.5 million cells for T47D. For each treatment (agRNA or negative control), we transfect four dishes for MCF7 cells and six dishes for T47D (*see* **Note 7**).
3. Check seeding 1 or 2 days later for evenness. If cells are seeded too unevenly, discard cells and try seeding again (*see* **Note 8**).
4. Two days after seeding, transfect with 25 nM agRNA using a lipid reagent that you know works best for your cell line (*see* **Note 9**). For our cells lines we use RNAi-Max® (Invitrogen).

3.4. Harvest (Crosslink in Dish)

1. Harvest cells when optimal knockdown is seen for your gene of interest. We harvest 3 days after transfection or 5 days after seeding.
2. Gntly wash cells with 10 mL PBS in dish. Scrape ~5 cm^2 of cells for qRT-PCR to check transfection.
3. Aspirate wash.
4. Add 15 mL of 1% formaldehyde in PBS.
5. Incubate for 10 min.
6. Quench by adding 1.5 mL of 1.25 M glycine to a final concentration 125 mM. Gently rock dish after adding glycine. Quench for 5 min.
7. Remove formaldehyde and glycine.
8. Harvest cells by scraping with rubber policeman (scrapper) in 7 mL of PBS and combining dishes of same treatment into one 15-mL conical tube.
9. Cells pelleted at 500×*g* (~1700 rpm) for 5 min at 4°C. Nuclei purification (optional).
10. Remove supernatant and add 4 mL cold hypotonic lysis buffer slowly while gently vortexing (1 mL at a time). Be sure to break pellet and set 5 min on ice.
11. Nuclei pelleted at 500×*g* for 5 min at 4°C (*see* Step 1 in **Section 3.5**).
    NOTE: The protocol can be stopped here and whole cells or nuclei can be stored at –80°C.
12. Lyse nuclei in 1 mL nuclear lysis buffer (Buffer B). For Buffer B, RNaseIn™ (Promega) and 1× EDTA-free Complete mini® protease inhibitors (Roche) are added just

before use. (NOTE: Your RNase inhibitor of choice can be substituted for RNaseIn™). Thoroughly break up pellet by pipetting.

13. Move to 1.5-mL eppendorf and sit on ice for 10 min.
14. Spin maximum speed for 10 min at 4°C.
15. Remove nuclear lysates to fresh tube.
16. Lysates can be stored at –80°C. We have successfully performed RIPs on lysates up to 1 year old. After 1 year, a considerable amount of precipitate will accumulate in the lysate and the amount of RNA recovered after immunoprecipitation drops significantly in some samples.

3.5. Immunoprecipitation

1. We do not sonicate our RIP samples. However, we have noticed a slight improvement in signal levels after a brief sonication, likely due to improved resuspension of complexes in the lysate. Sonication should be performed between Steps 11 and 12 in **Section 3.4**. Also, for most of our antibodies we have not seen a difference in whether or not we preclear the sample with IgG and/or beads so we usually omit that step. For different experiments, the omission of the preclearing step might not be tolerated.
2. 100 μL of lysate is diluted to 1 mL of IP buffer for each IP sample. For IP buffer, the RNaseIn™ and protease inhibitors are added just before use. One 20–50 μL lysate sample should be saved in –80°C for input control later. Also, 100 μL of lysate should be used for an IgG negative control. We do not reprecipitate from the same lysate – for our samples we have found that there is no complex left after one IP.
3. Samples are incubated with 2–4 μg of antibody overnight with rotation at 4°C. When first testing a new antibody, we use both 2 and 4 μg and check the results by quantitative real-time PCR, gel electrophoresis, and sequencing to be sure that that the antibody is pulling down the correct product reliably.
4. Recover complexes by adding 60–70 μL of Protein A/G Agarose beads equilibrated in IP buffer and rotate for 2 h. To equilibrate, spin down beads in suspension at low speed, remove storage buffer, and bring back to volume in IP buffer.
5. "Pulldown" by centrifugation at 300×*g* (~1000 rpm) for 2 min at 4°C.

3.6. Wash

1. Each sample is washed sequentially in 1 mL of each wash listed below for 5 min each. Complexes are pelleted each

time by centrifugation at $300\times g$ (~1000 rpm) for 1 min at room temperature. Wash buffer is completely removed by pipetting with clean tips each time and the last remains of buffer is pipetted using a smaller bore 200 μL pipette or gel loading tip. It is important to remove as much of the last wash possible before adding the next wash. Make sure to monitor for bead loss and ensure that beads are not lost resulting in loss of signal. Monitoring small amounts of bead loss can be done by slowly pipetting washes after removal along the side of a 50-mL tube or glass beaker in order to see the lost beads as they stick to the sides.

Washes:

a. Low-salt wash
b. High-salt wash
c. LiCl wash
d. TE, pH 8
e. TE, pH 8

3.7. Elution

1. Complexes are eluted by adding 400 μL freshly prepared Elution buffer.
2. Vortex and incubate for 15 min with rotation at room temp.
3. Spin at $8000\times g$ for 2 min.
4. Pipette supernatant into new labeled tubes.
5. Dilute 100 μL input samples to 300 μL with freshly prepared Elution Buffer (*see* **Note 10**). Treat input samples along with other samples for the rest of the protocol.
6. Add 16 μL of 5 M NaCl to final concentration of 200 mM.
7. Set at 65°C for at least 2 h to reverse crosslinking.
8. Add 16 μL of 1 M Tris–Cl, pH 6.5, 8 μL of 0.5 M EDTA, and 20 μg of Proteinase K to each sample.
9. Set at 42°C for 45 min.
10. Extract RNA by phenol:chloroform:isoamyl alchohol extraction.
    a. Samples should be at 400 μL.
    b. Add 500 μL phenol:chloroform:isomyl, vortex, and spin (>$11{,}000\times g$) for 12 min.
    c. Remove top clear layer (aqueous layer containing RNA) in a new 1.5-mL tube and do not include any cloudy layer between water and yellow phenol.
    d. To the aqueous layer, add 40 μg glycogen, 40 μL 3 M sodium acetate pH 5.2, and 800 μL of ethanol.
    e. Precipitate overnight at –80°C.

f. Next day, spin at maximum speed for 12 min. Pellet sizes vary and can be quite small.
g. Wash with 1 mL of 75% ethanol.
h. Spin 5 min at maximum speed.
i. Re-suspend pellets in 20 μL of DEPC-treated water. RIP product should be stored at –80°C. My experience has been that RIP product even at –80°C begins to deteriorate after about 2 weeks in storage and results become inconsistent. I general do not use samples more than 2 weeks old.

3.8. Detection

1. Use 10 μL of RNA sample and treat with 1 μL (2 U) of DNase I (Worthington) and 1 μL of 10× DNase I reaction buffer for 10 min at room temperature and deactivate for 10 min at 75°C.
2. Perform reverse transcription using ABI total cDNA library preparation kit.
3. After cDNA synthesis, we perform qRT-PCR using 4 mL of sample in a 20-mL total volume PCR reaction. Usually for rare noncoding transcripts, signals come up well outside the linear range for quantitative PCR so we do not believe that quantitation is reliable. However, for some more highly expressed transcripts, signals may come up in the linear range for qRT-PCR making quantitation a possibility. To maximize signal, it is desirable to keep amplicon sizes for PCR primers between 60 and 140 nucleotides long.
4. Samples are always run on a 3 or 4% agarose gel. Since signals are often very low, sometimes signals seen by qRT-PCR are only primer dimers and do not appear on the gel. It is important to validate product by gel.
5. Samples can be cut from the gel and sequenced to ensure the correct product is amplified.

4. Notes

1. This protocol has been adapted from that presented by the J. Lee lab at http://www.epigenome-noe.net/researchtools/protocol.php?protid=28.
2. Databases NCBI, Fantom4, DBTSS, and UCSC genome browser contain a wealth of information to develop leads concerning potential transcription start sites and noncoding RNAs. These databases can be accessed online at:

NCBI: http://www.ncbi.nlm.nih.gov/nucleotide/

Fantom 4: http://fantom.gsc.riken.jp/4/gev/gbrowse/hg18/

DBTSS: http://dbtss.hgc.jp/

UCSC genome

browser: http://genome.ucsc.edu/

3. The noncoding RNA transcripts that serve as substrates for agRNAs may be oriented in the sense or antisense direction with respect to the mRNA of the gene of interest. To search for RNA transcripts in either direction, primers should be tested targeting transcripts in either direction paired with primers to target either the 5′ or 3′ linker sequences to ensure all possible transcripts are detected.
4. We have found that RIP with anti-AGO antibodies is a straightforward technique once it is mastered. However, the technique is complex and can fail at any one of many steps. Therefore, researchers should approach the technique understanding that extreme care is necessary and expecting to invest a significant effort in optimization before useful results are achieved. This is not a technique that is likely to be successful on the first or second attempt.
5. For a first test, an additional negative control of untreated cells can be included to test if transfection itself is causing drastic changes in gene expression. Small changes in gene expression can be expected between untreated and mismatch or scramble treated cells simply due to the effects of slowed growth and stress caused by lipid-based transfections. These differences rarely are more than 20% and certainly should not exceed 40%. If differences are more drastic, alternative lipid reagents, transfections techniques, or even harvest conditions should be explored. After solid mismatch and scramble controls are identified, untreated samples cease to be a useful control.
6. Lipid-based transfections are inherently variable. Thus small changes in gene expression are generally difficult to reproduce reliably and may be due to off-target effects. For this reason we are typically only interested in gene silencing where gene expression is reduced by more than 50%. For gene activation, we are only interested in activation more than 2.5 or 3 fold. Smaller changes in gene expression may be due to a sequence specific agRNA effect but better transfection conditions should be explored to maximize the differences in gene expression to ensure success for subsequent more in depth analysis such as RIP and ChIP experiments.

7. The need for a large number of cells during the immunoprecipitation results in a large number of dishes at harvest depending on the number of treatments, which should be taken into account during the experimental design stage. RIP experiments in T47D cells for four treatments would result in 24 dishes at the time of harvest. We stagger harvests so that no more than nine dishes are handled at the same time during the crosslinking and scraping steps.
8. We have noticed that transfection efficiency depends dramatically on seeding density. When seeding large 15-cm dishes it is easy to "swirl" the media, which results in cells clustering in a ring around the dish. While some clustering of cells is unavoidable, this should be minimized to maintain high transfection efficiency. Ultimately, the efficiency of gene silencing or activation is the true test of effective seeding and spreading. If cells are seeded with extreme care, dramatic improvements in transfection efficiency can result.
9. Some lipid reagents allow for "reverse transfections" where the cells are seeded at the time of transfection. For these protocols, the 2 days after seeding before transfection are removed from the protocol. The cells are still harvested the same number of days after transfection that is optimal for that gene of interest and cell line. An added benefit of these protocols is that transfection efficiency seems much more consistent regardless of evenness of seeding density for "reverse transfected" protocols.
10. Since input samples have a much higher concentration of nucleic acids, it is not necessary to use a full 100 μL as input. The important point is to use the same amount of input for each treatment so results are comparable. We have used as little as 10 μL for input and I often use 50 μL. If input samples are too high a concentration this can inhibit detection by PCR. On the other hand, it is easy to dilute final volumes of input samples in order to get a more reliable PCR detection. It is, of course, desirable to use less input sample in order to be able to perform as many RIP experiments from a single harvest as possible.

Acknowledgments

This work was supported by grants from the National Institutes of Health (NIGMS GM77253 to DRC and NIBIB EB 05556 to JCS) and the Robert A. Welch Foundation (I-1244 to DRC).

References

1. Janowski, B. A., Hu, J., and Corey, D. R. (2006) Silencing gene expression by targeting chromosomal DNA with antigene peptide nucleic acids and duplex RNAs. *Nat Protoc* **1**, 436–443
2. Janowski, B. A., Huffman, K. E., Schwartz, J. C., et al. (2005) Inhibiting gene expression at transcription start sites in chromosomal DNA with antigene RNAs. *Nat Chem Biol* **1**, 216–222
3. Janowski, B. A., Huffman, K. E., Schwartz, J. C., et al. (2006) Involvement of AGO1 and AGO2 in mammalian transcriptional silencing. *Nat Struct Mol Biol* **13**, 787–792
4. Janowski, B. A., Younger, S. T., Hardy, D. B., Ram, R., Huffman, K. E, and Corey, D. R. (2007) Activating gene expression in mammalian cells with promoter-targeted duplex RNAs. *Nat Chem Biol* **3**, 166–173
5. Schwartz, J. C., Younger, S. T., Nguyen, N. B., et al. (2008) Antisense transcripts are targets for activating small RNAs. *Nat Struct Mol Biol* **15**, 842–848
6. Yue, X., Schwartz, J. C., Younger, S. T., Chu, Y., Gagnon, K. T., Elbashir, S., Janowski, B. A., and Corey, D. R. (2010) Regulation of transcription by small RNAs complementary to sequences downstream from the 3′ termini of genes. *Nat Chem Biol* **6**, 621–629
7. Jackson, A. L., Bartz, S. R., Schelter, J., et al. (2003) Expression profiling reveals off-target gene regulation by RNAi. *Nat Biotechnol* **21**, 635–637
8. Hannon, G. J., and Rossi, J. J. (2004) Unlocking the potential of the human genome with RNA interference. *Nature* **431**, 371–378
9. Persengiev, S. P., Zhu, X., and Green, M. R. (2004) Nonspecific, concentration-dependent stimulation and repression of mammalian gene expression by small interfering RNAs (siRNAs). *RNA* **10**, 12–18
10. Hammond, S. M., Boettcher, S., Caudy, A. A., Kobayashi, R., and Hannon, G. J. (2001) Argonaute2, a link between genetic and biochemical analyses of RNAi. *Science* **293**, 1146–1150
11. Liu, J., Carmell, M. A., Rivas, F. V., et al. (2004) Argonaute2 is the catalytic engine of mammalian RNAi. *Science* **305**, 1437–1441
12. Chu, Y., Yue, X., Janowski, B. A., Younger, S. T., and Corey, D. R. (2010) Argonaute proteins and modulation of progesterone receptor expression by promoter-targeted RNAs. *Nucl Acids Res* **38**, 7736–7748
13. Baroni, T. E., Chittur, S. V., George, A. D., and Tenenbaum, S. A. (2008) Advances in RIP-chip analysis: RNA-binding protein immunoprecipitation-microarray profiling. *Methods Mol Biol* **419**, 93–108
14. Nelson, P. T., De Planell-Saguer, M., Lamprinaki, S., et al. (2007) A novel monoclonal antibody against human argonaute proteins reveals unexpected characteristics of miRNAs in human blood cells. *RNA* **13**, 1787–1792
15. Landthaler, M., Gaidatzis, D., Rothballer, A., et al. (2008) Molecular characterization of human argonaute-containing ribonucleoprotein complexes and their bound target mRNAs. *RNA* **14**, 2580–2596
16. Denoeud, F., Kapranov, P., Ucla, C., et al. (2007) Prominent use of distal 5′ transcription start sites and discovery of a large number of additional exons in ENCODE regions. *Genome Res* **17**, 746–759

Chapter 21

Inhibition of Human Papillomavirus Expression Using DNAzymes

María Luisa Benítez-Hess, Pablo Reyes-Gutiérrez, and Luis Marat Alvarez-Salas

Abstract

Deoxyribozymes (DXZs) are catalytic oligodeoxynucleotides capable of performing diverse functions including the specific cleavage of a target RNA. These molecules represent a new type of therapeutic oligonucleotides combining the efficiency of ribozymes and the intracellular endurance and simplicity of modified antisense oligonucleotides. Commonly used DXZs include the 8–17 and 10–23 motifs, which have been engineered to destroy disease-associated genes with remarkable efficiency. Targeting DXZs to disease-associated transcripts requires extensive biochemical testing to establish target RNA accessibility, catalytic efficiency, and nuclease sensibility. The usage of modified nucleotides to render nuclease-resistance DXZs must be counterweighted against deleterious consequences on catalytic activity. Further intracellular testing is required to establish the effect of microenvironmental conditions on DXZ activity and off-target issues. Application of modified DXZs to cervical cancer results in specific growth inhibition, cell death, and apoptosis. Thus, DXZs represent a highly effective antisense moiety with minimal secondary effects.

Key words: DNAzyme, 10–23 DNAzyme, catalytic DNA, antisense, DNA structure, antisense DNA, papillomavirus, DNA, DNAzyme expression.

1. Introduction

Deoxyribozymes (DXZs) are small catalytic oligodeoxynucleotides capable of cleaving target RNA molecules in a sequence specific manner (1). These molecules represent a whole new generation of artificial catalytic nucleic acids originally obtained through the selective amplification of oligonucleotide combinatorial libraries by their ability to catalyze the *cis*-cleavage of a short RNA sequence (2). Like ribozymes, DXZs hybridize to the

J. Goodchild (ed.), *Therapeutic Oligonucleotides*, Methods in Molecular Biology 764,
DOI 10.1007/978-1-61779-188-8_21, © Springer Science+Business Media, LLC 2011

substrate RNA by Watson–Crick base pairing and cleave phosphodiester linkages resulting in the formation of RNA fragments containing 2′,3′-cyclic phosphate and 5′-hydroxyl ends (3, 4). DXZ catalysis and specificity have found many applications, but most attention has been focused on the knockdown of disease-associated genes (5). Because of its DNA nature, the use of DXZs as therapeutic oligonucleotides has the advantage of simple synthesis, cost-effectiveness, and robustness over other oligonucleotide technologies such as ribozymes and siRNAs (6).

The 10–23 motif is the most commonly used DXZ in therapeutic applications. Structurally, the 10–23 DXZ resembles a hammerhead ribozyme with a 15-nucleotide catalytic core flanked by two binding arms complementary to the target RNA. Cleavage is Mg^{++}-dependent and specifically produced at the RY (R, purine; Y, pyrimidine) junction (2, 7, 8) (**Fig. 21.1**).

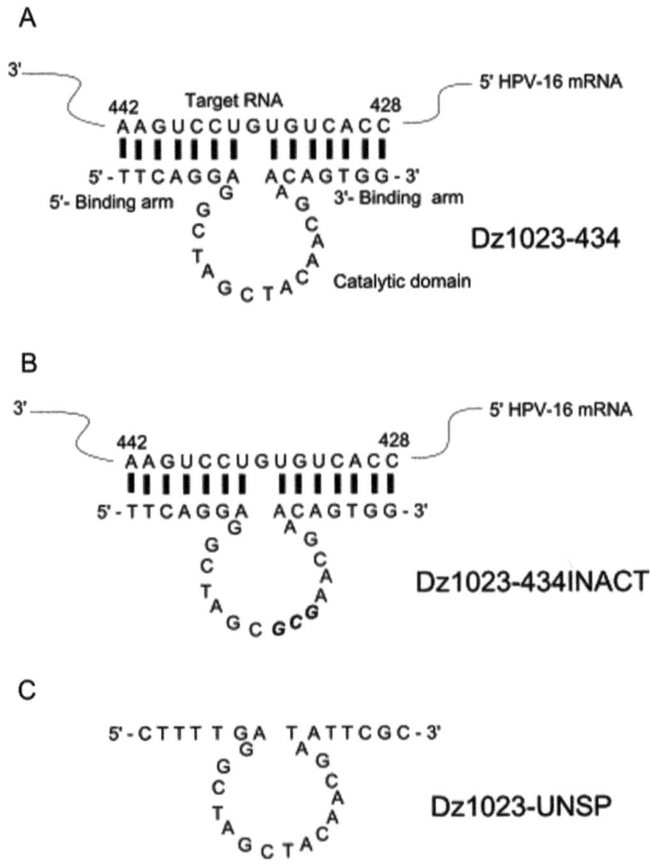

Fig. 21.1. Schematic representation of 10–23 deoxyribozymes (DXZs). (**a**) Dz1023-434 complementarity to HPV-16 E6/E7 mRNA nt 428-442 (Target RNA). The 10–23 catalytic domain and hybridization (binding) arms are indicated. Standard Watson–Crick pairs are indicated by lines. (**b**) Inactive Dz1023-434 control (Dz1023-434INACT); mutation changes (TAC for GCG) within the catalytic domain are indicated in bold italics. (**c**) Unspecific or scrambled control (Dz1023-UNSP).

The 10–23 DXZ is active at 37°C but requires nucleotide modifications to keep high activity under physiological conditions (9–11).

The relationship between cervical cancer and human papillomaviruses (HPVs) is associated with the expression of the viral early proteins E6 and E7, which are sufficient and necessary to acquire and maintain a transformed phenotype (12, 13). E6 and E7 protein products functionally neutralize cell cycle regulatory proteins, so that cell proliferation continues. Because they are retained and expressed in most cervical tumors there are often referred to as the hallmark of cervical cancer (14). The HPV types 16 and 18 are the most commonly HPVs found in cervical tumors worldwide and express E6 and E7 mRNA from a single promoter resulting in polycistronic mRNA containing both transcripts (15–17). The combination of E6 and E7 activities cause genomic instability, cell immortalization and transformation leading to cancer (12, 18, 19). Thus, targeting of either E6/E7 mRNA by antisense strategies would likely impede both E6 and E7 translation resulting in growth arrest and/or apoptosis leading to tumor growth suppression (20–23).

2. Materials

2.1. Oligonucleotide Purification Using Denaturing Polyacrylamide Gel Electrophoresis (D-PAGE)

1. Chemically synthesized single-stranded oligodeoxynucleotides (ssODNs), desalted and lyophilized. Store dry at 4°C.
2. Denaturing gel loading buffer: 80% formamide, 0.25% bromophenol blue, 0.25% xylene cyanol FF.
3. E buffer: 1 mM EDTA, 0.5 M ammonium acetate, 0.1% SDS. Store at room temperature.
4. 30% Acrylamide/*N*,*N*′-methylene-*bis*-acrylamide: 19:1 stock solution in deionized water (*see* **Note 1**). Filter through Whatman® 3 MM and store at 4°C.
5. Urea: ultrapure.
6. 5× TBE buffer stock: 0.45 M Tris–HCl, pH 8.3, 0.45 M boric Acid, 10 mM EDTA. Store at room temperature.
7. *N*,*N*,*N*′,*N*′-Tetramethylethylenediamine (TEMED). Store at 4°C.
8. Ammonium persulfate. Freshly prepare 10% solution in deionized water.
9. Vertical electrophoresis apparatus.
10. Silica-gel TLC plates with fluorescent indicator for visualizing DNA on gels by UV shadowing (*see* **Note 2**).

11. Syringe filters 0.45 μm.
12. 3-mL syringes.
13. UV lamp (230 nm).
14. Sephadex® G-50 pre-packed columns.

2.2. Labeling and Purification of DXZs and Target RNAs

2.2.1. DXZ Radiolabeling

1. T4 polynucleotide kinase (Promega). Store at 4°C.
2. 5× Forward reaction buffer: 250 mM Tris–HCl, pH 7.6, 50 mM $MgCl_2$, 25 mM DTT, 0.5 mM spermidine, 0.5 mM EDTA.
3. γ-[^{32}P]-ATP: 3000 Ci/mmol.
4. D-PAGE reagents and vertical electrophoresis apparatus.
5. Gel drier.
6. Liquid scintillation counter.
7. Liquid scintillation solution (Biofluor, Perkin-Elmer) and vials.

2.2.2. In Vitro Transcription (Run-off)

1. Linearized plasmid DNA templates with single-cut restriction enzymes (in the present example, *Not*I and *Eco*RI were used as they present a single cutting site right after the 3′ end of the target coding sequence): 1 μg (*see* **Note 3**)
2. RiboProbe® SP6 and T7 in vitro transcription systems (Promega Inc., Madison, WI)
3. Phenol saturated with water
4. Chloroform-isoamyl alcohol solution: 24:1
5. Ammonium acetate: 7.5 M, pH 7.5
6. Absolute and 70% (v/v) ethanol stored at –20°C and room temperature, respectively
7. α-[^{32}P]-UTP: 3000 Ci/mmol
8. D-PAGE reagents
9. Vertical electrophoresis apparatus
10. Radiographic film
11. Whatman® 3 MM filter paper
12. Variable mode imager scanner: Thypoon® 8600 with Image Quant® software

2.3. DXZ Cleavage Analysis

1. 1× DXZ buffer: 50 mM Tris–HCl, pH 7.5, 150 mM NaCl, and 2 mM spermidine. Filter through 0.22 μm and store at –20°C.
2. $MgCl_2$ stock solution: 1 M. Filter through 0.22 μm and store at room temperature.
3. Wet and dry ice.

4. 5× Stop buffer: 95% formamide, 20 mM EDTA, 0.05% xylene cyanol, 0.05% bromophenol blue.
5. D-PAGE reagents.
6. Vertical electrophoresis apparatus.
7. Liquid scintillation counter.
8. Liquid scintillation solution (Biofluor, Perkin-Elmer) and vials.

2.4. Cell Culture

1. C33-A (HTB-31) and SiHa (HTB-35) cervical cancer cell lines (American Type Culture Collection).
2. Dulbecco's modified Eagle Medium minimum essential growth medium (DMEM) supplemented with 5% of fetal bovine serum (FBS) (Invitrogen), 0.01 mg/mL gentamycin, 100 mg/mL penicillin, and 100 μg/mL streptomycin (Invitrogen).
3. CO_2 incubator.
4. 1× Phosphate-buffered saline (1× PBS): 2.7 mM KCl, 1.8 mM KH_2PO_4, 136 mM NaCl, 10 mM Na_2HPO_4, pH 7.4. Filter through 0.22 μm and store at room temperature.
5. FACS flow solution (BD Bioscience).
6. TEN buffer: 10 mM Tris–HCl, pH 7.5, 1 mM EDTA, 100 mM NaCl. Filter through 0.22 μm and store at room temperature.
7. Trypsin-Versene solution: Mix 0.5 mL of 2.5% trypsin stock (Invitrogen) with 25 mL of 1× PBS and 25 mL of versene solution (Invitrogen). Use freshly prepared.
8. Tissue-culture grade 100 mm and six-well plastic dishes.
9. Cell scrapers and spatulas.

2.4.1. DXZ Nuclease Resistance Assay

1. α-[^{32}P]-UTP: 3000 Ci/mmol
2. T4 polynucleotide kinase and RQ1 DNaseI (Promega)
3. TEN buffer: 10 mM Tris–HCl, pH 7.5, 1 mM EDTA, 100 mM NaCl
4. 10× RQI buffer: 40 mM Tris–HCl, pH 8.0, 100 mM $MgCl_2$, and 10 mM $CaCl_2$.
5. D-PAGE reagents and vertical electrophoresis apparatus
6. Liquid scintillation counter
7. Liquid scintillation solution (Biofluor, Perkin-Elmer) and vials.

2.4.2. DXZ Transfection

1. D-PAGE or HPLC-purified DXZ DNA. For cell uptake quantification use 5′-fluorescein-labeled DXZs resuspended in sterile deionized water (*see* **Note 1**).

2. Sterile 1.5-mL microtubes.
3. Basal DMEM growth medium (Invitrogen).
4. Lipofectin® transfection reagent (Invitrogen) stored at 4°C.
5. CO_2 incubator.
6. FACSCalibur® flow cytometer (BD Biosciences) with 488 nm excitation laser and band pass filters at 520/30 nm (FL1) and 542/85 nm (FL2).

2.4.3. Total RNA Extraction from Cultured Cells

1. Trizol® reagent (Invitrogen).
2. RNAsin® RNaseA inhibitor: 40 U/μL (Promega).
3. Chloroform:isoamyl alcohol: 24:1 v/v
4. Cell scrapers.
5. 75% 2-propanol.
6. 70% ethanol.
7. DNase/RNase-free water.
8. Microcentrifuge.
9. Vortex mixer.
10. Agarose gel electrophoresis reagents.
11. Horizontal electrophoresis apparatus.

2.4.4. Quantitative RT-PCR (QRT-PCR) Assays

1. Appropriate chemically synthesized DNA primer stocks (2 μM) stored in aliquots at −20°C. Primers may be purified by D-PAGE or HPLC to ensure length homogeneity.
2. 10 × Deoxynucleoside 5′-triphosphate (dNTP) mix: 10 mM each in deionized water, deoxyguanosine 5′-triphosphate (dGTP), deoxyadenosine 5′-triphosphate (dATP), thymidine 5′-triphosphate (TTP), and deoxycytidine 5′-triphosphate (dCTP). Store at −20°C.
3. 10 × PCR buffer: 100 mM Tris–HCl, pH 8.3, 500 mM KCl, 15 mM $MgCl_2$. Store at −20°C.
4. SuperScript® III Platinum® SYBR® Green One-Step qRT-PCR (Invitrogen).
5. SYBR Green® fluorescent dye 10,000× stock.
6. Rotor-Gene® RG-3000 real-time thermocycler (Corbett Research) with excitation (470 nm) and emission (510 nm) filters for the detection of SYBR Green®.
7. Thin-walled PCR microtubes: 50 and 100 μL.

2.5. Phenotypic Effects

2.5.1. Cell Viability Assays

1. Trypan blue stain solution: 0.4% in 1× PBS.
2. 15-mL Polystyrene conical tubes.
3. Inverted microscope with epifluorescence.
4. Neubauer cell counting chamber.

2.5.2. Cell Cycle/Apoptosis Analysis

1. Chilled 80% ethanol.
2. RNaseA working solution: 3 μg/mL freshly diluted in 1× PBS from a 100-μg/mL stock (Sigma-Aldrich). Store stock at –20°C.
3. 40 μm Nylon® mesh.
4. 5-mL Polystyrene round-bottom tubes.
5. 15-mL Conical polystyrene tubes.
6. Propidium iodide working solution: 0.33 μg/mL freshly diluted in deionized water from a 50-μg/mL stock. Store stock at –20°C.
7. FACSCalibur® flow cytometer (BD Biosciences).

3. Methods

3.1. Oligonucleotide Preparation

1. Prepare 18% polyacrylamide/7 M urea denaturing gels in 1× TBE buffer. In a clean and sterile beaker weigh 21 g urea and add 30 mL of the 30% acrylamide/bisacrylamide 19:1 stock solution and 5 mL 5× TBE. Dissolve by stirring and gentle warming for at least 1 h. Let the mixture reach room temperature and add 50 μL TEMED and 150 μL 10% ammonium persulfate. Mix well and pour mixture into the gel-casting cassette and attach comb. Leave polymerizing for at least 1 h at room temperature.
2. Lyophilized ssODNs and DXZs are resuspended in 30 μL of denaturing gel loading buffer and heated 5 min at 65°C. Remove comb and set vertical electrophoresis apparatus with 1× TBE buffer. Rinse wells with 1× TBE using a syringe and load the full sample volume. Electrophorese at 250 V for 3 h, most ssODNs migrate close to the xylene cyanol band. Wrap the gel in plastic wrap and ODNs are visualized by DNA shadowing using an UV lamp and a fluorescent TLC plate (*see* **Note 2**).
3. Full-length ssODNs and DXZs (usually the thickest and slowest migrating band) are excised with a scalpel, smashed with a 3-mL syringe, and poured in a sterile microtube. Smashed gels are eluted in 350 μL E buffer overnight at 4°C. Filter eluate using 0.45-μm syringe filters to exclude any polyacrylamide debris. The ssODNs are purified through Sephadex G-50 chromatography with deionized water and quantified by $UV_{260/280}$ absorbance.

3.2. Labeling and Purification of DXZs and Target RNAs.

3.2.1. DXZ Radiolabeling

1. Prepare 6% polyacrylamide/7 M urea denaturing gels in 1× TBE buffer. In a clean and sterile beaker weigh 21 g urea and add 10 mL of 30% acrylamide/bisacrylamide 19:1 stock and 5 mL 5× TBE. Add 50 μL TEMED and 150 μL 10% ammonium persulfate mix well and pour into the gel-casting cassette. Leave polymerizing for at least 1 h at room temperature.
2. Ten pmoles DXZ and 10 μCi γ-[^{32}P]-ATP are placed in a microtube with 10 U T4 polynucleotide kinase and 1× forward reaction buffer in a 20 μL final volume.
3. Labeling reactions are incubated at 37°C for 1 h.
4. Resuspend labeled DXZs in 30 μL of denaturing gel-loading buffer. Heat samples at 65°C for 5 min and load into 6% acrylamide/7 M urea gels.
5. Electrophorese at 250 V for 50 min. Wrap the gels in plastic wrap and expose to radiographic film (exposure time depends on label incorporation; normally 5 min exposure is recommended).
6. Develop radiographic film and place it below the gel. Mark the full-length band and excise from the gel with a scalpel. Transfer the gel band to a syringe, squash the gel and recover it in a microtube, and incubate in 350 μL of E buffer at 4°C overnight.
7. Filter the eluate using a 0.45-μm syringe filter to exclude polyacrylamide debris.
8. Extract eluate with phenol–chloroform (*see* **Note 4**).
9. Precipitate with absolute ethanol–ammonium acetate (*see* **Note 5**).
10. Rinse with 80% ethanol.
11. Resuspend in deionized water and quantify in Typhoon® 8600 fluorographic scanner or by scintillation counting (*see* **Note 6**).

3.2.2. In Vitro Transcription

1. Prepare 6% polyacrylamide/7 M urea denaturing gels in 1× TBE buffer.
2. Plasmids containing the target RNA sequence (pE6-GFP and pCR16HH) are linearized with a single-cut restriction enzyme (in the present example, *Not*I and *Eco*RI were respectively used) and cleaned by phenol–chloroform extraction and ethanol/ammonium acetate precipitation (*see* **Notes 4** and **5**).
3. Linearized plasmids are transcribed in vitro with SP6 or T7 RNA polymerases in the presence of 10 μCi α-[^{32}P]-UTP using Riboprobe® SP6/T7 in vitro transcription system as

described by the manufacturer. Labeled transcript production is optimized by the addition of 20- to 200-fold non-labeled UTP to the transcription reaction. DNA templates are removed by incubation with 20 U of RQI DNaseI by 30 min at 37°C.

4. Heat samples at 65°C for 5 min and incubate 5 min on ice to avoid any secondary structure that could modify electrophoretic mobility. Labeled transcripts are purified through preparative D-PAGE as described above.
5. Eluted transcripts are phenol–chloroform extracted and precipitated with ethanol/ammonium acetate. Resuspend labeled transcripts in 20 μL of deionized water and quantify in a scintillation counter (*see* **Note 6**). Adjust concentration to $5–20\times10^3$ cpm/μL and store at −70°C until use.

3.3. DXZ Cleavage Analysis

3.3.1. Importance of $[Mg^{2+}]$ in DXZ Cleavage

For therapeutic purposes, DXZs must show high target specificity and catalytic efficiency (CE). In a first step, biochemical parameters (such as temperature, $[Mg^{2+}]$, pH, etc.) are established by setting several cleavage reactions with changing conditions until an optimal cleavage (judged by the higher amount of cleaved product) is obtained. Most 10–23 DXZs are usually effective under physiological conditions but show a strong dependence on $[Mg^{2+}]$ (5), which can be limiting in some tissues (24). Thus, it is important to establish the $[Mg^{2+}]$ range for a given DXZ to determine its potential as therapeutic agent. To facilitate this step, the Dz1023-434a DXZ is incubated with a short HPV-16 E6 target with one accessible cleavage site (*see* **Note 3**). Specificity controls include an active DXZ with different specificity (Dz1023-434UNSP) and an inactive DXZ with a mutant catalytic domain (Dz1023-434INACT) that can hybridize with the target but is unable to cleave it (**Figs. 21.1** and **21.2**).

1. Prepare 6% acrylamide/7 M urea gel.
2. Set up cleavage reactions in 1.5-mL microtubes on ice by adding ^{32}P-labeled short E6 target ($5–20\times10^3$ cpm/μL), 2 μL 10× DXZ buffer and increasing $[Mg^{2+}]$ (0.01–20 mM) in a 20 μL final volume.
3. Add 1 pmol of the corresponding DXZ DNA on the wall of each microtube and briefly spin the tubes in a microcentrifuge.
4. Incubate 1 h at 37°C.
5. Stop reactions by adding 4 μL stop buffer to each microtube.
6. Heat reactions at 65°C for 5 min.
7. Load samples on 6% acrylamide/7 M urea gel.
8. Electrophorese at 250 V for 50 min.

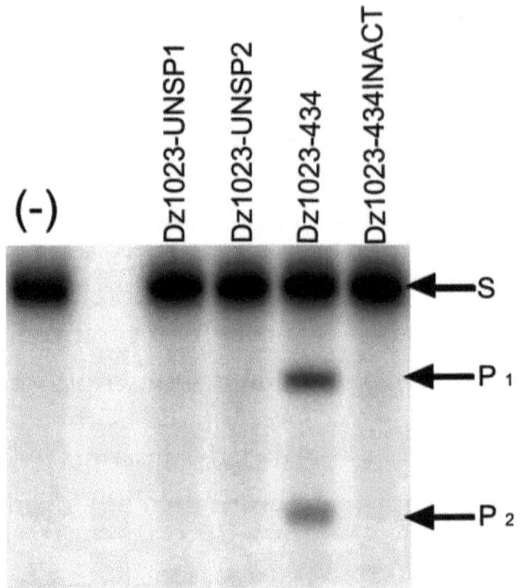

Fig. 21.2. DXZ cleavage activity on the short HPV-16 E6 target. Cleavage reactions were performed with the labeled short HPV-16 E6 RNA target (S) in absence of DXZs (–), with active DXZ (Dz1023-434), inactive DXZ control (Dz1023-INACT) and two unspecific DXZs controls (Dz1023-UNSP1 and Dz1023-UNSP2). The 5′ and 3′ cleavage products (P1 and P2, respectively) are indicated.

9. Dry gel and expose to radiographic film or fluorographic cassette.
10. Develop film or visualize in a fluorographic scanner, correspondingly.

3.3.2. DXZ Cleavage Kinetics

Kinetic analysis of DXZ cleavage is performed in a similar way to protein enzymes and ribozymes (20) (**Fig. 21.3**). Several cleavage reactions in optimal conditions are set for different concentrations of the target RNA *vs.* time. For the initial evaluation of DXZ cleavage on short HPV-16 E6 target transcripts use multiple turnover conditions (DXZ to target 1:30 molar ratio) (*see* **Note 7**). A similar protocol is used to evaluate mutant and modified DXZs (11).

1. Prepare 6% acrylamide/7 M urea gel.
2. Set up cleavage reactions in 1.5-mL microtubes on ice by adding the ^{32}P-labeled short or long E6 target (5–20 × 10^3 cpm/μL), 2 μL 10× DXZ buffer with Mg^{2+} in a 20 μL final volume.
3. Add 1 pmol DXZ DNA on the wall of each microtube and briefly spin the tubes in a microcentrifuge.
4. Incubate at 37°C for different periods of time (0–120 min).
5. Freeze tube in dry ice/ethanol to end reactions.
6. After the kinetic is complete, add 4 μL stop buffer to each microtube.

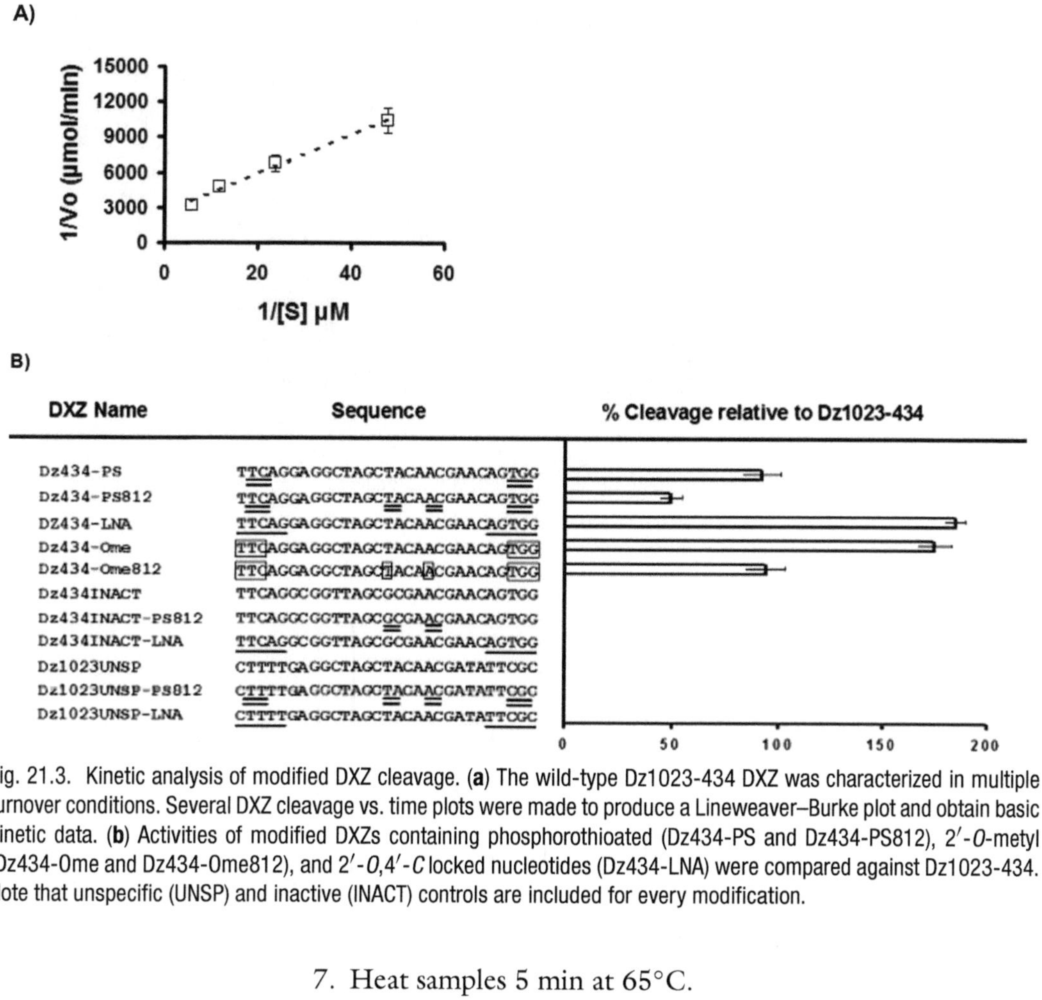

Fig. 21.3. Kinetic analysis of modified DXZ cleavage. (**a**) The wild-type Dz1023-434 DXZ was characterized in multiple turnover conditions. Several DXZ cleavage vs. time plots were made to produce a Lineweaver–Burke plot and obtain basic kinetic data. (**b**) Activities of modified DXZs containing phosphorothioated (Dz434-PS and Dz434-PS812), 2′-*O*-metyl (Dz434-Ome and Dz434-Ome812), and 2′-*O*,4′-*C* locked nucleotides (Dz434-LNA) were compared against Dz1023-434. Note that unspecific (UNSP) and inactive (INACT) controls are included for every modification.

7. Heat samples 5 min at 65°C.
8. Load samples on 6% acrylamide/7 M urea gel.
9. Electrophorese at 250 V for 50 min.
10. Dry gel and expose to radiographic film or fluorographic cassette.
11. Quantify cleaved and uncleaved target RNA (*see* **Note 8**).

3.4. DXZ Cleavage in Cultured Cells

To be used as therapeutic agents, DXZs require extensive test ing in cultured cells. Most relevant issues relate to cell uptake (**Fig. 21.4**), intracellular stability (nuclease resistance), cleavage of long transcripts (**Fig. 21.5a**) and evaluation of off-target effects (25). Native ssDNA is very short-lived in biofluids rendering low intracellular activity DXZs. Non-catalytic positions within the DXZ structure are often substituted with modified nucleotides to improve nuclease resistance with minimal impact on catalytic activity (26). Such modifications include phosphorothioates (27), 2′-*O*-methyl ribonucleotides (28), and lately 2′-*O*,4′-*C* locked nucleic acids (LNAs) (29). The effect of modified

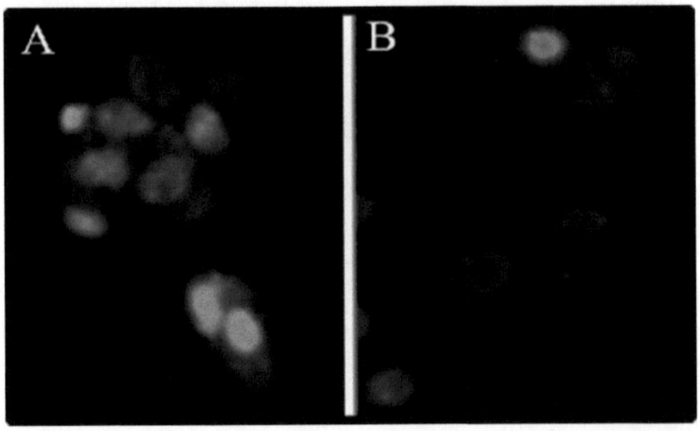

Fig. 21.4. Microphotography of SiHa cells transfected with fluorescein-labeled Dz1023-434DXZ (5 μM) using Lipofectin® reagent (**a**) or simple diffusion (**b**). Note the enhanced fluorescence produced by Lipofectin® transfection.

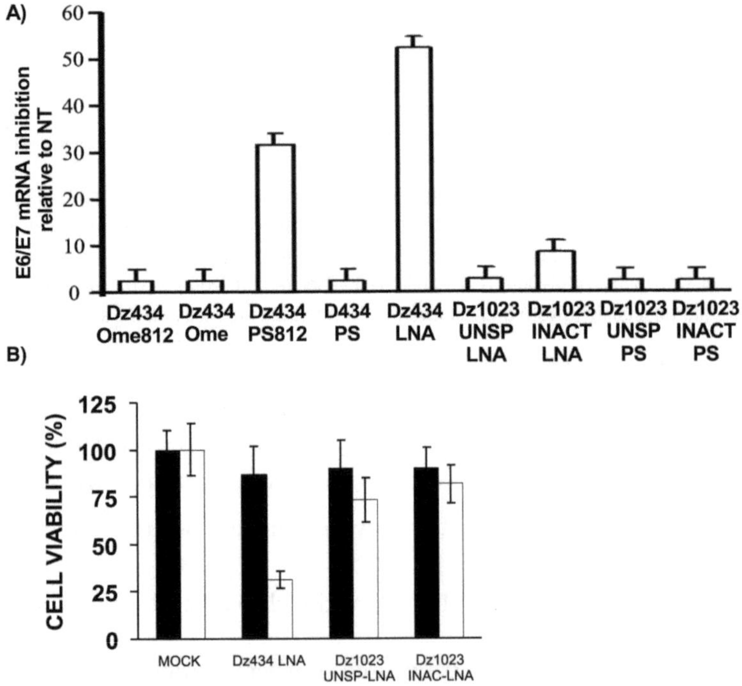

Fig. 21.5. Effects of Dz1023-434 in cultured cells. (**a**) SiHa cells were transfected with DXZ DNA (1 μM) from the indicated modification including unspecific (UNSP) and inactive (INACT) controls. Total RNA was extracted and RT-PCR analyzed for HPV-16 E6/E7 transcripts and corrected against β-actin mRNA. E6/E7 inhibition was plotted relative to non-treated (NT) cells. (**b**) Cell survival of SiHa (*white bars*) and C33-A (*black bar*) cells treated with 1 μM LNA-modified Dz1023-434 (Dz434-LNA). Note mock, unspecific (Dz1023UNSP-LNA), and inactive (Dz1023INACT-LNA) controls.

nucleotides on DXZ activity is first evaluated as described above and then successful DXZs are tested for nuclease resistance.

3.4.1. Nuclease Resistance Assay

1. Prepare 8% acrylamide/7 M urea gels.
2. In a fresh 1.5-mL microtube add 5–20×10^3 cpm radiolabeled DXZ, 1 μL 10× RQ1 buffer and 1 U of RQ1 DNaseI in 10 μL final volume (*see* **Note 9**).
3. Incubate for different time intervals at 37°C.
4. Load gels and electrophorese at 270 V for 70 min. Dry gels and expose to radiographic film or fluorographic scanner.
5. Select modified DXZs with better activity and nuclease resistance.

3.4.2. Cell Transfection with DXZ DNA

1. Seed 5×10^5 cells/well in 6-well dishes (*see* **Note 10**).
2. Incubate cells overnight and wash two times with sterile 1× PBS.
3. Prepare two 1.5-mL microtubes (tube A and B) for each DXZ transfection including not treated, mock, unspecific, and inactive controls.
4. Add 0.25 mL basal DMEM and 10 μL Lipofectin® reagent to microtube A (mixture A) (*see* **Note 11**).
5. Add 0.25 mL basal DMEM and different amounts of DXZ DNA to microtube B. For DXZ cell uptake experiments use 5′-fluorescein labeled DXZs.
6. Incubate for 15 min at room temperature.
7. Transfer all mixture A into microtube B. Mix gently.
8. Incubate at room temperature for 15 min.
9. Completely remove PBS and gently drip transfection mixture on the cell layer.
10. Incubate cells 2 h at 37°C in a CO_2 incubator with occasional tilting to avoid drying.
11. Add 2 mL of supplemented DMEM and incubate overnight at 37°C in a CO_2 incubator.
12. Aspirate transfection medium and add 2 mL of fresh supplemented DMEM.
13. For cell uptake analysis, incubate 2–4 h at 37°C in a CO_2 incubator.
14. Quantify transfection by flow cytometry.

3.4.3. Total RNA Extraction

1. Incubate cells 24–72 h postransfection at 37°C in a CO_2 incubator.
2. Transfected cells are washed twice with 1× PBS.

3. Completely remove PBS and add 1 mL Trizol® reagent.
4. Incubate 5 min at room temperature.
5. Transfer the lysate to a sterile 1.5-mL microtube and add 200 μL chloroform:isoamyl alcohol (24:1). Mix by inversion and keep 3 min at room temperature.
6. Centrifuge in a refrigerated microfuge at 16,000×*g* for 15 min at 4°C.
7. Transfer aqueous phase to a fresh 1.5-mL microtube and add 500 μL of 75% 2-propanol.
8. Incubate at –20°C for at least 30 min.
9. Centrifuge at 16,000×*g* for 30 min at 4°C. Discard supernatant and wash pellet with 1 mL 70% ethanol. Centrifuged at 16,000×*g* for 5 min at 4°C.
10. Discard supernatant and allow drying for 5 min at room temperature.
11. Resuspend RNA pellet in 100 μL DNase/RNAse-free water.
12. Quantify RNA using a spectrophotometer.

3.4.4. Quantitative RT-PCR

1. HPV-16 E6/E7 mRNA is amplified using target-specific primers and PCR conditions in the presence of Vista Green™ fluorescent dye using 100 μL PCR tubes (see **Note 12**). Fluorescence readings are set in the Real-Time PCR apparatus at 470 nm excitation and 510 nm emission.
2. The β-actin gene is used as internal control (30). HPV-16 E6/E7 mRNA levels are standardized against endogenous β-actin mRNA and plotted as a percentage of inhibition relative to controls. Standard quantification curves are made using serial dilutions of quantified DNA obtained from plasmids pCR16HH for HPV-16 E6/E7 and pBKS-actin for β-actin (*see* **Note 3**).
3. Analyze the relative abundance of HPV-16 E6/E7 transcripts using not treated, mock, unspecific, and inactive controls.

3.5. Phenotypic Effects

Inhibition of E6/E7 mRNA has many different effects at both cellular and molecular levels. Of particular interest for cancer therapeutics is the specific induction of cell death and/or apoptosis with low off-target effects. Because DNA itself has no toxicity, most DXZ off-target effects are due to the chemical modifications used and thus the inclusion of unspecific and inactive DXZ controls are very valuable to assert effect. Also, in many instances the transfection agent causes cell toxicity. Thus, it is necessary to estimate the effect of these variables by cell viability assays before

FACS an apoptosis analysis. Nevertheless, there are many procedures to evaluate cell viability, apoptosis, and cell cycle and we shall only refer to a couple of them (**Fig. 21.5b**).

3.5.1. Cell Viability

1. Cells are harvested 24–72 h post-transfection by trypsinization (*see* **Note 13**) and transferred to a 15-mL conical tube.
2. Centrifuge for 2 min at 650×*g*.
3. Discard supernatant and resuspend cell pellet in cold 1× PBS. Place on ice.
4. Take a 50-μL sample and add 50 μL of 0.4% Trypan blue stain solution.
5. Incubate 5 min at room temperature.
6. Count cells in a microscope with a Neubauer chamber. Blue-stained cells are not viable (*see* **Note 14**).

3.5.2. Cell Cycle and Apoptosis Analysis

1. Trypsinate 24–72 h transfected cells and transfer to a 15-mL conic tube.
2. Pellet cells by 5 min centrifugation at 1000×*g*. Discard supernatant.
3. Resuspend cells in FACS flow solution and pellet cells by 5 min centrifugation at 1000×*g*. Repeat twice.
4. Fix cells with 1 mL 80% ethanol. Re-suspend by gentle movement with the fingers and keep at –20°C for 24 h.
5. Pellet cells by centrifugation at 650×*g*.
6. Re-suspend in 300 μL 1× PBS and add 3 μL RNaseA solution and incubate at room temperature for 10 min.
7. Add 2 μL propidium iodide working solution in dark conditions.
8. Incubate 30–60 min at 4°C.
9. Pass cells through a 40-μm Nylon® mesh to avoid cell aggregates and transfer to 4-mL polyethylene tubes.
10. Analyze cell cycle/apoptosis in a flow cytometer.

4. Notes

1. Deionized water has a resistivity of 18.2 MΩ-cm and total organic content of less than five parts per billion.
2. Alternatively, staining with 0.02% ethidium bromide may be used to visualize ssODNs.
3. The pE6-GFP plasmid contains the HPV-16 E6 gene cloned in the pGreenLantern 2 vector (Invitrogen) (21). It

is used to produce target transcripts for in vitro DXZ cleavage assays from a SP6 promoter. The pCR16HH plasmid contains the HPV-16 E6/E7 cloned in the pCR3.1 vector (Invitrogen). It is used to produce target transcripts for in vitro DXZ cleavage assays from a T7 promoter (31). The pBKS-actin plasmid contains the β-actin cDNA cloned in the pBKS vector (Stratagene) (32). Other target RNAs can be produced from a vector of choice using the appropriate single-cut restriction endonucleases.

4. Phenol–chloroform extraction. Add half a volume of phenol saturated with deionized sterile water and vortex. Add half a volume of chloroform:isoamyl alcohol (24:1) and vortex. Centrifuge at 16,000×*g* for 5 min. The top aqueous phase is transferred to a fresh microtube for ethanol precipitation.
5. DNA precipitation by ethanol–ammonium acetate. Add half a volume of 7.5 M ammonium acetate, pH 7.5, and two volumes of cold (–20°C) absolute ethanol, mix and pellet DNA by 30-min centrifugation at 16,000×*g*.
6. Radioactive gel quantification may be simplified by using a fluorographic scanner. Nevertheless, quantification through scintillation counting can also be used. For such purpose, cleavage percentage is calculated from the formula:

$$\text{Cleavage percentage} = [\text{cleaved RNA dpm}/(\text{cleaved RNA dpm} + \text{non} - \text{cleaved RNA dpm})] \times 100$$

The dpm are calculated from cpm using the counting efficiency. For K_m and V_{max} determination, many DXZ cleavage assays must be performed with different target RNA concentrations to produce cleavage rate *vs.* substrate molar concentration plots. Molar concentration of the ^{32}P-U-labeled RNA is obtained converting dpm to μCi using the equivalency 1 μCi = 2.22×10^6 dpm. The number of U pmoles in the sample is calculated through the formula (33):

$$\text{pmoles of } U \text{ in sample} = \text{pmoles of UTP added} \times [\mu\text{Ci in transcript}/\ (\mu\text{Ci UTP added} \times \text{decay factor})]$$

Decay factor is the correction for the half-life of ^{32}P at the moment of the experiment. The number of pmoles of U in the sample is then converted to pmoles of RNA including the number of U residues in the transcript sequence:

$$\text{pmoles of transcript RNA} = \text{pmoles of } U \text{ in sample}/(\text{number of } U \text{ per transcript RNA})$$

7. Due to the long size of the full HPV-16 E6/E7 transcript (>1 kb), it is impractical (if not impossible) to use multiple turnover conditions. Therefore, for long transcripts we use single turnover conditions (DXZ to target 100:1 molar ratio) were observed. Rate constants (k_{obs}) are obtained from a curve fitted to the data (least squares) using the equation $y = x(1 - e^{-kt})$, where y is the fraction reacted at time t and x is the fraction reacted at $t = \infty$, and k is the k_{obs}.
8. Results are analyzed by plotting the molar amount of the cleaved target *vs.* time (min). The slopes (μM/min) represent the initial velocity (V_o). Further plots of $1/V_o$ vs. 1/[target RNA] are used for Lineweaver–Burk analysis. Kinetic parameters K_m, V_{max}, and K_{cat} are obtained from the graph (Y intercept $= 1/V_{max}$), K_m is the slope (K_m/V_{max}), and K_{cat}($K_{cat} = V_{max}/[Dz]$). The CE $= K_{cat}/K_m$.
9. Additionally, nuclease resistance assays may be performed with crude protein lysates of the target cells instead of DNaseI using 1× RQ1 buffer.
10. For DXZ intracellular testing it is necessary to use both cell lines expressing the target RNA and off-target controls of similar or identical lineage lacking the target RNA. The cervical carcinoma cell line SiHa contains and expresses HPV-16 E6/E7 genes (34). C33-A cells are derived from a cervical adenocarcinoma and contain no HPV DNA (35). Alternatively, cells transfected with the target gene may be used.
11. In our hands, Lipfectin® reagent yielded the best results (>90% transfection efficiency) with the cell lines used. Other transfection reagents may be used but transfection conditions must be optimized.
12. One μg of total RNA was used for quantitative RT-PCR amplification with SuperScript® III Platinum® SYBR® Green One-Step qPCR kit (Invitrogen). Previously validated primers E6U and E7L and primers targeting HPV-16 E6/E7 and Actin-*forward* and Actin-*reverse* targeting β-actin gene are used as described (21). Briefly, PCR of HPV-16 E6/E7 was conducted by reverse transcription for 30 min at 45°C, denaturing for 2 min at 94°C, and 30 cycles of denaturing for 1 min at 94°C, hybridization for 45 s at 45°C and polymerization for 1 min at 72°C. For the β-actin gene, PCR modifications were done at the

denaturing (45 s at 94°C) and hybridization steps (45 s. at 60°C).

13. To trypsinate cells wash twice with 1× PBS. After final aspiration add 2 mL Trypsin-Versene solution followed by a 5-min incubation at 37°C. Closely monitor for detached cells to avoid over-digestion. Once the monolayer is starting to detach, stop trypsin by addition of 5 mL supplemented DMEM and homogenize by pipetting up and down.

14. Cell viability determination must be performed within 30 min after Trypan blue addition.

References

1. Baum, D. A., and Silverman, S. K. (2008) Deoxyribozymes: useful DNA catalysts in vitro and in vivo. *Cell Mol. Life Sci.* **65**, 2156–2174.
2. Santoro, S. W., and Joyce, G. F. (1997) A general purpose RNA-cleaving DNA enzyme. *Proc. Natl. Acad. Sci. USA* **94**, 4262–4266.
3. Santoro, S. W., and Joyce, G. F. (1998) Mechanism and utility of an RNA-cleaving DNA enzyme. *Biochemistry* **37**, 13330–13342.
4. Bhindi, R., Fahmy, R. G., Lowe, H. C., Chesterman, C. N., Dass, C. R., Cairns, M. J., Saravolac, E. G., Sun, L. Q., and Khachigian, L. M. (2007) Brothers in arms. DNA enzymes, short interfering RNA, and the emerging wave of small-molecule nucleic acid-based gene-silencing strategies. *Am. J. Pathol.* **171**, 1079–1088.
5. Dass, C. R., Choong, P. F., and Khachigian, L. M. (2008) DNAzyme technology and cancer therapy: cleave and let die. *Mol. Cancer Ther.* **7**, 243–251.
6. Alvarez-Salas, L. M., and DiPaolo, J. A. (2007) Molecular approaches to cervical cancer therapy. *Curr. Drug Discov. Technol.* **4**, 208–219.
7. Faulhammer, D., and Famulok, M. (1997) Characterization and divalent metal-ion dependence of in vitro selected deoxyribozymes which cleave DNA/RNA chimeric oligonucleotides. *J. Mol. Biol.* **269**, 188–202.
8. Cairns, M. J., King, A., and Sun, L. Q. (2000) Nucleic acid mutation analysis using catalytic DNA. *Nucleic Acids Res.* **28**, E9.
9. Yuan, B. F., Xue, Y., Luo, M., Hao, Y. H., and Tan, Z. (2007) Two DNAzymes targeting the telomerase mRNA with large difference in Mg2+ concentration for maximal catalytic activity. *Int. J. Biochem. Cell Biol.* **39**, 1119–1129.
10. Takamori, K., Kubo, T., Zhelev, Z., Rumiana, B., Ohba, H., Doi, K., and Fujii, M. (2005) Suppression of bcr/abl chimeric gene by conjugate DNA enzymes in human cells. *Nucl. Acids Symp. Ser. (Oxf.)* **49**, 333–334.
11. Reyes-Gutierrez, P., and Alvarez-Salas, L. M. (2009) Cleavage of HPV-16 E6/E7 mRNA mediated by modified 10-23 deoxyribozymes. *Oligonucleotides* **19**, 233–242.
12. Pirisi, L., Yasumoto, S., Feller, M., Doniger, J., and DiPaolo, J. A. (1987) Transformation of human fibroblasts and keratinocytes with human papillomavirus type 16 DNA. *J. Virol.* **61**, 1061–1066.
13. Durst, M., Dzarlieva-Petrusevska, R. T., Boukamp, P., Fusenig, N. E., and Gissmann, L. (1987) Molecular and cytogenetic analysis of immortalized human primary keratinocytes obtained after transfection with human papillomavirus type 16 DNA. *Oncogene* **1**, 251–256.
14. Alvarez-Salas, L. M., Benitez-Hess, M. L., and DiPaolo, J. A. (2003) Advances in the development of ribozymes and antisense oligodeoxynucleotides as antiviral agents for human papillomaviruses. *Antivir. Ther.* **8**, 265–278.
15. Broker, T. R. (1987) Structure and genetic expression of papillomaviruses. *Obstet. Gynecol. Clin. North Am.* **14**, 329–348.
16. Smotkin, D., Prokoph, H., and Wettstein, F. O. (1989) Oncogenic and nononcogenic human genital papillomaviruses generate the E7 mRNA by different mechanisms. *J. Virol.* **63**, 1441–1447.
17. Schneider-Gadicke, A., and Schwarz, E. (1986) Different human cervical carcinoma cell lines show similar transcription patterns

of human papillomavirus type 18 early genes. *EMBO J.* **5**, 2285–2292.
18. DiPaolo, J. A., Popescu, N. C., Alvarez, L., and Woodworth, C. D. (1993) Cellular and molecular alterations in human epithelial cells transformed by recombinant human papillomavirus DNA. *Crit. Rev. Oncog.* **4**, 337–360.
19. Duensing, S., and Munger, K. (2002) The human papillomavirus type 16 E6 and E7 oncoproteins independently induce numerical and structural chromosome instability. *Cancer Res.* **62**, 7075–7082.
20. Alvarez-Salas, L. M., Cullinan, A. E., Siwkowski, A., Hampel, A., and DiPaolo, J. A. (1998) Inhibition of HPV-16 E6/E7 immortalization of normal keratinocytes by hairpin ribozymes. *Proc. Natl. Acad. Sci. USA* **95**, 1189–1194.
21. Alvarez-Salas, L. M., Arpawong, T. E., and DiPaolo, J. A. (1999) Growth inhibition of cervical tumor cells by antisense oligodeoxynucleotides directed to the human papillomavirus type 16 E6 gene. *Antisense Nucl. Acid Drug Dev.* **9**, 441–450.
22. Butz, K., Denk, C., Ullmann, A., Scheffner, M., and Hoppe-Seyler, F. (2000) Induction of apoptosis in human papillomaviruspositive cancer cells by peptide aptamers targeting the viral E6 oncoprotein. *Proc. Natl. Acad. Sci. USA* **97**, 6693–6697.
23. Jiang, M., and Milner, J. (2002) Selective silencing of viral gene expression in HPV-positive human cervical carcinoma cells treated with siRNA, a primer of RNA interference. *Oncogene* **21**, 6041–6048.
24. Romani, A. M., and Scarpa, A. (2000) Regulation of cellular magnesium. *Front Biosci.* **5**, D720–D734.
25. Alvarez-Salas, L. M. (2008) Nucleic acids as therapeutic agents. *Curr. Top Med. Chem.* **8**, 1379–1404.
26. Santoro, S. W., Joyce, G. F., Sakthivel, K., Gramatikova, S., and Barbas, C. F. (2000) RNA cleavage by a DNA enzyme with extended chemical functionality. *J. Am. Chem. Soc.* **122**, 2433–2439.
27. Cieslak, M., Niewiarowska, J., Nawrot, M., Koziolkiewicz, M., Stec, W. J., and Cierniewski, C. S. (2002) DNAzymes to beta 1 and beta 3 mRNA down-regulate expression of the targeted integrins and inhibit endothelial cell capillary tube formation in fibrin and matrigel. *J. Biol. Chem.* **277**, 6779–6787.
28. Fokina, A., Novopashina, D., Meschaninova, M., Vorobjeva, M., Zenkova, M., Francois, J. C., and Venyaminova, A. (2008) Effective cleavage of structured RNAs by tandems of 10-23 DNAzymes with 3′-modified oligo(2′-O-methylribonucleotide)-effectors. *Nucl. Acids Symp. Ser. (Oxf.)* **52**, 525–526.
29. Vester, B., Hansen, L. H., Lundberg, L. B., Babu, B. R., Sorensen, M. D., Wengel, J., and Douthwaite, S. (2006) Locked nucleoside analogues expand the potential of DNAzymes to cleave structured RNA targets. *BMC Mol. Biol.* 7, 19.
30. Guapillo, M. R., Marquez-Gutiérrez, M. A., Benitez-Hess, M. L., and Alvarez-Salas, L. M. (2006) A bacterial reporter system for the evaluation of antisense oligodeoxynucleotides directed against human papillomavirus type 16 (HPV-16). *Arch. Med. Res.* **37**, 584–592.
31. Marquez-Gutierrez, M. A., Benitez-Hess, M. L., DiPaolo, J. A., and Alvarez-Salas, L. M. (2007) Effect of combined antisense oligodeoxynucleotides directed against the human papillomavirus type 16 on cervical carcinoma cells. *Arch. Med. Res.* **38**, 730–738.
32. Aquino-Jarquin, G., Benitez-Hess, M. L., DiPaolo, J. A., and Alvarez-Salas, L. M. (2008) A triplex ribozyme expression system based on a single hairpin ribozyme. *Oligonucleotides* **18**, 213–224.
33. DeYoung, M. B., Siwkowski, A., and Hampel, A. (1997) Determination of catalytic parameters for hairpin ribozymes. *Methods Mol. Biol.* **74**, 209–220.
34. Baker, C. C., Phelps, W. C., Lindgren, V., Braun, M. J., Gonda, M. A., and Howley, P. M. (1987) Structural and transcriptional analysis of human papillomavirus type 16 sequences in cervical carcinoma cell lines. *J. Virol.* **61**, 962–971.
35. Yee, C., Krishnan-Hewlett, I., Baker, C. C., Schlegel, R., and Howley, P. M. (1985) Presence and expression of human papillomavirus sequences in human cervical carcinoma cell lines. *Am. J. Pathol.* **119**, 361–366.

INDEX

J. Goodchild (ed.), *Therapeutic Oligonucleotides*, Methods in Molecular Biology 764,
DOI 10.1007/978-1-61779-188-8, © Springer Science+Business Media, LLC 2011

MIX
Papier aus verantwortungsvollen Quellen
Paper from responsible sources
FSC® C105338

If you have any concerns about our products,
you can contact us on
ProductSafety@springernature.com

In case Publisher is established outside the EU,
the EU authorized representative is:
Springer Nature Customer Service Center GmbH
Europaplatz 3, 69115 Heidelberg, Germany

Printed by Libri Plureos GmbH
in Hamburg, Germany